Climate Change and Adaptation Planning for Ports

As key links in transportation and supply chains, the effect of climate change on seaports has broad implications for the development prospects of the global economy. However, the picture is very uncertain because the impacts of climate change will be felt very differently around the world, both positively and negatively. This book addresses the need for quality theoretical analysis, highly innovative assessment methodologies and insightful empirical global experiences so as to identify the best international practices, planning and appropriate policies to effectively adapt to, develop resilience to and, indeed, benefit from the impacts posed by climate change on transportation and supply chains.

With contributions from reputable scholars, inter-governmental organizations and senior professionals in transportation and supply chains coming from five continents, this book investigates climate change, adaptation strategies and planning of transportation and supply chains from different angles. It encourages innovative and effective solutions in developing adaptation measures and resilience by ports, transportation and supply chains to climate change. This book will be essential for anyone interested in understanding climate change and its impacts on transportation and supply chain infrastructures, as well as the effective ways to adapt to such impacts.

Adolf K. Y. Ng is Professor of Transportation and Supply Chain Management at the Asper School of Business of the University of Manitoba, Canada.

Austin Becker is Assistant Professor of Coastal Planning, Policy and Design at the University of Rhode Island (URI), USA.

Stephen Cahoon is the Director of the Sense-T NICTA Logistics Lab at the University of Tasmania, Australia.

Shu-Ling Chen is Senior Lecturer at the Australian Maritime College of the University of Tasmania, Australia.

Paul Earl is Senior Scholar at the Asper School of Business of the University of Manitoba, Canada.

Zaili Yang is Professor of Maritime Transport at Liverpool John Moores University, UK.

Routledge Studies in Transport Analysis

Maritime Logistics Value in Knowledge Management
Eon-Seong Lee and Dong-Wook Song

The Management of Maritime Regulations
Hristos Karahalios

Public Private Partnerships in Transport: Trends and Theory
Edited by Athena Roumboutsos

Climate Change and Adaptation Planning for Ports
Edited by Adolf K. Y. Ng, Austin Becker, Stephen Cahoon, Shu-Ling Chen, Paul Earl and Zaili Yang

Climate Change and Adaptation Planning for Ports

Edited by Adolf K. Y. Ng, Austin Becker, Stephen Cahoon, Shu-Ling Chen, Paul Earl and Zaili Yang

Routledge
Taylor & Francis Group
LONDON AND NEW YORK

First published 2016
by Routledge
4 Park Square, Milton Park, Abingdon, Oxon OX14 4RN

and by Routledge
605 Third Avenue, New York, NY 10017

Routledge is an imprint of the Taylor & Francis Group, an informa business

British Library Cataloguing in Publication Data
A catalogue record for this book is available from the British Library

Library of Congress Cataloging in Publication Data
Catalog record for this book has been requested

ISBN: 978-1-138-79790-1 (hbk)
ISBN: 978-1-315-75681-3 (ebk)

Typeset in Times New Roman
by Saxon Graphics Ltd, Derby

Contents

List of figures

List of tables

About the editors

Adolf K. Y. Ng is Professor in Transportation and Supply Chain Management at the Asper School of Business of the University of Manitoba, Canada. He obtained his DPhil from University of Oxford (St. Antony's College) UK, in geography, and excels in the research and teaching of port management, transport geography, intermodal transportation, climate change and transportation infrastructure planning, port-focal logistics and global supply chains. He is a productive researcher, which includes the publication of three scholarly books (*Climate Change and Adaptation Planning for Ports*, *Port-Focal Logistics and Global Supply Chains*, and *Port Competition*), more than 40 journal papers, and other forms of publications. He has received numerous accolades around the world, such as Fulbright Scholar Program (USA), Endeavour Research Fellowship (Australia), Universités Parisiennes Fellowship (France), Rh Award for Outstanding Contributions to Scholarship and Research in Interdisciplinary Studies, and Associates' Achievement Award for Outstanding Business Research (both Canada), and has provided key strategic advices to intergovernmental organizations, such as the United Nations (UN), European Commission (EC), and the African Development Bank (AFDB). He is currently a council member of the International Association of Maritime Economists (IAME), the co-editor of *Journal of Transport Literature*, associate editor of *The Maritime Economist*, and editorial board member of several reputable transportation and logistics journals. He has led in organizing many international conferences, such as the International Forum on Shipping, Ports and Airports (held in Hong Kong and Chengdu, China, 2008–13), and The Warming of the North: Challenges and Opportunities for Arctic Transportation, Supply Chain Management, and Economic Development (held in Ottawa, Canada, March 2015).

Austin Becker is Assistant Professor of Coastal Planning, Policy, and Design at the Department of Marine Affairs of the College of the Environment and Life Sciences of the University of Rhode Island. He obtained his PhD in Environment and Resources at Stanford University. His teaching and research on sustainable design, port planning and policy, GIS, and climate policy contribute to the untangling of complex planning and policy problems involving uncertainty, mismatched incentive structures, and climate change. He uses urban coastal areas, and seaports in particular, as a lens through which to study tools that enhance stakeholder participation in long term planning. He is a regularly invited speaker at international expert group meetings, as well as numerous conferences in the US. He earned a Master of Marine Affairs and Master of Environmental Science and Management at URI, and a BA at Hampshire College. He maintains a 500-ton captain's license for ocean-going vessels.

Stephen Cahoon is the Director of the Sense-T NICTA Logistics Lab at the University of Tasmania. He is a member of Tasmanian, national and international transport and logistics related industry committees and currently supervises ten PhD candidates in fields related to port management, shipping and supply chain management. He is currently a council member of the International Association of Maritime Economists (IAME), associate editor of the *Asian Journal of Shipping and Logistics*, and editorial board member of the *Journal of Marine Science and Technology* and the *Journal of Engineering for Maritime Environment.*

Shu-Ling Chen is Senior Lecturer at the Australian Maritime College, University of Tasmania. She received her PhD in Maritime Studies at Cardiff University. Her main research interests and publications focus on port management and strategy, port governance, dry port development, quality management of supply chains and shipping business strategies.

Paul Earl is Associate of the Transport Institute, Senior Scholar in the Asper School of Business, and Associate of the Desautels Centre for Private Enterprise and the Law at the University of Manitoba Law School. He holds a BA Sc. in Civil Engineering and MA Sc. in transportation and economics, both from the University of Toronto, and an interdisciplinary PhD (history, agricultural economics) from the University of Manitoba. His current research is on Canadian transportation issues, and on the western Canadian grain industry.

Zaili Yang is Professor of Maritime Transport at Liverpool John Moores University, UK. His research interests are system safety, security and risk-based decision-making modelling, especially their application in marine and supply chain systems. He has received more than £2 million in external grants from the EU and the UK research council to support his research. His research findings have been published in more than 120 technical papers in risk and supply chain areas. He has won several awards for his research work, including a Stanley Grey Fellowship by the IMarEST in 2004–06.

About the contributors

Michele Acciaro is Assistant Professor of Maritime Logistics at Kühne Logistics University (KLU) since January 2013. Previously he held the position of Senior Researcher – Green Shipping at the Research and Innovation department of DNV-GL in Oslo, and of Deputy Director at the Center for Maritime Economics and Logistics (MEL) in Rotterdam.

Jean-Christophe Amado is a climate change expert with a focus on strategy, risk management, planning and evaluation. He has developed numerous climate change strategies and plans with executives and senior-level managers, and has authored high-profile risk assessments, tools and guidance for clients in the public, energy, transportation, financial services, food and retail sectors. He is the author of many articles, book chapters and public reports, and was recognized in 2012 as an Aspen Institute/National Geographic Environment Forum Scholar for his leadership on climate change.

Carlos Andrade has a PhD in Physical Oceanography and an MSc. in Meteorology. He is a member of the Colombian Academy of Sciences, Captain (ret) in the Colombian Navy, a former Director of the Colombian Center for Investigations in Oceanography and Hydrography (CIOH), a former Chief of the Colombian Navy Diving and Salvage Department and General Manager of Oceanic Explorations of Colombia (EXOCOL) SAS in Cartagena of Indies, Colombia.

Austin Becker is Assistant Professor of Coastal Planning, Policy, and Design at the Department of Marine Affairs of the College of the Environment and Life Sciences of the University of Rhode Island. He obtained his PhD in Environment and Resources at Stanford University. His teaching and research on sustainable design, port planning and policy, GIS, and climate policy contribute to the untangling of complex planning and policy problems involving uncertainty, mismatched incentive structures, and climate change. He uses urban coastal areas, and seaports in particular, as a lens through which to study tools that enhance stakeholder participation in long term planning. He is a regularly invited speaker at international expert group meetings, as well as numerous conferences in the

US. He earned a Master of Marine Affairs and Master of Environmental Science and Management at URI, and a BA at Hampshire College. He maintains a 500-ton captain's license for ocean-going vessels.

Benjamin Brooks has a degree in environmental science and PhD in human factors. He currently works on research in areas such as innovation in high-risk environments, climate change and decision-making. He works with regulators, private companies, port authorities and emergency management agencies.

Stephen Cahoon is the Director of the Sense-T NICTA Logistics Lab at the University of Tasmania. He is a member of Tasmanian, national and international transport and logistics related industry committees and currently supervises ten PhD candidates in fields related to port management, shipping and supply chain management. He is currently a council member of the International Association of Maritime Economists (IAME), associate editor of the *Asian Journal of Shipping and Logistics*, and editorial board member of the *Journal of Marine Science and Technology* and the *Journal of Engineering for Maritime Environment*.

Shu-Ling Chen is Senior Lecturer at the Australian Maritime College, University of Tasmania. She received her PhD in Maritime Studies at Cardiff University. Her main research interests and publications focus on port management and strategy, port governance, dry port development, quality management of supply chains and shipping business strategies.

Claude Comtois is Deputy Director of the Interuniversity Research Centre on Enterprise Networks, Logistics and Transportation and Professor at Montreal University. Visiting professorships include more than 15 foreign universities. His teaching and research are centred on shipping and ports. He is the author or co-author of over 100 scientific publications and 250 communications. He has supervised or participated in more than 40 research projects. He currently supervises projects on the competitiveness of port systems, configuration of ocean shipping networks, adaptation measures of maritime transport to environmental changes and Arctic shipping.

Richenda Connell, CTO and co-founder of Acclimatise, is an expert in climate risk assessment and management. She has more than 20 years' experience working on decision-making and environmental issues, on climate change adaptation, air pollution and environmental impact assessment. She was Expert Reviewer for the IPCC Working Group II Fourth Assessment Report (AR4). She sits on the Steering Committee of the Urban Climate Change Research Network (UCCRN). From 1999 to 2005, she was Senior Scientist then Technical Director at the UK Climate Impacts Programme (UKCIP). Prior to UKCIP, she worked as an air quality consultant at Environmental Resources Management. She holds a doctorate in atmospheric chemistry from the University of Oxford.

Maria Inés Cusano has a PhD in Logistics, Transport, and Territory from the Italian Centre of Excellence for Integrated Logistics (Genoa). She is now a post-doctoral research fellow at the Department of Economics of the University of Genoa. Her scientific research is mainly related to maritime economics and regional economics with a focus on environmental issues related to transport activity and on the effects of climate change on coastal cities.

Paul Earl is Associate of the Transport Institute, Senior Scholar in the Asper School of Business, and Associate of the Desautels Centre for Private Enterprise and the Law at the University of Manitoba Law School. He holds a BA Sc. in Civil Engineering and MA Sc. in transportation and economics, both from the University of Toronto, and an interdisciplinary PhD (history, agricultural economics) from the University of Manitoba. His current research is on Canadian transportation issues, and on the western Canadian grain industry.

Miguel Esteban is Project Associate Professor at the Graduate School of Frontier Science, University of Tokyo, Japan. He received his PhD in Coastal Engineering from Yokohama National University in Japan in 2007, and then continued his work with post-doctoral fellowships at the United Nations University Institute of Advanced Studies (UNU-IAS) and Kyoto University. Subsequently he also worked as Associate Professor at Waseda University.

Claudio Ferrari, PhD in Transport Economics at the Department of Economics (Genoa), is currently Associate Professor at the University of Genoa. Leader of many national and international research projects on transport-related problems, his scientific research is mainly related to transport economics and transport planning, with a focus on port efficiency studies and on transport network integration.

John Higginbotham is Senior Fellow at the Norman Paterson School of International Affairs and Head of the Arctic Program and Senior Fellow at the Center for International Governance Innovation. He has served in senior Canadian diplomatic posts in Washington, DC, Hong Kong and Beijing, and as Assistant Deputy Minister in Transport Canada (where he coordinated Canada's Asia-Pacific Gateway and Corridor Initiative), Foreign Affairs and International Trade, and the Canada School of Public Service.

John Hunter has 40 years of post-qualification civil engineering experience, gained mainly in the design and construction of marine civil engineering works and heavy foundations in the UK, North and South America, Africa, South-East Asia and the Middle East.

Andrew Hunting is continuing his PhD on 'modelling the effects of a changing climate on infrastructure', which is funded by the Australian Research Council. A particular area of investigation is chloride ingress with results presented in Asia and Europe.

Michael C. Ircha is Professor Emeritus of Civil Engineering at the University of New Brunswick. He is also Senior Advisor to the Association of Canadian Port Authorities and an Adjunct Research Professor at Carleton University. He has taught port management courses in various countries, including the World Maritime University and the Shanghai Maritime University.

Paul Kapelus is the Director of Synergy Global Consulting, and has worked at the interface of business and society for the past 23 years. He works in the natural resources sectors, with mining, oil, gas, agriculture and forestry companies. He is a Young Global Leader of the World Economic Forum and a Desmond Tutu Leadership Fellow.

Geraldine Knatz is Professor of the Practice of Policy and Engineering, a joint appointment between the University of Southern California Price School of Public Policy and the Sonny Astani Department of Civil and Environmental Engineering at the USC Viterbi School of Engineering. She served as Executive Director of the Port of Los Angeles from 2006 to January 2014. She was the first woman to serve in this role and made a significant impact through the creation and implementation of the San Pedro Bay Ports Clean Air Action Plan, an aggressive plan that reduced air emissions by combined port operations of over 70 per cent over five years, which is recognized around the world for its innovation and success. She was also Managing Director of the Port of Long Beach, where she also led a number of environmental initiatives, including the Green Port Policy and Truck Trip Reduction Program. She is past president of the American Association of Port Authorities and past president of the International Association of Ports and Harbors. She serves as Gov. Brown's appointee on the Ocean Protection Council. She has received numerous awards, including Outstanding Women in Transportation (*Journal of Commerce*, 2007), Woman Executive of the Year (*Los Angeles Business Journal*, 2007), Compass Award (Women's Leadership Exchange, 2008), honorary PhD, Maine Maritime Academy (2009), and the Peter Benchley Ocean Award (Blue Frontier Campaign, 2012). In 2014, she was named a member of the National Academy of Engineering in recognition of her international leadership in the engineering and development of environmentally clean urban seaports.

Daniel Kong is a lecturer with the Department of Civil Engineering at Monash University. Prior to joining Monash, he worked at RMIT University where his research focused on exploring the structural resilience of seaport structures in a changing climate and infrastructure management projects.

Diana Liverman is Co-Director of the Institute of the Environment at the University of Arizona where she is also Regents Professor of Geography and Development. Her work focuses on the human dimensions of climate change, especially climate impacts, adaptation and policy in Latin America.

Darryn McEvoy is a principal researcher in climate change adaptation, specialising in urban resilience. He currently acts as chief investigator to a number of projects in Australia and the Asia Pacific region. As well as leading the climate resilient seaports project, other project examples include the development of a climate adaptation plan for RMIT University, community resilience to flooding in Queensland, reviewing the natural disaster risk assessment for Tasmania, and conducting vulnerability assessments and adaptation planning for UN-Habitat in Honiara (Solomon Islands) and Port Vila (Vanuatu). He is a member of the international scientific committee for 'Urbanization and Global Environmental Change' and a global adviser for the Global Compact Cities Program.

James McGregor is an economist with over 20 years of experience in global private and public sectors helping make better-informed decisions, particularly in light of political and non-commercial risks. He possesses expertise within the oil and gas, mining, power, retail, agriculture, agrifood, fisheries, national parks and public policymaking sectors.

Steven Messner has led many studies in the past ten years concerning climate change resiliency and adaptation, including the *California Adaptation Strategy*, the *San Diego Regional Focus 2050 Adaptation Study*, and the *Port of San Diego's Climate Mitigation and Adaptation Plan.* He is currently leading a regional study on climate change and economic resiliency in California.

Will Morgan manages technical illustration requirements for WorleyParsons' UK Infrastructure group, providing support to a team of over 100 professional staff. With over ten years of experience, he uses CAD and graphic design packages to deliver bespoke illustration work, with quality, clarity and consistency at the core of his approach.

Jane Mullett is a research fellow with the Climate Change Adaptation Program, in the Global Cities Research Institute at RMIT University. Her work concentrates on the resilience of critical infrastructure and also how climate change can be best communicated. She was lead researcher on three resilient seaports projects (in Australia and the Pacific) and is currently working on an assessment of community resilience to road network failure in Australia. She has also worked on the impacts of extreme heatwaves in Melbourne.

Adolf K. Y. Ng is Professor in Transportation and Supply Chain Management at the Asper School of Business of the University of Manitoba, Canada. He obtained his DPhil from University of Oxford (St. Antony's College) UK, in geography, and excels in the research and teaching of port management, transport geography, intermodal transportation, climate change and transportation infrastructure planning, port-focal logistics and global supply chains. He is a productive researcher, which includes the publication of three scholarly books (*Climate Change and Adaptation Planning for Ports*, *Port-Focal Logistics and Global*

Supply Chains, and *Port Competition*), more than 40 journal papers, and other forms of publications. He has received numerous accolades around the world, such as Fulbright Scholar Program (USA), Endeavour Research Fellowship (Australia), Universités Parisiennes Fellowship (France), Rh Award for Outstanding Contributions to Scholarship and Research in Interdisciplinary Studies, and Associates' Achievement Award for Outstanding Business Research (both Canada), and has provided key strategic advices to intergovernmental organizations, such as the United Nations (UN), European Commission (EC), and the African Development Bank (AFDB). He is currently a council member of the International Association of Maritime Economists (IAME), the co-editor of *Journal of Transport Literature*, associate editor of *The Maritime Economist*, and editorial board member of several reputable transportation and logistics journals. He has led in organizing many international conferences, such as the International Forum on Shipping, Ports and Airports (held in Hong Kong and Chengdu, China, 2008–13), and The Warming of the North: Challenges and Opportunities for Arctic Transportation, Supply Chain Management, and Economic Development (held in Ottawa, Canada, March 2015).

Winfried Osthorst has been working as Professor for Political Management at the University of Applied Science Bremen, Germany, since 2008. His focus is on regional environmental conflicts and regional development; one topic of which is the work on climate adaptation of the regional ports in north-west Germany.

José Daniel Pabón is Associate Professor at the research group 'Weather, Climate and Society', Department of Geography, National University of Colombia, focusing on climate modelling and climate change scenarios.

Olivia Palin is a climate risk management expert at Acclimatise, where she leads the firm's work in Latin America and the Caribbean. She is active in mainstreaming climate change into decision-making processes and development planning, on Small Island Developing State (SIDS) issues in general and on global private sector supply chains. She holds an MSc in Environmental Change and Management from Linacre College, University of Oxford, and a BA in Geography from Jesus College, University of Oxford.

Ben Pope is an economist specializing in natural resources, especially in the context of the extractive industries. He has extensive experience incorporating environmental and social considerations into financial models to help decision-makers address the concerns of a variety of stakeholders and facilitate sustainable use of our limited resources.

Spencer Samsom entered the transportation industry as an owner operator. Studying supply chain management at the Asper School of Business of the University of Manitoba, Canada, solidified his interest in the industry. With a specific interest in the capital requirements of fleet management, Spencer currently serves as President of a transport equipment-remarketing firm.

Sujeeva Setunge is the Deputy Head of School, Civil Environmental and Chemical Engineering at RMIT University, responsible for the Civil Engineering teaching and research. She is currently leading a number of research projects delivering outcomes to the industry on management of civil infrastructure assets including buildings, bridges, storm water pipe networks, water mains and seaports using advanced reliability based methods. Her more recent research is directed at resilience of infrastructure under natural hazards with a major focus on critical road structures. She is on the editorial board of the journal of disaster resilience of the built environment and is a co-chair of the board of the centre for pavement excellence (Asia Pacific). She had the lead responsibility for the engineering component of the research on resilient seaports.

Hope Sherwin has spent the past 15 years working with the private sector, civil society and governments to promote corporate responsibility towards people and planet. Most recently she led the Middle East business of Synergy Global Consulting, specializing in the oil and gas, mining and infrastructure sectors. She has worked with clients across the Middle East, Europe and Africa.

Tomoya Shibayama is Professor of Civil and Environmental Engineering at Waseda University, Tokyo, Japan. He is also an Emeritus Professor at Yokohama National University. He received his doctoral degree in engineering from the University of Tokyo. Formerly, he was Associate Professor at the University of Tokyo, and at the Asian Institute of Technology and was a Professor at Yokohama National University.

Brian Slack is Distinguished Professor Emeritus at Concordia University, Montreal, Canada. Throughout his career his research has been focused on ports and shipping. He is presently engaged in two main areas: the influence of time in container shipping, and the impact of climate change on water levels and inland shipping.

Greg Smith has extensive experience in bulk shipping and port operations having served in senior management positions with a major capital city port, a major bulk export port and various bulk shipping companies. He holds an MBA (AMC) and Master of Sustainable Management (Sydney University).

Tiffany C. Smythe is an ocean and coastal policy scholar and practitioner. She is a Coastal Management Extension Specialist with the Coastal Resources Center and Rhode Island Sea Grant at the University of Rhode Island Graduate School of Oceanography, and Adjunct Professor of Marine Affairs at the University of Rhode Island. Work included herein was completed while a Research Fellow at the US Coast Guard Academy.

Ben Staley joined WorleyParsons' Technical Illustration and Graphics department in August 2007. He has knowledge of a wide range of industry-standard software packages, together with visualization skills and an understanding of the role of illustration in environmental consultancy. He is experienced at creating customized graphical solutions that meet clients' needs.

Vladimir Stenek is Senior Climate Change Specialist at the International Finance Corporation, the private sector lending arm of the World Bank Group. He leads IFC's Climate Risk and Adaptation Program, which provides investment and operational solutions to the challenges of climate change impacts.

Hiroshi Takagi is Associate Professor at the Graduate School of Engineering, Tokyo Institute of Technology, Japan. After graduation from Yokohama National University in 1999, he has been working as a researcher, engineer and officer at Yokohama National University, Penta-Ocean Construction Co. Ltd. and Japan International Cooperation Agency (JICA).

Alessio Tei has a PhD in Logistics, Transport and Territory from the Italian Centre of Excellence for Integrated Logistics (Genoa). He is Post-Doctoral Research Fellow at the Department of Economics of the University of Genoa and Adjunct Professor at the Naval Academy (Leghorn). His scientific research is mainly related to transport economics and regulation, maritime and port economics with a focus on the effects of transport on regional development and on the efficiency of transport infrastructure.

Alexei Trundle is a research associate at the RMIT University Climate Change Adaptation Program, an interdisciplinary research unit which integrates climate science and adaptation planning approaches into both student learning and research projects. Recent work has focused on adaptation in urban contexts in the Asia-Pacific, including participatory action research in Port Vila (Vanuatu), Honiara (Solomon Islands), Hue (Viet Nam) and Satkhira (Bangladesh), as well as sector-specific capacity building exercises with seaport stakeholders and Port Authorities across Australia, Fiji and Papua New Guinea.

Jerome Verny is Scientific Director of MOBIS, the international research institute for transport and supply chains at NEOMA Business School (France). He has authored numerous international and national publications on transport and logistics. In 2009, he was named Young Researcher of the Year by the International Transport Forum (ITF-OECD). His research interests mainly cover the fields of economic and transport geographies.

Chengpeng Wan is currently working towards a PhD in the School of Energy and Power Engineering at the Wuhan University of Technology (WUT), People's Republic of China. He received his BSc in Marine Engineering from WUT in 2012. His major research interests include formal maritime safety assessment;

risk-based multiple-attribute decision-making and risk management of container supply chains.

Tianni Wang is an MSc candidate in Supply Chain Management at the Asper School of Business of the University of Manitoba, Canada, under the supervision of Prof. Adolf K. Y. Ng. Her research focuses on the areas of transportation and logistics, climate change adaptation and supply chain management. She earned her bachelor degree in logistics management from the Xiamen University of Technology as an outstanding graduate in 2013. She has obtained a Quality Management Technologist certificate from the Chinese Institute of Industrial Engineers, and the certificate of Junior Logistics Management Talent from the Logistics Association of the Republic of China.

Richard Washington is Professor of Climate Science at the School of Geography and the Environment and Fellow of Keble College, University of Oxford, UK. He has degrees from the University of Natal and University of Oxford and taught at the University of Natal and University of Cape Town. His doctorate was on African rainfall variability and change. He took up a university lectureship position and fellowship at Keble College in 1999, a readership in 2006 and was made professor in 2010. He has served as the World Climate Research Program representative to the International Council for Science Southern Africa as well as membership of several external steering committees including AFRICANNESS (African Earth System Science) and the Stockholm Environment Institute, Oxford. He was one of 12 members of the UK Natural Environment Research Council (NERC) Climate Science Strategy Panel for the forthcoming 2007–12 NERC Science Strategy. He has taught several World Climate Program 'Climate Information and Prediction Services (CLIPS)' workshops. He has served on the panel of judges for the Best Research Paper (SA Society for Atmospheric Scientists) from 2003 to 2009.

Stewart Wright is an oceanographer and environmental expert who has led environmental teams in the UK and Canada over a 15-year career. He has managed Environmental Assessments for coastal and offshore development across the world, with specific interest in climate change adaptation, oceanographic modelling, underwater noise and the integration of environment into engineering design.

Zaili Yang is Professor of Maritime Transport at Liverpool John Moores University, UK. His research interests are system safety, security and risk-based decision-making modelling, especially their application in marine and supply chain systems. He has received more than £2 million in external grants from the EU and the UK research council to support his research. His research findings have been published in more than 120 technical papers in risk and supply chain areas. He has won several awards for his research work, including a Stanley Grey Fellowship by the IMarEST in 2004–06.

Di Zhang is Associate Professor of National Engineering, Research Center for Water Transport Safety, Wuhan University of Technology, China. He received his BSc in Navigation Technology, MSc in Traffic Information Engineering and Control, and PhD in Vehicle Operation Engineering from Wuhan University of Technology in 2005, 2008 and 2011 respectively. With financial support from the China Scholarship Council, he was a full-time researcher at Liverpool John Moores University (LJMU), from October 2010 to September 2011. His major research interests include risk assessment and decision science applied in marine systems. He has authored two book chapters, 15 refereed journal papers and over 20 refereed conference papers. He is Associate Fellow of the Royal Institute of Navigation (AFRIN).

Foreword

Mitigating climate change through reduction of greenhouse gases is now a mainstream activity of ports. There are numerous publications and cases studies on how ports have been successful in reducing the production of greenhouse gases from their operations. The World Port Climate Initiative (WPCI) of the International Association of Ports and Harbors was created by the 55 largest ports to share practices to reduce greenhouse gas production and to spur similar action among the maritime industries on an international scale. Yet, most ports, including those in the WPCI, have focused more on greenhouse gas reduction than adaptation to climate change. This is not surprising. Reducing greenhouse gases can be easier and cost-effective for a port and its customers while also providing the public relations value of being seen as a "green" or sustainable port operation. Climate change adaptation is harder to do. It takes longer to accomplish. It can be very costly. It involves long-term strategies and long-term investment not likely to provide an immediate public relations benefit.

This is the first book of its kind to focus on climate change adaptation for ports. The editors illustrate, through examples from every continent, how a global problem is tackled by solutions for adaptation that are regionally defined. But this book is not a compendium of best practices. In fact, the editors are to be applauded for including cases where the efforts to create a climate change adaptation plan have failed. Many ports operate as landlords leasing their waterfront to port terminal operators under long-term concessions. Several of the authors illustrate the challenges ports face in incentivizing a concession holder to take actions to adapt to sea level rise when the length of their concession may only be 20 or 30 years. Why would a private company make investments in the waterfront property that would benefit the future holder of the concession? In other cases, very long-term concessions that may extend for over 50 years and written prior to concerns about global climate change, further limit a seaport authority's options. These stories serve as a warning for ports. The ports that become the most resilient to climate change in the future are the ones where climate change policy is considered in future concession agreements.

Ports sit on the frontlines of global climate change. Many ports already address sea level rise in their design of new piers and infrastructure. Ports are fortunate in typically having technical and engineering expertise to address these issues head on. But as the editors show, sea level rise is only one of multiple challenges ports face due to climate change. Adaptation to climate change is not the same as preparing for the next major storm event. The melting permafrost that supports railroad tracks in Canada, the extreme temperatures experienced by dockworkers in Australia, the lower water levels of the US Great Lakes and the Gatun Lake in the Panama Canal: all these examples illustrate the wide range of impacts on port infrastructure, operations, and worker health and safety. The challenge for all ports is to figure out how to proactively address climate adaptations today rather than waiting until they are forced to react in a crisis situation. Remarkably, the efforts of many ports to "green" their operations over the past decade have had the unintended consequence of making the ports less resilient. Reducing the production of greenhouse gases by replacing the use of diesel-powered equipment with electrically powered equipment is one example where the vulnerability of the port may be increased in the case of a severe weather event involving the loss of electrical power.

As many of the authors point out, the port is just a node in a global supply chain. A port that is well-positioned to address climate change and able to bounce back into operation after an extreme weather event could still have its operations effectively shut down by failures in other sectors of the supply chain. While this book provides an excellent review of efforts by various seaport authorities to minimize the impacts of climate change on their operations, it also points out the value of a comprehensive policy at a national or international level to ensure supply chain resiliency.

The editors are globally recognized as scholars in the fields of maritime transport, geography, economics, logistics, and coastal management, each with experience in multiple regions of the globe. They have pulled together a themed but extremely diverse collection of cases that showcase the breadth of climate change impacts ports might experience. More importantly, they showcase how ports, policymakers and academics are enhancing decision-making tools, stakeholder input, and strategies to adapt to climate change. The book will serve as a valuable resource for port personnel, governmental agency staff, educators, and students. While not providing all the answers to how a port manages adaptation to climate change, the stories are so varied and the struggles so clearly outlined, that the book will serve as a catalyst for creative solutions.

Geraldine Knatz, PhD, CEO Port of Los Angeles (retired), Professor of the Practice of Policy and Engineering, University of Southern California

Preface

> All wise rulers … have to consider not only present difficulties but also future, against which they use all diligence to provide; for these, if they be foreseen while yet remote, admit of easy remedy, but if their approach be awaited, are already past cure, the disorder having become hopeless … in its beginning it is easy to cure, but hard to recognize; whereas, after a time, not having been detected and treated at the first, it becomes easy to recognize but impossible to cure.
>
> Niccolò Machiavelli in *The Prince*, published in 1513; translated into English in 1910.

Only a century ago most people who had travelled to a continent other than the one that they were born/lived in would repeatedly brag about such a once-in-a-lifetime experience to their next generations. Nowadays, moving around the globe, whether it is for business, searching for a new destiny, or simply temporarily escaping from the sometimes spirit-shattering work pressure, has become a fact of life for many of us. At home, we get used to buying imported goods labeled "Made in China" from megastores with (in most cases) reasonable prices, while eating decent Korean food doesn't mean that you have travelled to Korea; and indeed, professors in Canadian universities would find that most of their classes are now filled with students originating from anywhere other than Canada. How life has changed, even within such a short period of time!

Many would say that this is what it means to live in a globalized era: the convenience brought by technological innovation, cheaper goods flocking in from around the world *via* international trade and transportation, and more regular interactions between people from different races and backgrounds. Ultimately, as they say, we are all striving for a better future. But being a global citizen is far from easy – we are now guaranteed to face even more uncertainties, and have to make riskier decisions on a daily basis: we might have more opportunities coming from different parts of the world, such as jobs and easier travel, but can we get used to the cultural shock in a completely strange place easily? To what extent can we increasingly work with people with completely different backgrounds, values, and mindsets, while not toppling our own established habits and beliefs? Even if you decide to stay in your home country, these problems could prevail as your city/company is recruiting more and more "aliens," and you may soon find that life is becoming harder as the place that you live is changing so fast, it is so diversified that you can hardly keep pace. Whether deciding to move or to stay, one thing is certain: with such rapid and diversified changes taking place which are mostly outside our own control, whether we are able to adapt to such changes will be crucial to our destiny in this contemporary world.

This is what adaptation is all about: the more uncertain our world has (will) become, the more proactive and resilient we should be to relieve the negative impacts, and turn risks and perils into opportunities. It sounds easy, but achieving it is a much taller order. How can we know what "uncertainty" in the future really means, and be proactive about it? Even if we can recognize that things might not seem right, does that imply they are inevitable? Even if they are, does it really matter to us? Even if it matters, why not leave it for the next generation to solve? And who would appreciate the efforts and commitment, especially if the precautionary measures do not seem to pay off (yet)? As a matter of fact, there are just so many reasons (or excuses, depending on how you look at it) that can discourage us from doing anything. The first editor of this book was once told by his friend, who was a director of the port authority of a major port in North America, that "for most industry people, the future means tomorrow." This shows the difficulty in even trying to get a reasonably effective solution to the challenge. To be fair, this isn't just a reflection of industrial professionals, but fundamentally human nature: so often we are not only reluctant (due to inertia, institutional setting, or simply indolence) to change and adapt, but our limited knowledge and vision, not helped by the real aspects of life, have jeopardized us being proactively adaptive to new, uncertain risks and challenges.

But all those who downplay the importance of adaptation have missed a crucial point: adaptation is the wormhole between the past and the future. If one could sense that an exogenous change might cause some impacts, it means that the current settings might soon be (if not already been) obsolete, and require a rethink, no matter whether it is just a minor re-adjustment or a toppling of the status quo. Adaptation offers the opportunity to review what has happened, and actively prepare for betterment in the future. On the contrary, resistance, or an indifferent attitude, deprives us of any chance of seeing the current setting improve, leaving it to gradually decline. In fact, adaptation and progression belong to both sides of the same coin. While survival isn't only reserved for the strongest or the most intelligent, any societies, sectors, or industries that fail to respond to challenges, or keep on taking short-term actions that try to "mainstream" problems into existing economic, societal, policy, and/or institutional frameworks, are like those who choose to turn a blind eye to the time bombs inside their homes, praying that they won't explode for as long as possible. With the ruthlessness of neoliberal ideology shaping how the global economy works nowadays, we are under no illusion that what (not) to adapt would always involve the painful process of prioritizing objectives and the (re-)allocation of (largely financial) resources under a highly intensified competitive environment. But if the future really only means tomorrow for industry people due to frontline obligations, scholars and policymakers should then assume the role of protectors, and undertake the responsibility to safeguard the safety and wellbeing of the next generations.

And this is exactly what this book is all about: encouraging our society to work together to safeguard our future generations for the long-term future. Throughout the production process, we always tried to "think outside the box," maintained an interdisciplinary and multi-hierarchical approach, and ensured that our (including the contributing authors') ideas would pose direct impacts to the benefits of long-term human welfare. The composition of the editorial team, consisting of scholars from three continents, reflected such a desire. The authors were selected carefully, and we made sure that every single chapter would enrich the overall lessons learnt on the general topic of ports' adaptation to climate change impacts through a vigorous peer-review process and editorial guidance. Thus, this book would not have come to fruition without the generous support from many parties, not least the comments from the anonymous reviewers in enhancing the quality of the chapters. Of course, we are grateful for the support of our publisher, Routledge, for this initiative. A special thank you is also given to Huiying Zhang who assists in handling part of the production process. Finally, the first editor would like to reserve a few lines to thank his family's unconditional support—notably continuously tolerating the 'family kid' to stay thousands of miles away from them in the name of contribution to society and, of course, his own career advancement. This book is dedicated to them.

The efforts, contents and suggestions in this book might not be perfect. But perhaps imperfectness is exactly what we need—the strive to perfection sustains further interests and inputs from existing and new personnel, in terms of both research findings and professional experiences, and this process will only encourage more discussion and collaboration for this purpose. If this is what is required to develop effective adaptation strategies, plans, and actions for the wellbeing of future generations, we strongly believe that this book will go on to play a significant role in making it happen.

Adolf K. Y. Ng, Austin Becker, Stephen Cahoon,
Shu-Ling Chen, Paul Earl, and Zaili Yang, June 10, 2015

1 Time to act

The criticality of ports in adapting to the impacts posed by climate change

Adolf K. Y. Ng, Austin Becker, Stephen Cahoon, Shu-Ling Chen, Paul Earl and Zaili Yang

1. Setting the scene

Nowadays, ports are key nodes in determining the smoothness of transportation and cargo flows along global and supply chains (Ng and Liu, 2014). Thus, not surprisingly, climate change impacts to ports (which can include seaports, river ports, inland ports, logistics and distribution centres, etc.) would have broad implications for the development prospects of the global economy and human welfare. However, there are many uncertainties, given that the impacts of climate change can significantly diversify in different countries and regions, both positively and negatively. There is an urgent demand for quality theoretical analysis, highly innovative assessment methodologies, and insightful global experiences so as to identify the best international practices, planning strategies, and the most appropriate policies so as to effectively adapt to, develop resilience, and benefit from, the impacts posed by climate change on ports, transportation and global supply chains.

In the past years, considerable research has focused on mitigating climate change. Adaptation research, and the development of effective solutions to deal with the impacts posed by climate change, remain rather scarce. During the United Nations Conference on Trade and Development (UNCTAD)'s *Ad Hoc* Expert Meeting on Climate Change and Adaptation (held in Geneva, Switzerland, 29–30 September 2011), participants generally agreed that:

> Substantial input and the sharing of both global and local experiences are required so as to better understand the issue of adaptation to climate change. Furthermore, reliable data, information and experiences on this issue is seriously inadequate, if not unavailable altogether.
>
> (UNCTAD, 2012)

This topic has begun to generate substantial interests to scholars, policymakers and industrial practitioners. Nevertheless, it is still in its embryonic stage, and needs collaborative research and the sharing of global and local experiences, from both developed and developing countries and regions. In this context, any efforts on adaptation must not simply be regarded as to 'avert the negative effects' and/

or 'minimize costs'. Quite the contrary, effective adaptation, and the development of resilience, can create new economic, business and social opportunities, thus improving wellbeing for future generations. A good example is the recent warming of the Arctic region, where adaptation may generate considerable opportunities for northern ports, maritime transportation, exploitation of natural resources within the region, and catalyze the economic and social development of the aboriginal communities (e.g. the Inuit in Alaska, Greenland and northern Canada). However, the chance of transforming risks into opportunities would diminish if we continue to employ a 'business as usual' approach *via* incremental adjustments, or simply trying to 'mainstream' identified risks into existing economic, planning and institutional frameworks. New, innovative approaches should be developed not only for minimizing uncertainties and clarifying targets and priorities, but also catalyzing a transformative adaptation process (Moser and Boykoff, 2013). This will allow the threatened areas (e.g. ports and port infrastructures) to develop long-term plans and actions that can be applied into a normative, historically contingent and economically sensible context, and fundamentally enhance capacities and resilience to (diversified) climate change impacts.

Given the likely irreversible trend of climate change and its increasingly serious and visible impacts on different parts of the world, as well as both the international and local nature of adaptation and resilience, publications need to offer substantial appeal to different ports and transportation infrastructures, now and in the near future. With the highly diversified impacts posed by climate change ranging from the damaging power of hurricanes on the port of New York/New Jersey (e.g. Hurricane Katrina in 2012) to the lowering of water level along the port of Montreal and the St. Lawrence-Great Lakes (SLGLS) navigation system, such publications require theoretical and methodological innovation, and real-world case studies from the developed and developing worlds. Understanding such, with contributions from reputable researchers, inter-governmental organizations and senior professionals in transportation and supply chains coming from five continents, this book investigates climate change, adaptation strategies and planning of transportation and supply chains from different angles (academic, professional and geographical), but with a unified focus surrounding ports. By encouraging innovative and effective solutions to develop adaptation measures, it serves as an ideal companion to anyone interested in understanding climate change, its impacts and risks on transportation and supply chain infrastructures, effective ways to adapt to such impacts, and perhaps more importantly, how the global economy and human welfare may benefit from such opportunities.

We believe that this book has achieved three major objectives. First, obviously, it greatly enhances our understanding of ports adapting to climate change risks. As illustrated by the forthcoming chapters, the contributors analyse ports in the context of transportation and supply chains, both theoretically and through empirical evidence. Second, it discusses, elaborates, and in some cases, provides, international/regional best practices aiming to successfully establish long-term resilience and adaptation to climate change risks. Third, it establishes and consolidates an international and sustainable partnership and research consortium to shape the strategic directions to build resilient ports in the face of the climate change challenge, especially in their dynamic relationship with transportation and supply chains. Throughout the process, we requested all contributors (especially academic ones) to 'think outside the box' and, as much as possible, In addition to editorial guidance and suggestions, each of the chapters has been peer-reviewed by at least two relevant experts.

2. Uniqueness and targeted audience

We make no attempts to claim that this book is the first publication to address the issues as stated above. Throughout the past decades, both academic and policy researchers have published research addressing the adaptation of infrastructures, including transportation infrastructures, to the risks and impacts posed by climate change (e.g. Asariotis and Benamara, 2012; Driesen, 2010; Moser and Boykoff, 2013; National Research Council of the US (NRC), 2010; Palutikof et al., 2013; Pappis, 2010). Having said so, we strongly believe that this book is unique, and highly complementary to the aforementioned publications, as few books or monographs directly address ports' adaptation to climate change and its impacts. For instance, Palutikof et al. (2013) largely focused on urban infrastructures, food and water security, while Driesen (2010) addressed how neoliberalism affected the US government in developing climate policy in the past decades. On the other hand, the report by the NRC paid attention to the problem of climate change in general, and offered a theoretical discussion on how the impacts of climate change should be approached and adapted. Others address the issue of 'adaptation of ports to climate change', but most of them only consist of a section/some chapter(s) (and often play backseat roles), indicating that 'ports' adaptation to climate change' stands as an issue that should not be overlooked, rather than highlighting its significance as a topic that must be focused upon. This book is a pioneer that systematically contributes to climate change and adaptation planning, with a focus on the dynamic relationships between ports, transportation and supply chains, while simultaneously paying substantial attention to the 'soft' components (e.g. planning, management, economics, policy). The current volume provides both insightful theoretical and methodological analysis as well as empirical experiences on climate change adaptation from different continents contributed by reputable researchers, policymakers and industrial practitioners from nearly all corners of the globe.

We believe that researchers, especially those in geography, planning, environmental studies, economics, transportation, supply chain management, marine policy and management, will find this book very interesting, as will senior policymakers and industrial practitioners. Indeed, it will appeal to anyone who is interested in climate change, its impacts and risks on transportation and supplychain infrastructures, and the related economic activities that are likely to be affected. It will be a quality companion to policymakers and industrial practitioners who would like to learn about the interrelationship between climate change, transportation and supply chains, and will serve as a solid platform for further research, planning and development of appropriate policies and effective industrial practices in addressing this increasingly critical issue. Finally, with its rich contents, this book serves as an ideal introduction for any postgraduate or undergraduate students who pursue courses and/or research in transportation, supply chains, public policy and regional development (either as complete courses or modules within related courses), e.g. sustainable transportation; supply chain management; climate change, resilience and adaptation planning; port operations, management and policy; climate change adaptation in the built environment; coastal management; transport geography and economics; globalization; environmental management; public policy and regional development; development economics; methodologies in transportation and supply chain research; Arctic development.

3. Structure

The rest of this chapter briefly introduces the contents. This book is divided into three main sections reflecting a balance between the theoretical, methodological and empirical chapters, namely: (1) approaches to the climate change challenge for ports in transportation and supply chains; (2) global experiences and the identification of best practices; and (3) adaptation in the Arctic: melting ice and northern ports, transportation and supply chains.

In the **approaches to the climate change challenge for ports in transportation and supply chains**, the chapters develop theoretical and methodological frameworks for researchers, policymakers and industrial practitioners to develop international best practices in addressing the impacts posed by climate change in the long-term future. A critical analysis provides an understanding of the uncertainties that climate change would pose, while the section further discusses the impacts of climate change, and the ways to address them, from different theoretical angles. It provides constructive, practical solutions to enhance the quality of adaptation planning to climate change. Also, it consists of proposed methodologies to evaluate how transportation and supply chain stakeholders perceive climate change risks under financial constraints, how to assess the effectiveness of adaptation planning, as well as the physical and engineering solutions in adapting to the impacts posed by climate change on ports and supply chains.

In Chapter 2, Messner, Becker and Ng consider the roles of adaptation practitioners and institutions in ports' adaptation planning, with a focus on stakeholder dynamics and approaches. They argue that effective planning processes depend on stakeholder input, effective management of the inputs and guidance from practitioners. Also, they offer recommendations and examples for successful adaptation planning at ports, notably the use of the 'Theory of Change' as an alternative to the current planning paradigm. Chapter 3 is a methodological chapter with Yang, Cahoon, Chen, Ng and Becker building a fuzzy logic method to assess stakeholders' perception of climate change impacts to ports. They attempt to introduce a reliable, quantitative assessment tool to help policymakers and industrial practitioners to develop effective adaptation strategies, plans and actions.

Adaptation to and planning for climate change require global, regional and local perspectives. The next section, **global experiences and the identification of best practices**, bridges these perspectives by providing empirical research from around the world. The chapters consist of the experiences of adaptation strategies, planning and actions to climate change impacts from both developed and developing countries and regions, in Asia, Europe, North America and Oceania.

In Chapter 4, Slack and Comtois study the impacts of climate change on the SLGLS navigation system in Canada. It is a good example reflecting the diversified nature of climate change impacts, where the lowering of water level (rather than rising water level as experienced by some seaports) poses substantial challenges to SLGLS and the ports located within the system (e.g. port of Montreal). They offer insight on the adaptation measures that are being considered to ensure the commercial viability of ports and shipping on the system, and discuss the costs and operational effectiveness of the adaptation measures in place. In Chapter 5, Wang, Samsom, Ng and Earl focus on CentrePort Canada (the major inland port in Manitoba) and the Hudson Bay Railway (connecting the inland areas of Saskatchewan and Manitoba with the port of Churchill along Hudson Bay) in the province of Manitoba in Canada. They investigate the adaptation planning process of these facilities, and the major barriers tackled during the process. The chapter strengthens the close interaction between ports, logistics and supply chain systems as demonstrated by Ng and Liu (2014). In Chapter 6, Smythe investigates the recovery process of the port of New York/New Jersey after the damage done to port infrastructures by Hurricane Sandy in 2012. She reports the key lessons learned, including the successes and challenges, and argues that future port resilience research should respond to port stakeholders' needs for decision support and planning tools.

Turning our attention to Europe, in Chapter 7 Osthorst reviews a recent adaptation project that has taken place in the port of Bremerhaven in Germany, called the 'Northwest 2050'. By analyzing the port's vulnerability, and identifying adaptation demands of different segments of the port sector's value chains, he argues that adaptation can take different forms that are dependent on strategic responses to central challenges. He emphasizes that adaptation to climate change impacts by ports requires policy integration, and should link port development issues to the issue of sustainability. In Chapter 8, Cusano, Ferrari and Tei discuss how climate change is perceived with respect to the future development of Italian ports. They probe the discrepancy between the 'publicized' and 'actual' interventions, and how national legislation reduces the attractiveness for private port terminal operators to invest in adaptation measures. They also argue that there are few policy tools that local port authorities can use, and a general lack of integrated, long-term strategies or plans for the issue.

Given the economic and political influences of Asia nowadays and in the foreseeable future, the adaptation plans and actions of Asian ports cannot be understated. In Chapter 9, Esteban, Takagi and Shibayama describe the significant consequences of a potential increase in typhoon strength and sea level rise to Japanese port infrastructures, with emphasis on the ports along Tokyo Bay. Japan is regularly prone to tropical cyclones, some of which cause widespread damage, damaging breakwaters and other port installations, disrupting operations and having substantial effects on its economy. They conduct an economic analysis of climate change impacts on the port areas in Tokyo Bay, and strive to quantify the magnitude of the problem. In Chapter 10, Zhang, Wan and Yang study the deficiencies of green port development in China through developing an assessment model in measuring green port development, and applying it to the port of Tianjin. They discuss some corresponding countermeasures so as to improve the ability of ports to cope with climate change impacts.

In this case, Latin America is equally important, especially with the recent expansion of the Panama Canal that would no doubt pose considerable impacts on maritime transportation and supply chains. In Chapter 11, Stenek, Amado, Connell, Palin, Wright, Pope, Hunter, McGregor, Morgan, Staley, Washington, Liverman, Sherwin, Kapelus, Andrade and Pabón explore the major risks and opportunities for the port sector, and examines specific implications on Terminal Maritimo Muelles El Bosque (MEB) in Cartagena, Colombia. They found that MEB is generally in a relatively good position (compared to nearby ports) in adapting to climate change impacts. However, they also identify materially significant impacts in timeframes that are relevant for investment decisions, and propose adaptation investments that should be incorporated in the MEB's financial model. In Chapter 12, Acciaro gives an account on the climate change assessment risk for the Panama Canal, and illustrates how such risk can be mitigated through adequate adaptation plans. He discusses various climate change related risks in the Panama Canal and quantifies the economic loss that would result from the unlikely event of the canal's closure. The results suggest that climate change poses potentially a huge challenge to the Panama Canal area, and as such might affect the Panamanian economy and the efficiency of global supply chains.

The global coverage of this book is completed with two chapters focusing on the Southern Hemisphere. In Chapter 13, Cahoon, Chen, Brooks and Smith explore the climate change risks to which Australian ports are exposed, and discuss existing and potential adaptation strategies used by them. They investigate how climate change may impact on a coal supply chain from the coalmines through to the port terminal, and highlight the importance of a coordinated approach so as to adapt to challenges posed by climate change. In Chapter 14, McEvoy, Mullett, Trundle, Hunting, Kong and Setunge report on the conceptual framing of a USAID-funded decision support toolkit tailored to the port authorities of Pacific island states (with them focusing on Suva and Port Moresby in Fiji and Papua New Guinea, respectively), as well as its development and contents. They argue that the knowledge gained from the toolkit showcases important lessons and guidance for others seeking to translate climate information for more informed adaptation planning not only for ports, but also for other critical infrastructure assets.

The last section is a special section on **the adaptation in the Arctic: melting ice and northern ports, transportation and supply chains**. The melting of ice routes will certainly impact maritime transportation and supply chains in the

Arctic. As stated earlier, adaptation to climate change impacts could generate new opportunities, and this section illustrates how the global society might benefit from climate change through effective adaptation plans and actions. The chapters are contributed by several speakers who have participated in a conference entitled *The Warming of the North: Implications for Arctic Transportation, Supply Chain Management, and Economic Development*, organized by Adolf K. Y. Ng, Paul Earl and several other members and institutions of the University of Manitoba (held in Ottawa, Canada; 2–3 March 2015). Most of the expenses of the conference, and thus the chapters in this section, are funded by the Social Sciences and Humanities Research Council of Canada (SSHRC)'s Connection Grants (43620).1

In Chapter 15, Ircha and Higginbotham examine the capability of the Northwest Passage (NWP) in supporting increased shipping, including ports, ship technology, search-and-rescue capability, navigational aids, and governance and regulatory concerns. Despite the melting ice, they point out the many factors that militate against any real growth of shipping along the NWP in the foreseeable future. In this case, they argue that Canada's northern strategy must be renewed so as to clarify and consolidate the many initiatives being undertaken by federal and territorial departments, agencies and private-sector partners involved in the Arctic's maritime affairs. They also argue that Canada should play more proactive roles in the Arctic Council so as to maintain its focus on responsible marine, resource, economic and community development, and promote shared goals among the Arctic states. In Chapter 16, Verny focuses on the performance of Russian ports along the Northeast Passage (NEP). He notes the problem of navigability along the NEP as a fundamental limitation on the development of ports located in this area. In this context, he notes that the extent to which the Russian authorities could adopt a combination of a proactive policy and public–private partnership, with foreign investments to be pivotal in deciding the future of Russian Arctic ports.

Finally, the volume ends with Chapter 17 summarizing the main messages brought out by different chapters, and contextualizing the contributions of this book within the broader framework of adaptation planning. The various chapters herein serve as a useful barometer to help gauging how the global port community is working towards the long-term goals of resilience and adaptation, from problem identification, to strategy implementation and revision, and ultimately to transform challenges into opportunities. More importantly, understanding the complexity of ports and adaptation to climate change, we should put much more emphasis on the criticality of a partnership approach between researchers, policymakers and industrial practitioners in solving this problem successfully. This view complements Xiao et al. (2015)'s latest findings who argue that collaborations and partnerships between stakeholders become increasingly important when the risks posed by climate change to ports and port infrastructures become more visible and explicit in the near future, of which we believe that it is already taking place.

Note

1 Other members and institutions involved in organizing the conference include R. McLachlin, A. Phillips, E. Tyrchniewicz, D. Duval, K. Chmelnytzki, S. VanDeKeere, Department of Supply Chain Management of the Asper School of Business of the University of Manitoba, and the University of Manitoba Transport Institute.

References

Asariotis, R. and Benamara, H. (Eds.) (2012). *Maritime Transport and the Climate Change Challenge*. New York, NY: Earthscan.

Driesen, D. M. (Ed.) (2010). *Economic Thought and US Climate Change Policy*. Cambridge, MA: MIT Press.

Moser, S. and Boykoff, M. T. (2013). 'Climate change and adaptation success'. In: Idem (Eds.) *Successful Adaptation to Climate Change: Linking Science and Policy in a Rapidly Changing World*. Abingdon: Routledge, pp. 1–33.

National Research Council of the US. (2010). *Adapting to the Impacts of Climate Change*. Washington, DC: National Academies Press.

Ng, A.K.Y. and Liu, J. J. (2014). *Port-Focal Logistics and Global Supply Chains*. Basingstoke: Palgrave Macmillan.

Palutikof, J., Boulter, S. L., Ash, A. J., Smith, M. S., Parry, M., Waschka, M. and Guitart, D. (Eds.) (2013). *Climate Adaptation Futures*. Hoboken, NJ: Wiley-Blackwell.

Pappis, C. P. (2010). *Climate Change, Supply Chain Management and Enterprise Adaptation: Implications of Global Warming on the Economy*. Hershey: IGI Global.

UNCTAD. (2012). *Ad Hoc Expert Meeting on Climate Change Impacts and Adaptation: A Challenge for Global Ports – Main Outcomes and Summary of Discussions*. Geneva, Switzerland: UNCTAD (ref. UNCTAD/DTL/TLB/2011/3).

Xiao, Y., Fu, X., Ng, A.K.Y. and Zhang, A. (2015). 'Port investments on coastal and marine disasters prevention: economic modeling and implications'. *Transportation Research Part B: Methodological* 78: 202–221.

2 Port adaptation for climate change

The roles of stakeholders and the planning process

Steven Messner, Austin Becker and Adolf K. Y. Ng

1. Introduction

As stewards for many activities and resources, port authorities will face challenging decisions on how to manage risks, such as increases in sea level rise (SLR) and flooding, resulting from global climate change. In particular, they must consider the concerns and objectives of the wide variety of stakeholders who depend on the port's functionality. Indeed, given the importance of ports in the global economy and supply chains (Ng and Liu, 2014), the development of effective solutions, including local case experiences and their application to other ports around the world (UNCTAD, 2012), to enable ports to adapt to climate-change risks effectively is a top priority.

This chapter considers the roles of port stakeholders and institutions[1] in adaptation planning for ports through an investigation of the adaptation planning process undertaken at the port of San Diego (PSD) in the USA, with a focus on stakeholder dynamics.[2] Through document reviews, semi-structured interviews with stakeholders, and first-hand experiences of the authors,[3] the chapter provides a background on the roles of stakeholders relevant to adaptation planning at ports, as well as explaining the process of an adaptation plan's development. It argues that effective planning processes depend on stakeholder input, effective management of the input, and guidance from adaptation practitioners.[4] Also, it offers recommendations and examples for successful adaptation planning at ports, notably the application of the Theory of Change (ToC) (Bours, 2014) as an alternative to the current planning paradigm. The rest of the chapter is structured as follows. Section 2 provides a theoretical discussion on climate change and port's adaptation. It is followed by the case study in Section 3. Finally, the conclusion can be found in Section 4.

2. Climate change and adaptation at ports

2.1 The port as an institution in the context of climate change

The primary function of a port has traditionally been the transfer of cargo and/or passengers between a waterway and the shore (Talley, 2009), but today's ports are more than a system of ship channels, wharves and connections to truck and train transport. Centralized authorities or private operators generally serve as port managers. Whether publicly or privately operated, the port authority or management agency carries out institutional planning functions for development projects. In addition to their vital role in global maritime commerce, port authorities often oversee adjacent or interrelated activities, such as airports, bridges, tunnels, industrial parks, natural lands, public marinas and public recreational facilities (beaches, parks). Ports thus fulfill an expanded role as a cultural element of the community, embedded within and held accountable for the goals of a larger society (Messner et al., 2013; Burroughs, 2005). As stewards for many or all of these activities, port authorities will face challenging decisions to manage increasing risks as SLR advances and storm intensities increase with global warming (IPCC, 2013), threatening port properties, operations and critical utilities not previously impacted. For example, decisions to protect existing port operations from SLR could have adverse effects on adjacent beaches and tourism income. Similarly, adjacent natural lands that support wildlife, provide beneficial ecosystem functions, and serve to dampen the effects of storms, could become entirely inundated without a coordinated planning effort to manage coastal resiliency at a regional level.

2.2 Port stakeholders

The major stakeholders of the port include the port operator (sometimes a public port authority) and the firms engaged in the transfer of cargo or passengers. However, a broader set of stakeholders share a variety of goals and missions with respect to the long-term functioning of the port, including business success, facilitating trade, economic growth, and public goods and services. Adaptation to risks posed by climate change and building resilience to storm events at ports will most likely be carried out by those individuals and groups who will be directly affected (i.e. at risk), and who may assume responsibility for implementing and sustaining the adaptation measures over time (NRC, 2010). This more limited group of port stakeholders is referred to here as the *port stakeholder cluster*.[5]

The *port stakeholder cluster* (de Langen, 2004) may be divided into two major categories: *internal* and *external* stakeholders. Stakeholders that are part of the port management organization (e.g. the port operator, shareholders, managers and employees) may be considered as *internal stakeholders*. A diverse array of actors and organizations fall into the broader category of *external stakeholders*. These include *economic/contractual stakeholders* that are involved in certain port operations such as stevedoring companies, shipping agencies, insurers, ship repairers, port tenants and others. *Public policy stakeholders* include government agencies responsible for transport and economic affairs, as well as environmental agencies, planning departments, and emergency management agencies. *Community/environmental stakeholders* consist of community groups, neighboring residents, general (tax-paying) public, environmental groups and others. *Academic and technical consultants* can also play roles in port planning development, though they may not be stakeholders in the true sense of the word. These may include academic or consulting organizations or non-governmental groups that conduct independent work and offer knowledge of climate-change science, the interpretation and presentation of that science, and interaction techniques with diverse stakeholder groups. There is increasing use of the term 'adaptation practitioners' (Bours et al., 2013) to recognize individuals who have developed specialized knowledge in this emerging field. This chapter uses this term (adaptation practitioners) to describe individuals who have assisted work on adaptation planning, and considers their role along with community and port cluster stakeholders. Indeed, the first author of this chapter served as an adaptation practitioner in his consultant role for PSD's adaptation planning process. This will be further discussed in the next section.

3. Case study: Port of San Diego

3.1 The development of an adaptation plan

In 2010, PSD began development of a stakeholder-inclusive adaptation planning process in conjunction with a greenhouse gas (GHG) emission-reduction plan that addressed all of the properties in their jurisdiction around San Diego Bay. Like most of the ports in the State of California, the port authority holds authority for planning on parcels bordering the water. In San Diego, activities on these parcels include commercial activities (e.g. shipping, hotels, restaurants, commercial fishing), public recreational use (e.g. shoreline parks, boating), and natural habitat.

The PSD's adaptation planning process aimed to identify, assess and develop strategies to help both reduce GHG emissions and address local vulnerabilities to climate change (Hooven et al., 2011). The adaptation plan addressed the need to better understand climate change impacts and prioritized key vulnerabilities within port tidelands that will require adaptation strategies in order to retain operational, recreational and natural resources functions at the port. The California Adaptation Strategy (State of California, 2010), an ongoing collaboration with the *Local Governments for Sustainability USA Sea Level Rise Adaptation Strategy* (ICLEI, 2012), and the California SLR Interim Guidance (CO-CAT, 2010) all informed the planning process. Based on its location and activities, the process found SLR to be the primary impact of climate change expected to affect PSD.

A four-step assessment methodology was used to facilitate the process, described in the following section.[6]

Step 1: Develop GIS model to map SLR. To help port stakeholders visualize the range of impacts of SLR on the future activities at the port, PSD's consulting team, an environmental planning consulting firm,[7] collected geographic information system (GIS) data to address the impact of SLR for areas within PSD's jurisdiction and in San Diego Bay. These datasets included location-specific information on PSD's jurisdictional boundaries, port-specific land use types, coastal and marine habitats and protected areas, the ten geographic planning areas used in PSD's Master Plan (referred to hereafter as Planning Districts), the stormwater conveyance system, elevation monuments (control points), LIDAR-derived elevation contours, San Diego Bay bathymetry and aerial photography. Using the GIS model the consulting team then calculated areas potentially subjected to inundation under each SLR scenario for all port lands in the GIS using the predicted SLR elevation (Figure 2.1).

Projected inundation levels in 2050 (2.85m above global mean sea level) and in 2100 (3.85m above global mean sea level) were consistent with available science and local SLR projections (Gersberg, 2010; CO-CAT, 2010) and included a 100-year flood exceedance. The model included recorded local events such as storm surges, wave run-up, El Niño events and circulation patterns.

Step 2: Assess likelihood of SLR inundation levels in years 2050 and 2100. The 'likelihood' of inundation represents the susceptibility of port functions to flooding as projected in the 2050 and 2100 SLR scenarios. For this analysis, 'likelihood' did not measure the probability that an area as a whole would be inundated in 2050 or 2100. Rather, 'likelihood' measured the extent of each port function (commercial, safety, recreation, habitat) inundated by a SLR scenario within a specific Planning District. This definition of likelihood was based on an assumption that the greater the inundation extent for each port function determined in the GIS analysis, the more likely that port function will be impacted on a regular basis.

Step 3: Assessing the consequences of SLR. The 'consequences' of inundation represent the severity of impacts resulting from SLR in the 2050 and 2100 scenarios. The consulting team then developed impact-severity determinations.

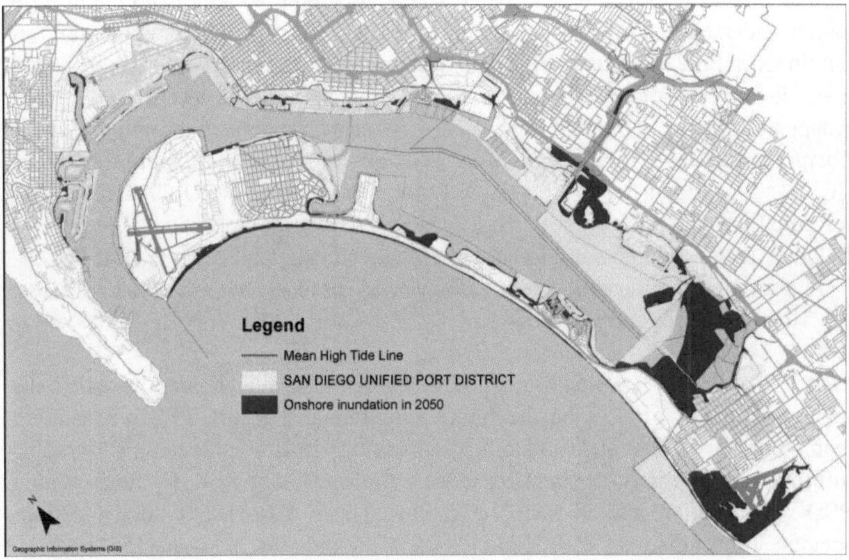

Figure 2.1 GIS model – example 2050 scenario.
Source: PSD public workshop presentation, February 2012

The consulting team's experts all had adaptation-planning expertise: one had more experience with port infrastructure, one with climate science, and one with natural-resource management. Consequence ratings (ranging from 1 to 5) were assigned by comparing the port function's percentage of inundation within each Planning District to the percentage of a port function's inundation port-wide. Therefore, a higher consequence rating was assigned to those port functions impacted by SLR to a greater extent both within the Planning District and port-wide. Conversely, those port functions that may be impacted to a great extent in a single planning district but less so port-wide were assigned a relatively lower consequence rating.

The three experts determined consequence ratings individually, then results were compared. There was a high degree of agreement in the individual rankings, but in cases of divergence, the team discussed the reasons further to find consensus.

Step 4: Determine overall risk to port functions. The risk assessment process described in Steps 1–3 was performed for each of the ten Planning Districts PSD uses around the San Diego Bay. Risk summary tables were developed for each Planning District to incorporate the results of the GIS analysis identifying the percentage of each port function (commercial, safety, recreation, and habitat) potentially inundated. In addition, the tables provide details of the land uses, habitats, roads and infrastructure that might be affected by SLR. Finally, PSD and its consulting team developed an initial list of adaptation strategies developed for each port function to respond to inundation.

The risk levels within individual planning districts increase between 2050 and 2100 as SLR increases. Planning districts with high and very high risk in 2050 support mainly commercial activity and natural habitat. The risk to commercial activities is mainly associated with water-based uses such as marinas, berthing areas along marine terminals, waterfront businesses and/or operations with infrastructure immediately adjacent to the shore. Habitats most at risk in 2050 include eelgrass and beach dunes. In 2100, increased inundation associated with SLR was predicted to be widespread throughout the port's jurisdiction and in and around San Diego Bay.

3.2 Stakeholders involved in the adaptation plan

PSD made a significant effort to develop stakeholder input groups and cultivate input during Steps 1–4. Between 2010 and 2013, at least seven stakeholder meetings related to the development of the adaptation plan had been held, with opportunities for input at each meeting. Three of these meetings were focused on scoping and developing the GIS model in Step 1. The remaining four meetings were distributed between Steps 2–4. Since the adaptation planning effort encompassed most parcels around the San Diego Bay (excluding the US Navy's operations), planning representatives from the five cities surrounding the Bay provided input to the process. Figure 2.2 shows a more complete list of stakeholders to the process.

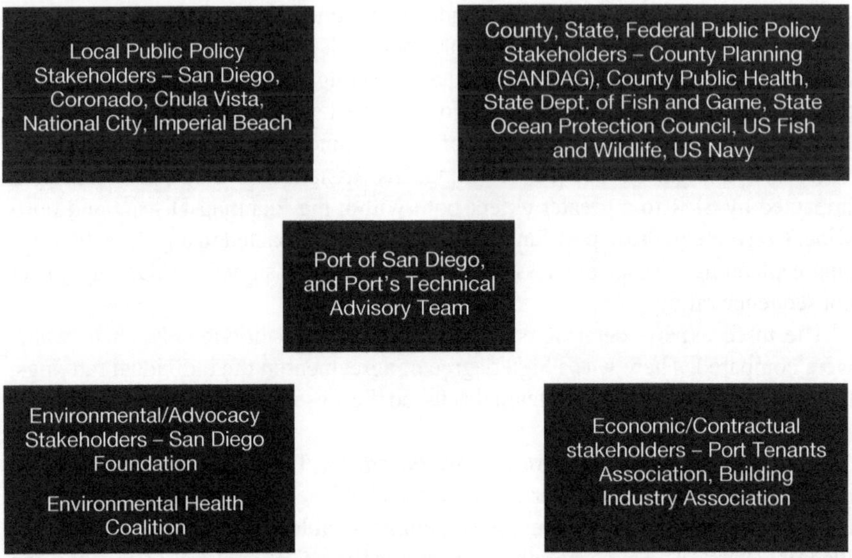

Figure 2.2 PSD adaptation process stakeholders.
Source: Authors

The most active stakeholders included several (but not all) of the cities, the Port Tenants Association, and the local environmental groups. Several comment letters from the Port Tenants Association, who perceived business risks associated with both inaction and overreaction, and letters from the Environmental Health Coalition, with perceived risks to their community from inaction, delayed the planning process for weeks or months at a time while the issues were incorporated to the extent possible in the adaptation plan.

Another active member of the advisory group was the San Diego Foundation, a local non-profit community development organization that has been involved with regional adaptation planning since 2005. It initiated the earliest adaptation planning studies in the county and has been a very effective facilitator of communications between institutional agents within San Diego County, including the port.

3.3 The influences of practitioners on the process of adaptation planning

A technical advisory team, consisting of researchers from the University of California at San Diego (UCSD) – Scripps and San Diego State University and the environmental consulting team led by Messner, presented specialized information needed to carry out a risk assessment involving different property types (commercial, public use, natural). Thus, the technical advisors, called adaptation practitioners here, influenced the process, but did not act as 'stakeholders', as they had no direct stake in specific outcomes of the process, e.g. whether the outcomes were considered more business friendly, resident friendly, or environmentally friendly. However, adaptation practitioners do have an interest in a successful planning process being executed and being recognized for their work in facilitating the project.

The following is an example of how influence occurred when business and environmental stakeholders disagreed over the risk assessment procedures and how risk levels were defined. Environmental groups presented a novel evaluation framework that would have skewed the risk assessment of natural habitats so that almost all impacts identified would fall into the 'very high' risk category, thus warranting the implementation of specific protection strategies. To resolve this dispute, the environmental consulting team developed a risk evaluation matrix with a strong basis on previous completed studies to handle the risk evaluation more objectively. Figure 2.3 gives an example of the matrix used to evaluate all the impacts.

Another example of adaptation practitioner influence in the adaptation planning process occurred when disagreement arose over how the consultants chose the inundation projections to use in planning for 2050 and 2100. Some stakeholders favored a 'worst case' scenario projection that included the maximum annual high tide levels, coupled with a very severe storm surge level from the recent past, along with a higher-end estimate of overall SLR in 2050 and 2100. Local environmental activists, including the Environmental Health Coalition, expressed concern that risks would be generally underplayed, given uncertainty in SLR projections and future storm events (CO-CAT, 2010). Although some level of conservatism in the evaluation of inundation is usually employed, the technical advisory team first developed the maximum surge plus tide height from historical records, and then obtained the latest research on anticipated storm surge increase with climate change. This approach satisfied both the environmental activists and the business community (notably the Port Tenants Association) who were initially concerned about portraying high degrees of risk when considerable uncertainty over the range of SLR still exists.

The successful adaptation practitioner will also need to understand that the adaptation planning process is new and so not well-defined. Expectations from clients are often quite high and, combined with limited budgets, there is a premium on communication of concrete and realistic expectations from the consultant to the client and stakeholders. These examples provide evidence that adaptation

		CONSEQUENCE				
		1	2	3	4	5
LIKELIHOOD	5	Medium	High	Very high	Very high	Very high
	4	Medium	Medium	High	Very high	Very high
	3	Low	Medium	Medium	High	Very high
	2	Low	Low	Medium	Medium	High
	1	N/A	Low	Low	Medium	Medium

Figure 2.3 SLR inundation risk rating (risk = likelihood × consequence).

Source: Messner et al. (2013)

practitioners can influence a decision-making process and thus must be careful to behave impartially, though subjectivity in decision making can rarely be entirely eliminated. For example, in the above risk analysis table, experts will have different opinions on the consequence of an impact, e.g. whether it is a 'two' or a 'three'. This subjectivity can be mitigated through process transparency and good documentation, but not eliminated.

3.4 Challenges in adopting the adaptation plan

To further explore stakeholders' perception on the adaptation planning process, the authors conducted semi-structured, in-depth interviews with ten relevant PSD personnel and stakeholders (hereinafter called 'interviewees') in late 2012. The interviewees included consultants, environmental interest groups, the institutional agents within the County of San Diego and the San Diego Foundation. They were queried on the overall impression of the planning process; the uniqueness of climate adaptation planning; the public outreach and accountability components of the process; the nature, organization and stakeholders involved; the existing conditions; PSD's vulnerabilities to SLR; risk analysis, goals and the prioritization for actions; and the establishment of adaptation strategies.[8]

To date, the adaptation portion of PSD's climate plan had not been adopted, in contrast to the emission reduction plan, which PSD adopted in December 2013. The mitigation plan called for sixteen specific emission reduction measures, a majority of which would have some negative financial impact on port tenant operations. The emission reduction plan received considerable input from the local environmental stakeholders as well as from port tenants. Often the differences were pronounced with environmental stakeholders pushing for faster action and measurable results. PSD's environmental planning consultant helped to resolve a number of these disputes by applying their expertise with a number of cost-effectiveness studies[9] and proposing less controversial strategies that achieved similar emission reductions across the broad mix of port businesses – maritime, surface transportation, buildings, and stationary combustion sources.

In contrast, the (not-yet-adopted) adaptation plan included about 50 measures and strategies. Many of these did not include quantifiable outcomes, but rather called for clarification of uncertainties, such as further research or developing more goals, strategies and plans. In addition, some measures overlapped with emission reduction measures and adaptation occurred as a co-benefit (e.g. the use of trees for shading and cooling [adaptation] as well as reducing energy use and increasing carbon sequestration [mitigation], energy efficiency [mitigation] to also reduce peak load stress during heat waves [adaptation]).

Not surprisingly, in the absence of any regulations, the interviewees generally had few concrete ideas about how the adaptation strategies would/should be implemented, or how implementation would/should be tracked and monitored. Some noted that, apart from measures aimed at public parks that would be PSD's responsibilities, much of the implementation would be left to other port stakeholders (notably port tenants – who pay leases to the port which serves as

PSD's major income). They reported that port tenants were mainly concerned with the financial implications of adaptation, and thus largely maintained a wait-and-see attitude. Expecting to share the cost of adaptation, these (private) stakeholders exerted considerable pressure on the port to empower them to decide how to define and implement adaptation strategies. Thus, planners hesitated to assign more direct responsibilities. Quoting an interviewee, the broad list of options allowed 'maximum flexibility' for port stakeholders to carry out strategies.

A factor that explains the adoption of a controversial GHG emission reduction plan, but the delay in adopting the adaptation plan, is a lack of perceived urgency among some key stakeholders around adaptation. Environmental groups and local residents who are the most vulnerable (i.e. in known flooding areas or near the shore) often see the climate problem with a higher sense of urgency and feel that actions are warranted sooner rather than later. However, business groups and residents with lower perceived risk tended to be more concerned with their economic viability over five-year planning periods, and less concerned about issues beyond this timeframe.

At the time the PSD adaptation plan was suspended in early 2013, there was considerable guidance, but no requirements from the state or federal governments concerning adaptation planning. In contrast, the State of California adopted many regulations concerning GHG emission reductions and these were often cited as the justification in moving forward the port's GHG emission-reduction plan.

As is the case with many port authorities, the PSD has the power (or shares the power) to adopt requirements on their tenants pertaining to land use, environmental protection, and economic incentives. In the adoption of their GHG mitigation plan, the port and the environmental consultants frequently relied on specific or parallel requirements from the State of California. However, requirements (as opposed to guidance) from the state could not be cited in the adaptation planning process and, as mentioned, eventually the planning was suspended.

Since the suspension of the adaptation planning process in early 2013, the state, later in the same year, adopted requirements for the first time for port authorities to develop adaptation plans responding to the threats from SLR and flooding.[10] The plans to the state are due in 2019. The PSD adaptation plan (or at least part of it) could perhaps be used eventually to fulfill the new requirements, but as of the beginning of 2015, the PSD adaptation planning process has not been formally reactivated.

3.5 An alternative adaptation-planning paradigm

The PSD case provides a good example of a well-researched planning effort with notable efforts made to obtain stakeholder input. Ultimately, the adaptation planning process was suspended. According to interviews at the time, stakeholders were unclear on fundamental planning issues such as how this plan would be implemented, who would actually implement it, and who would pay the accompanied costs. In addition, local residents viewed climate change more in terms of the long-term impacts to their lives, whereas businesses focused on their viability over the next several years. The new requirement from the State of California to prepare a plan by 2019 (see Section 3.4) will ensure that some kind of adaptation plan is eventually adopted; notwithstanding the uncertainties and different perspectives among stakeholders that led to the suspension of the plan.

To address these uncertainties and differing perspectives, some researchers suggest the Theory of Change (ToC) as an alternative approach to designing and evaluating climate change adaptation plans (Bours et al., 2014), international development programs (Valters, 2014) and other complex planning efforts (James, 2011). Anderson defined ToC as 'an explanation of how a group of stakeholders expects to reach a commonly understood long-term goal' (Anderson, 2005, p. 3).

ToC processes first develop a desired outcome from process stakeholders, and then reverse engineer a plan to achieve it. This contrasts with the forward-looking process outlined in the four steps discussed above. Instead of diving into the scientific details of the SLR mapping and risk assessment from the outset, as the port's consulting team did, stakeholders in a ToC process start by defining a sufficiently detailed long-term goal or goals, and work backwards in time up to the present, laying out each step in a single or multiple decision pathway (the 'causal' pathway). For each step, stakeholders outline intermediate outcomes, quantifiable metrics, 'indicators' and assumptions. The pathways are depicted in a diagram called a 'change map' (Figure 2.4), accompanied by a narrative. ToC is designed to be an iterative process; in other words, the strategy should be reviewed regularly and modified to reflect emerging conditions and new knowledge.

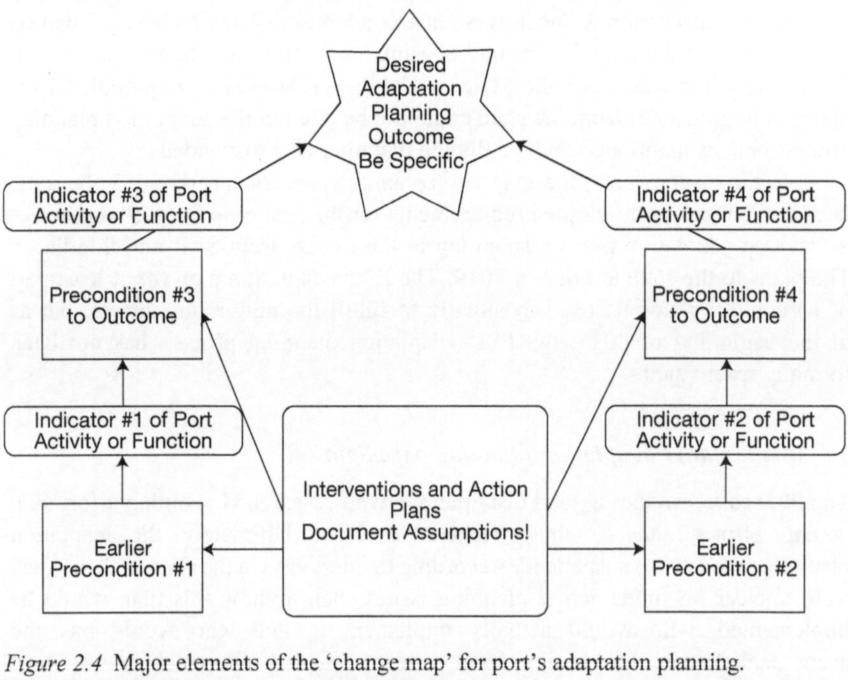

Figure 2.4 Major elements of the 'change map' for port's adaptation planning.
Source: revised format based on Anderson (2005)

Key steps (called tasks) in adaptation planning using the ToC approach are as follows.

Task 1: Identify the goal or 'long-term outcome'. This should be a clear and specific statement of what the group is working toward. A very high-level statement (e.g. 'improved resilience to SLR') will probably not be useful since the backward mapping process could be too ambiguous to get started. The long-term goal should be clear enough to initiate a reverse planning process and offer some direction for adaptation strategy prioritization. For example something like, 'preserving economic vitality of the port while minimizing impacts on local residents' could provide a starting point for the ToC process. The involvement of community groups should take place within this early stage. In the PSD adaptation planning example, this task was included in Step 1, but only briefly as the ICLEI (2012) SLR adaptation effort for San Diego Bay was in progress at the time and discussions of goals had occurred to some extent in that process.

Task 2: Develop a pathway of change (the 'change map'). This step involves identifying and sorting 'preconditions related to the ultimate outcome of interest into a pathway of change that moves linearly and chronologically toward the long-term goal' (Anderson, 2005, p.12). As noted above, the pathway of change is designed for backwards mapping. Stakeholders identify specific sub-steps or intermediate outcomes along the way. For each intermediate outcome, stakeholders consider the necessary conditions that need to exist (the 'preconditions') for this outcome to be realized.

In the PSD planning example, this task was largely left for the fourth and final step after risks were fully evaluated and strategies developed. This highlights a fundamental difference between forward-looking, linear planning and ToC. In ToC, a good amount of time is spent on mapping out planning steps before the individual, detailed strategies are developed. One of the shortcomings in the port's consulting team was a lack of experience in alternative planning processes that could have redirected the approach to something like ToC in the planning process when stakeholder resistance was first experienced and to have served as a focal point for further stakeholder consensus building. One of the advantages of this approach versus the PSD planning example, is that stakeholders get an earlier sense of implementation challenges, and how those can be overcome and by whom. When waiting until the final step, as in the PSD example, to consider implementation, more time to develop a consensus on issues such as implementation responsibilities and cost sharing can be spent.

Task 3: Operationalize outcomes. For this task, stakeholders identify 'indicators' or milestones against which progress can be measured. This involves selecting the indicators and 'thresholds' that can describe or measure progress along the decision pathway. Indicators are not necessarily the specific metrics used to measure progress; rather they are the key concepts underlying the milestones along the pathway towards achieving the long-term goal. Thresholds are related to the indicators and identify the point at which an outcome is met and could, for instance, be a certain percentage of properties being protected against sea-level rise in a specified manner, port operational efficiency, to name but a few. In contrast, the PSD and other forward-looking planning processes start by looking at details like metrics of SLR adaptation (cost impacts, acres potentially flooded, etc.) and move directly to action plans after metrics are explored.

Task 4: Define interventions. At this point stakeholders and practitioners can start to plan activities, i.e. to develop action plans. These may be broad-ranging, from concrete strategies such as building structures differently, to changes in insurance coverage, new planning horizons, or other options.[11]

Task 5: Articulate assumptions. Clearly laying out the assumptions used to develop the change map is critical. Since ToC is designed to be an iterative process involving future updates, clearly specifying the assumptions will be essential in developing effective updates. Anderson (2005) asserted that assumptions should address questions that come up when a theory of change is being critiqued. For example, if certain stakeholder groups did not participate in the initial mapping exercise, clearly note why, if they were invited, etc. As another example, if business groups are given a wide range of options for adaptation strategies, clearly note why the group believed the flexibility was warranted (e.g. the uncertainty of future impacts).

Valters (2014) reviewed potential problems with ToC and cites 'a considerable danger that the approach will privilege a linear cause and effect narrative of change'. Rather than expanding discussion of long-term outcomes and the uncertainties involved in getting there, it can become a more complex version of a linear planning effort.

ToC stresses an emphasis on establishing the desired outcomes first. The consulting team, lacking background in alternative planning processes, was not able to help the port recognize the importance of this. In contrast, a forward-looking plan was attempted from the start in the PSD process, with many (50), often complex, strategies developed to potentially defend against any significant risk. The risk matrix developed for the strategies by the consulting team helped to prioritize thinking, but did not ultimately map out when certain actions should be taken in the future. ToC, if implemented carefully, could potentially overcome the stated obstacles.

Conclusion

Regardless of the planning approach taken, institutions such as port authorities will be challenged to deal with the longer-term and more uncertain nature of adaptation planning and will need to often 'break out' of their existing planning paradigms to be successful. Adaptation practitioners with specialized knowledge in helping institutions manage these uncertainties can add much needed assistance towards successful planning.

Despite a rather well-researched adaptation plan with considerable technical expertise from the consulting team applied, and despite efforts made to encourage stakeholder input from a cross-section of the communities affected by the port, the adaptation component of the PSD plan stalled, at least for the time being. Reasons include a lack of urgency among some stakeholders, lack of upfront goal setting and consensus building, and the lack of a focal point that could effectively hold the strings together and forge consensus building. Instead of the forward-looking, linear planning process in the PSD example, the reverse planning steps in the ToC could be used to overcome a number of the obstacles with emphasis on preliminary process mapping. Early process mapping could inform the adaptation planning effort with a clearer focal point and clearer roles for stakeholders. The iterative nature of ToC allows for remapping over time when key preconditions and thresholds are not met, rather than invalidating the entire plan. In contrast, forward-looking, linear path planning is often confronted by plan invalidation when key economic (e.g. recession) or political change occurs.

Adaptation at ports will likely include the adoption of protective measures before it is too late, but not before they are needed, as in the case of raising piers and wharves to accommodate a working range for loading and unloading of cargo, passengers, etc. Intermediate milestones (e.g. various triggering SLR heights), inherent to the ToC process, will be much more important to define for SLR planning at ports, in contrast to planning for increased heat which can typically be done well in advance without regrets. ToC offers one potential way to revise the current planning paradigm in institutions to address some of the issues encountered by the PSD adaptation planning process.

Notes

1 As discussed later in the chapter, port authorities typically carry out institutional planning functions.
2 A portion of this chapter is adapted from Becker et al. (2014).
3 Defined later in the chapter as technical advisors to the adaptation planning process with specific expertise in the field. This expertise can come from academia (i.e. SLR, precipitation forecasts, public health studies), consultants, or a specialist within the institution.
4 For example, Messner actually led the consulting team in the development of the Port of San Diego's climate adaptation plan.
5 For more details on stakeholders, see Becker and Caldwell (forthcoming).
6 A detailed description of this process can be found in Messner et al. (2013).
7 Messner led this consulting team. See Section 2.
8 Confidentiality was guaranteed to all the interviewees. Their identities, and any linkage between comments and identities, would remain anonymous and unidentifiable throughout the interview and study process.
9 These included California-specific studies from seven cities, three counties, a port (Long Beach), an airport (San Francisco), as well as state reference materials and studies from the International Association of Ports and Harbors. A full list of references is available from Appendix C of the adopted Port Climate Plan.
10 California State Assembly Bill No. 691 was adopted and approved by the Governor on 5 October 2013.
11 For further details on strategies, see Becker and Caldwell (forthcoming).

References

Anderson, A. A. (2005). *The Community Builder's Approach to Theory of Change: A Practical Guide to Theory Development*. New York, NY: The Aspen Institute Roundtable on Community Change.

Becker, A. and Caldwell, M. (forthcoming). 'Stakeholder perceptions of seaport resilience strategies: a case study of Gulfport (Mississippi) and Providence (Rhode Island)', *Coastal Management* (in press).

Becker, A., Matson, P., Fischer, M. and Mastrandrea, M. (2014). 'Towards seaport resilience for climate change adaptation: Stakeholder perceptions of hurricane impacts in Gulfport (MS) and Providence (RI)', *Progress in Planning* (in press, doi: 10.1016/j.progress.2013.11.002).

Bours, D., McGinn, C. and Pringle, P. (2013). 'Monitoring and evaluation for climate change adaptation: A synthesis of tools, frameworks and approaches', *SEA Change CoP*, Phnom Penh and UKCIP, Oxford.

Bours, D., McGinn, C., and Pringle, P. (2014). 'The Theory of Change approach to climate change adaptation programming', *SEA Change CoP*, Phnom Penh and UKCIP, Oxford.

Burroughs, Richard (2005). 'Institutional change in the Port of New York', *Maritime Policy and Management,* 32 (3), 315–28.

CO-CAT (California Climate Action Team) (2010). *State of California Sea Level Rise Interim Guidance Document*. Developed by the Sea Level Rise Task Force of the Coastal and Ocean Working Group, accessible at: http://opc.ca.gov/webmaster/ftp/pdf/agenda_items/20110311/12.SLR_Resolution/SLR-Guidance-Document.pdf (accessed 16/4/2015).

de Langen, P. W. (2004). *The Performance of Seaport Clusters; A Framework to Analyze Cluster Performance and an Application to the Seaport Clusters of Durban, Rotterdam and the Lower Mississippi*. Doctoral thesis, Erasmus University Rotterdam, Netherlands.

Gersberg, R. (2010). 'Description of methods for creating the inundation maps for sea level rise scenarios in San Diego Bay' (personal e-mail communication with Dr Emily Young of the San Diego Foundation).

Hooven, C., Hirsch, J., White, M., Daugherty, D., Messner, S., Moran, L. and Kim, D. (2011). 'Port of San Diego Climate Mitigation and Adaption Plan', Air and Waste Management Association, Greenhouse Gas Strategies in a Changing Climate Conference (San Francisco, CA).

ICLEI (Local Governments for Sustainability USA) (2012). 'Sea level rise adaptation strategy for San Diego Bay', Local Government for Sustainability USA (ICLEI), accessible at: http://www.icleiusa.org/climate_and_energy/Climate_Adaptation_Guidance/san-diego-bay-sea-level-rise-adaptation-strategy-1/san-diego-bay-sea-level-rise-adaptation-strategy.

IPCC (Intergovernmental Panel on Climate Change) (2013). *Climate Change 2013. The Physical Science Basis. Working Group I Contribution to the Fifth Assessment Report of the Intergovernmental Panel on Climate Change. Summary for Policymakers*. T. F. Stocker and D. Qin, Intergovernmental Panel on Climate Change (IPCC).

James, C. (2011). *Theory of Change Review: A Report Commissioned by Comic Relief*, accessible at: http://www.theoryofchange.org/wp-content/uploads/toco_library/pdf/James_ToC.pdf (accessed 14/4/2015).

Messner S., Moran L., Reub, G. and Campbell, J. (2013). 'Climate change and sea level rise impacts at ports and a consistent methodology to evaluate vulnerability and risk', *WIT Transactions on Ecology and the Environment*, 169 (doi: 10.2495/13CP0131).

Ng, A.K.Y. and Liu, J. J. (2014). *Port-Focal Logistics and Global Supply Chains*. Basingstoke: Palgrave Macmillan.

NRC (National Research Council) (2010). *America's Climate Choices: Adapting to the Impacts of Climate Change*. Washington, DC: America's Climate Choices.

State of California (2010). *2009 California Climate Adaptation Strategy: A Report to the Governor of the State of California in Response to Executive Order S-13-2008*. Sacramento, CA: Natural Resources Agency.

Talley, W. K. (2009). *Port Economics*. New York, NY: Routledge.

UNCTAD (2012). *Ad Hoc Expert Meeting on Climate Change Impacts and Adaptation: A Challenge for Global Ports: Geneva, Palais des Nations, 29–30 September 2011: Main Outcomes and Summary of Discussions*. Geneva, Switzerland: UNCTAD.

Valters, C. (2014). *Theories of Change in International Development: Communication, Learning, or Accountability?* The Asian Foundation and the Justice and Security Research Program, International Development Department, London School of Economics and Political Science, London, UK (JSRP paper no. 17).

3 Analyzing risks posed by climate change on ports

A fuzzy approach

Zaili Yang, Stephen Cahoon, Shu-Ling Chen, Adolf K. Y. Ng and Austin Becker

1. Introduction

With most of the world's traded cargoes carried by ships (Ng and Liu, 2014), ports provide crucial linkages connecting hinterlands and other components along global supply chains. As noted by Ng (2006), many interactive attributes affect the competitive positions of ports; these include climate change impacts, such as sea level rise (SLR) and extreme weather conditions (e.g. temperature, precipitation and strong winds), all of which have broad implications for the development prospects of the global economy with substantial economic costs. Hence, the inability of ports to adapt to climate change could negatively affect the wellbeing of global and local economies.

The risks posed by climate change can be enormous, resulting in catastrophic consequences influencing human lives and activities. This was exemplified by Hurricanes Katrina and Sandy's impacts on the North American Atlantic coastline in the last several years (Becker et al., in press), while in Australia, the operations of ports, logistics, and supply chains were also significantly affected (Ng et al., 2013). In this regard, Schaeffer et al. (2012) warned that, due to climate change, by 2100, the sea levels in some parts of the world may be as much as 80 cm higher than nowadays. Nevertheless, tackling climate change is a two-fold process that includes both mitigation and adaption. According to the United Nations Framework Convention on Climate Change (UNFCCC), adaptation to climate change is "the adjustment in natural or human systems in response to actual or expected climatic stimuli or their effects, which moderates harm or exploits beneficial opportunities," while mitigation is "a human intervention to reduce the sources or enhance the sinks of greenhouse gases (GHG)" (UNFCCC, 2012).

Due to a lack of global action to reduce emissions, it is likely too late to avoid all deleterious effects posed by climate change (Applegate, 2010). Thus, the development of appropriate adaptation measures to climate change is now not a choice but a necessity to ensure that maritime stakeholders (including port service suppliers, users, and public bodies) understand the risks posed by climate change on ports, and work hand-in-hand to undertake appropriate adaptation planning, measures, and strategies (Becker et al., 2013; Becker and Caldwell, 2015). Techniques to assess risk use both qualitative and quantitative approaches to address the differences between identified climate change adaptation responses (Willows and Connell, 2003; Füssel, 2007; Wilby et al., 2009; Apel et al., 2009; Klinke and Renn, 2002). However, given that the dynamics between climate change and ports can significantly differ between regions, there are still many uncertainties. As noted by the USA's National Research Council's (NRC) report on climate change, research on reducing the uncertainties of decision-making when dealing with climate change impacts on human welfare remains scarce (NRC, 2010a). Moreover, the exact nature of that risk remains ambiguous and subject to interpretation from different maritime stakeholders. Decision-makers urgently need new analysis methods. Issues related to sources of uncertainty, influencing factors, barriers to adaption, and enablers of adaptation require qualitative assessment while evaluation of the risks is undertaken using a range of different quantitative assessment approaches. Wilby et al. (2009) reviewed the various methods and classified them into three approaches: methods requiring limited resources (sensitivity analysis, change factors, climate analogues, and trend extrapolation); statistical methods (pattern-scaling, weather generation, and empirical downscaling); and techniques requiring significant computing resources (dynamical downscaling and coupled climate models [ocean–atmosphere/Global Climate Model]). Thus, the selection of approaches to be used in a particular situation will vary according to circumstance. When statistical risk data and evidence-based information are absent, subjective judgments from domain experts are often used to assess the impact of climate change ports as the initial perception of the stakeholders, which can be updated further and developed when new evidence is collected. Compared to numerical values, linguistic variables are often used to describe subjective judgments. However, such linguistic descriptions define risk-assessment parameters to a discrete extent so that they can at times be inadequate. Fuzzy set theory is well-suited to model such subjective linguistic variables and deal with discrete problems (Yang et al., 2010).

Understanding such, this chapter employs a fuzzy approach to conduct a quantitative analysis on the perception of maritime stakeholders of the risks posed by climate change on ports. The literature review reveals that risk quantification normally relies on the analysis of three parameters{--}*occurrence likelihood*, *consequence severity* and the *probability of consequence*{--}defined as the probability that damage consequences happen given the occurrence of the event (Yang et al., 2008). To investigate the risks posed by climate change, the *probability of consequence* is heavily influenced by (and thus referred to) the timeframe in which the consequence could occur. The model is therefore based on three variables related to the potential effects posed by climate change on ports: *timeframe*, *likelihood*, and *severity of consequence*. The calibration of the model is demonstrated through a pioneer survey towards several stakeholders who are closely linked to a port located along the North American West Coast1 (hereinafter called "the studied port"). Risk input data is collected with respect to each parameter using some predefined linguistics terms. By doing so, a risk analysis model can be developed to simulate maritime stakeholders' perception of the risks posed by climate change. With reference to the risk estimation results, appropriate adaptation measures to climate change impacts on ports will also be developed.

The rest of the chapter is structured as follows. Section 2 provides a literature review on previous research on climate change (including similar research on shipping and ports). The risk model is established and presented in Section 3. In Section 4, the data collected from the studied port for the pilot study is introduced, while the model's application and the analytical results can be found in Section 5. Finally, Section 6 illustrates the concluding remarks.

2. Literature review

There is no shortage of research investigating climate change, including its risk and challenges, especially SLR (Sanchez-Arcilla et al., 1996; Liu 1997; Jevrejeva et al., 2012; Schaeffer et al., 2012), vulnerability of coastal areas (Nicholls and Hoozemans, 1996; El-Raey, 1997; Shea and Dyoulgerov, 1997; McGinnis and McGinnis, 2011; Hanak and Moreno, 2012) and the construction of coastal defenses via the marine eco-systems (Chemane et al., 1997; Tobey et al., 2010). In this regard, many illustrate the urgency for mitigation and adaptation plans on economic activities and sectors. Simultaneously, many researchers also investigate the relationship between climate change and the built environment. Notable examples include Rosenzweig et al. (2011) who studied New York City's climate adaptation plan so as to protect coastal infrastructure, and Hanson et al. (2011) who measured the exposure of major port cities to the risks and challenges posed by climate change.

However, while recognizing adaptation as an integral component (Wheeler et al., 2009; Posas, 2011), research is (still) dominated by mitigation, notably the measurement and control of GHG and pollutants (Scott et al., 2004; Peters et al., 2009; Yang et al., 2012), with shipping and ports being no exception (Corbett et al., 2009; Eide et al., 2009; Ng and Song, 2010; Psaraftis and Kontovas, 2010; Eide et al., 2011; Geerlings and van Duin, 2011; Villalba and Gemechu, 2011; Berechman and Tseng, 2012). Such trends were similar in the non-academic literature. Among the climate change actions plans developed in 34 states in the USA, only 15 (none specifically in ports) include elements of any significance for adaptation (NRC, 2010b). On the other hand, the American Bar Association's book on the USA's climate change policy in 2007 was illustrative, with only two out of nearly 800 pages addressing adaptation (Verchick, 2010). This should not be surprising, given the much wider availability of international and bi-lateral protocols and regulations on mitigating climate change (Keohane and Victor, 2010). Moreover, previous research, which often focuses on the engineering and physical technicalities, put emphasis on investigating the risks of climate change rather than planning and decision-making. Quoting NRC's report on climate change:

> In the context of climate change, a better understanding of human behavior and of the role of institutions … is as fundamental to effective decision making as a better understanding of the climate system … research investments in the behavioral and social sciences can provide this knowledge, but such investments have been lacking … [while] barriers and institutional

factors … have also constrained progress in these areas … However, improving our understanding of the flexibility and efficacy of current institutions and integrating this body of knowledge with existing work on international treaties, national policies, and other governance regimes remains a significant research challenge.

(NRC, 2010b, pp. 75–77)

This is not to deny that some relevant research has been undertaken. For example, Preston et al. (2011) studied 57 climate adaptation plans around the world{--}paying special attention to the "quality" of their planning processes; Osthorst and Manz (2012) looked at the changing relationship between stakeholders and their surrounding regions while developing climate adaptation strategies in northern Germany. Simultaneously, Maunsell (2008), Stenek et al. (2011), and Becker et al. (2012) studied the potential impacts and challenges imposed by climate change on ports, and all of them emphasized the necessity of developing adaptation plans in ports. Finally, Mullett and McEvoy (2011) applied an integrated assessment approach to analyze the potential risks faced by Australian ports, and used the findings to work closely with port stakeholders in developing effective adaption strategies.

However, as Preston et al. (2011) indicated, many such studies are desktop research based on information as laid down in the plans only and not supported by empirical study. Although most of them address the impacts of climate change, including on ports, and highlight the necessity for adaptation planning and management, few actually go further and develop quantitative, longitudinal measurement techniques in assessing risk-based climate change adaptation, of which the effect of adaptation plans on risk reduction can be effectively measured in a quantitative manner. It is clear that a research gap has yet to be filled, requiring further research to be investigated.

3. The model

Many risk assessment approaches, such as Qualitative Risk Assessment (QRA) approach, identified as deductive risk assessments, have been widely used and are easily conducted based on historical data. Such historical data is not always available, however, and its collection is often time-consuming, expensive, and/or includes many uncertainties. Consequently, these approaches may not be well-suited to deal with climate change risk analysis, in which a high level of uncertainty in data exists. For example, the severity of consequences of SLR can be dynamic and thus sometimes very difficult to be evaluated precisely. Linguistic assessments have been used to cope with such imprecision. Fuzzy set theory (FST) is well-suited to model such subjective linguistic variables and deal with discrete problems (Yang et al., 2010). According to the theory, linguistic variables can be characterized by fuzzy numbers, in which their membership functions to a set of categories are defined to describe the degrees of the linguistic variables.

Fuzzy logic, based on FST, accommodates such linguistic terms through the concept of partial membership. In FST, everything is a matter of degree. Therefore, any existing element or situation in risk assessment could be analyzed and assigned a value (a degree) indicating how much it belongs to a member of the linguistic variables used to describe risk parameters. Furthermore, the membership functions can be defined as curves to describe how each point in the input and output space is mapped onto a membership value (or degree of membership) between 0 and 1. Due to the advantage of simplicity, straight-line membership functions (i.e. triangular and trapezoidal) have been commonly used to describe risks in safety assessment (Wang, 1997). For example, a fuzzy number $\tilde{M}$ is a special fuzzy set $\tilde{M} = \{(x, \mu_{\tilde{M}}(x)), x \in R\}$ where x takes its values on the real line $R: -\infty < x < +\infty$ and $\mu_{\tilde{M}}(x)$ is a continuous mapping from R to the close interval [0, 1]. A triangular fuzzy number can be defined by a triplet (a, b, c) and its membership function is mathematically expressed as (Cheng et al 1999):

$$\mu_{\tilde{M}}(x) = \begin{cases} 0 & x < a \\ \dfrac{x-a}{b-a} & a \le x \le b \\ \dfrac{c-x}{c-b} & b \le x \le c \\ 0 & x > c \end{cases}$$

The addition, multiplication and reciprocal operations of the triangular fuzzy numbers are expressed below (Kwong and Bai, 2003; Chen and Chen, 2005):

1 Fuzzy number addition ⊕
(a1, b1, c1) ⊕ (a2, b2, c2) = (a1+a2, b1+b2, c1+c2) (1)
2 Fuzzy number multiplication ⊗
(a1, b1, c1) ⊗ (a2, b2, c2) = (a1a2, b1b2, c1c2) (2)

Consequently, the fuzzy membership functions in climate risk assessment, consisting of five sets of overlapping triangular or trapezoidal curves, are generated using the linguistic categories identified in knowledge acquisition and the fuzzy Delphi method (Bojadziev and Bojadziev, 1995). The typical linguistic variables and their membership functions for the three risk parameters may be defined with reference to Yang et al., (2008) and characterized as shown in Tables 3.1 to 3.3. The linguistics terms are suggested by domain experts through the *Ad Hoc* Expert Meetings of the United Nations Conference on Trade and Development (UNCTAD) (see Section 4). Although it is possible to have some flexibility in the definition of the five sets of membership functions to suit different specific situations, the support from multiple experts appropriately chosen is necessary to ensure realistic and non-biased membership functions (Kuusela et al., 1998).

Table 3.1 Timeframe

Linguistic terms	*Description*	*Fuzzy numbers*
Very Short (VS)	Less than 1 year	(0.7, 0.9, 1, 1)
Short (S)	Approximately 5 years	(0.5, 0.7, 0.9)
Medium (M)	Approximately 10 years	(0.3, 0.5, 0.7)
Long (L)	Approximately 15 years	(0.1, 0.3, 0.5)
Very Long (VL)	More than 20 years	(0, 0, 0.1, 0.3)

Table 3.2 Severity of consequence

Linguistic terms	*Description*	*Fuzzy numbers*
Catastrophic (CA)	Very severe economic loss and/or disruption on the facilities/systems/services requiring a very long period and very high cost of recovery	(0.7, 0.9, 1, 1)
Critical (CR)	Severe economic loss and/or disruption on the facilities/systems/services requiring a long period and long cost of recovery	(0.5, 0.7, 0.9)
Major (MA)	Significant economic loss and/or disruption on the facilities/systems/services requiring certain length of time and cost of recovery	(0.3, 0.5, 0.7)
Minor (MI)	Some economic loss and/or disruption on the facilities/systems/services requiring some time and cost of recovery	(0.1, 0.3, 0.5)
Negligible (NE)	A bit of disruption on the facilities/systems/services, and possibly with some economic loss, but with no real impacts on the continuance of services, nor does it require significant time and cost of recovery	(0, 0, 0.1, 0.3)

Table 3.3 Likelihood

Linguistic terms	*Description*	*Fuzzy numbers*
Very High (VH)	It is very highly likely that the stated effect will occur, with a probability around 90% of at least 1 such incident within the indicated timeframe	(0.7, 0.9, 1, 1)
High (H)	It is highly likely that the stated effect will occur, with a probability around 70% of at least 1 such incident within the indicated timeframe	(0.5, 0.7, 0.9)
Average (A)	It is likely that the stated effect will occur, with a probability around 50% of at least 1 such incident within the indicated timeframe	(0.3, 0.5, 0.7)
Low (L)	It is unlikely that the stated effect will occur, with a probability around 30% of at least 1 such incident within the indicated timeframe	(0.1, 0.3, 0.5)
Very Low (VL)	It is very unlikely that the effects will occur, with a probability around 10% of at least 1 such incident within the indicated timeframe	(0, 0, 0.1, 0.3)

If *T, C* and *L* represent respectively "*Timeframe*", "*Severity of consequence*" and "*Likelihood,*" the fuzzy safety score *R* can be defined by using the following fuzzy set manipulation.

$$R = T \otimes C \otimes L \tag{3}$$

where the symbol "⊗" represents fuzzy multiplication operation in the fuzzy set theory. The membership function of *R* is thus described by:

$$\mu_R = \mu_T \otimes \mu_C \otimes \mu_L \tag{4}$$

where μ_T, μ_C, and μ_L can be presented by any form of triangular or trapezoidal fuzzy numbers with reference to the defined linguistics variables in Tables 3.1 to 3.3. For example, if one believes that the *L* of a climate threat is "high" with 50 percent confidence and "very high" with 50 percent confidence, then μ_L can be calculated as (0.6, 0.8, 0.85, 0.95).

If multiple (i.e. *n*) assessors/experts provide their judgments, an average value $\frac{1}{n}\sum_{i=1}^{n} e_i$ is used based on an assumption that the experts have the same importance in the assessment in Eq. (5). If the assumption becomes too strong to be accepted in special cases, a weighted value $\sum_{i=1}^{n}(e_i \times w_i)$ can be used, where w_i represents the normalized weight the ith expert has in the assessment.

μ_R is a fuzzy number which needs to be "defuzzified" so as to prioritize the risk level it indicates. There are many defuzzification methods available. Many defuzzification algorithms have been developed, of which a centroid approach (Mizumoto, 1995) may be well-suited to modeling the fuzzy expressions of climate risks. When μ_R is a triangular number, $\mu_R = (a_1, a_2, a_3)$, its defuzzification can be presented in Eq (3). When it is a trapezoidal fuzzy number $\mu_R = (a_1, a_2, a_3, a_4)$, its defuzzified value can be calculated by Eq. (6).

$$\mu_R^* = \frac{\int_{a_1}^{a_2} \frac{x-a}{a_2-a_1} x dx + \int_{a_2}^{a_3} \frac{a_3-x}{a_3-a_2} x dx}{\int_{a_1}^{a_2} \frac{x-a_1}{a_2-a_1} dx + \int_{a_2}^{a_3} \frac{a_3-x}{a_3-a_2} dx} = \frac{1}{3}(a_1 + a_2 + a_3) \tag{5}$$

$$\mu_R^* = \frac{\int_{a_1}^{a_2} \frac{x-a}{a_2-a_1} x dx + \int_{a_2}^{a_3} x dx + \int_{a_3}^{a_4} \frac{a_4-x}{a_4-a_3} x dx}{\int_{a_1}^{a_2} \frac{x-a_1}{a_2-a_1} dx + \int_{a_2}^{a_3} x dx + \int_{a_3}^{a_4} \frac{a_4-x}{a_4-a_3} dx} = \frac{1}{3} \frac{(a_4+a_3)^2 - a_4 a_3 - (a_1+a_2)^2 + a_1 a_2}{(a_4+a_3-a_2-a_1)} \tag{6}$$

Consequently, the higher μ_R^* *is*, the higher the risk level of the threat it indicated.

4. Pilot study: data collection

To demonstrate the feasibility of the model, a pilot study on a port located along the North American West Coast (hereinafter called "the studied port") was undertaken. The necessary data was collected through a pioneer questionnaire survey (see Appendix), duly completed by three stakeholders from the studied port in 2012. They included two consultants appointed to develop the studied port's climate plan, and a senior official from a health and environmental group which was directly involved in the development of the studied port's climate plan (hereinafter called "experts"). Simultaneously, they possessed diversified interests and perceptions about how ports should adapt to climate change and its impacts. This would also provide good insight to researchers and industrial practitioners on how the fuzzy model developed in the last section can be applied in practice. Also, there is a consensus among the groups of experts and authors to assign the same weight to each interviewee.

To identify the potential threats caused by climate change to the studied port, the questionnaire contains two parts: Environmental Drivers due to climate change (ED) and Potential Threats (PT) of ED on the studied port. In total, three EDs and 13 PTs were identified and presented. The EDs and PTs were developed through intensive documental reviews, notably making references to the outcomes of the *Ad Hoc* Expert Meetings organized by UNCTAD in 2011, entitled *Climate Change Impacts and Adaptation: A Challenge for Global Ports* (UNCTAD, 2012). About 40 transport and environmental scholars and industrial experts from around the world, including two authors of this chapter, participated in this meeting. During the meeting, they identified the major threats and impacts posed by climate change, and provided suggestions on the development of adaptation strategies and solutions. Hence, this study is a response to the meeting's call for providing a better information and analytical tool, so as to help stakeholders and policymakers in port planning, development, and operations to make better decisions in adapting to climate change and its impacts.

5. Pilot study: analytical results

It is essential to analyze the risk levels of the associated PTs so that the effectiveness of adaption measures can be quantified and measured and that distraction of resources can be rationalized based on a cost benefit (i.e. risk reduction) analysis. Using the fuzzy risk modeling developed earlier, the risk of the PT#1 "high waves that can damage the studied port's facilities" under the ED#1 "Sea level rise" can be evaluated as follows.

First, the data collected from three experts are aggregated into an average value with respect to *T*, *C*, and *L*. For example, the three experts' judgments of the "timeframe" of the PT#1 are "Long," "Long," and "Long," respectively. With reference to the membership of "Long" in Table 3.1, $\mu_T = (0.1, 0.3, 0.5,)$. Similarly, μ_C and μ_L can be obtained as $\mu_C = (0.3, 0.5, 0.7)$ and $\mu_L = (0.5, 0.7, 0.73, 0.9)$.

Next, using Eq (4), the membership of the fuzzy risk score of PT#1 is calculated as follows.

$$\mu_R = \mu_T \otimes \mu_C \otimes \mu_L = (0.015, 0.105, 0.11, 0.315)$$

Using Eq (6), the defuzzified value μ_R^* of PT#1 can be measured as follows.

$$\mu_R^* = \frac{\int_{a_1}^{a_2} \frac{x-a_1}{a_2-a_1} x dx + \int_{a_2}^{a_3} x dx + \int_{a_3}^{a_4} \frac{a_4-x}{a_4-a_3} x dx}{\int_{a_1}^{a_2} \frac{x-a_1}{a_2-a_1} dx + \int_{a_2}^{a_3} x dx + \int_{a_3}^{a_4} \frac{a_4-x}{a_4-a_3} dx} = \frac{1}{3} \frac{(a_4+a_3)^2 - a_4 a_3 - (a_1+a_2)^2 + a_1 a_2}{(a_4+a_3-a_2-a_1)} = 0.146$$

In a similar way, the safety degrees of all 13 PTs can be obtained and is presented in Table 3.4.

Table 3.4 Safety degrees of all PTs

Environmental drivers (EDs)	*Potential threats (PTs) of ED on the studied port*		*Safety of PTs*
	PT#	*Descriptions*	
Sea level rise	1	High waves that can damage the studied port's facilities	0.146
	2	Port installations (like cranes and warehouses) in the studied port get flooded	0.088
	3	Transport infra- and super-structures in the studied port get flooded	0.135
	4	Coastal erosion at or adjacent to the studied port	0.185
	5	Deposition and sedimentation along the studied port's channels	0.148
Storm surge intensity and/or frequency	6	Waves that can damage the studied port's facilities	0.216
	7	Flooding within the studied port due to storm surge	0.310
	8	Downtime in the studied port operation due to high winds	0.163
	9	High wind damage to port installations (like cranes and warehouses) in the studied port	0.191
	10	Coastal erosion at or adjacent to the studied port	0.265
	11	Deposition and sedimentation along the studied port's channels	0.199
Changing quality and quantity of agricultural and seafood production	12	Reduce the competitiveness of the studied port dedicated to such products	0.064
	13	Negatively affect the economic wellbeing of surrounding communities which largely depend on the studied port	0.072

The analytical results in Table 3.4 indicate that the potential threat (PT#7) of "Flooding within the studied port due to storm surge" has the highest risk in the studied port, followed by PT#10 "Coastal erosion at or adjacent to the studied port," while PT#12 "Reduce the competitiveness of the studied port dedicated to such products" and PT#13 "Negatively affect the economic wellbeing of surrounding communities which largely depend on the studied port" have shown relatively high safety levels.

Conclusion

This chapter has developed a new risk model analyzing the perception on the risks posed by climate change on ports. This new approach will enable maritime stakeholders, such as port planners, port managers, and policymakers to quantitatively evaluate the risks caused by climate change. It can be used not only to identify the PTs of highest risk levels but also, as a benchmark, to effectively measure the effectiveness of adaptation measures. It therefore offers a contribution to port planners, port managers, and policymakers in developing appropriate adaptation strategies. The model can be used either as a stand-alone technique, or as part of an integrated risk control method in climate change and port adaption for prioritizing critical impacts and evaluating the cost-effectiveness of adaptation measures. In turn, it helps in drawing up a state-of-the-art measurement tool regarding vulnerability and risk assessment for ports and climate change, both at an individual port level and at a large scale. The results in the case study only represent the perception of the stakeholders on climate risks in the investigated port. They may be varied when investigating the ports in other regions. Therefore use of a large scale of data being collected from different regions through a project "Climate change and port adaptation{--}a global survey," led by a co-author of this paper, will help discover an interesting finding on the variation of climate risk perceptions among ports across regions and countries in future. Furthermore, the model can be further developed using fuzzy rule-based methods in order to simplify the associated calculation process. Finally, new studies capable of combining the risk assessment method in the current work with a new cost model for optimal selection and implementation of adaptation measures are needed. Empirical research based on ports that have suffered disruptions in their operations due to climate change events can be conducted to further test the effectiveness of climate change measures adopted in those ports.

Acknowledgments

This study was supported by an EU-funded project ENRICH (612546), Marie Curie IRSES, 2013–2017, and the University of Manitoba's University Research Grant Program (42227).

Note

1 The identity of the port is not released due to confidentiality.

References

Apel, H., Aronica, G. T., Kreibich, H. and Thieken, A. H. (2009). 'Flood risk analyses – How detailed do we need to be?', *Natural Hazards*, 49 (1): 79–98.

Applegate, J. S. (2010). 'Embracing a precautionary approach to climate change.' In: Driesen, D. M. (Ed.), *Economic Thought and US Climate Change Policy*. Cambridge, MA: MIT Press, pp. 171–196.

Becker, A. and Caldwell, M. (2015). Stakeholder perceptions of seaport resilience strategies: A case study of Gulfport (Mississippi) and Providence (Rhode Island)', *Journal of Coastal Management*, 43 (1): 1–34.

Becker, A., Inoue, S., Fischer, M. and Schwegler, B. (2012). 'Climate change impacts on international seaports: Knowledge, perceptions, and planning efforts among port administrators', *Climatic Change*, 110 (1): 5–29.

Becker, A., Matson, P., Fischer, M. and Mastrandrea, M. (in press). 'Towards seaport resilience for climate change adaptation: Stakeholder perceptions of hurricane impacts in Gulfport (MS) and Providence (RI)', *Progress in Planning*. doi: 10.1016/j.progress.2013.11.002.

Berechman, J. and Tseng, P. H. (2012). 'Estimating the environmental costs of port related emissions: The case of Kaohsiung', *Transportation Research D: Transport and the Environment*, 17: 35–38.

Bojadziev, G. and Bojadziev, M. (1995). *Fuzzy Sets, Fuzzy Logic, Application*. Singapore: World Scientific,.

Chemane, D., Motta, H. and Achimo, M. (1997). 'Vulnerability of coastal resources to climate changes in Mozambique: A call for integrated coastal zone management', *Ocean and Coastal Management*, 37 (1): 63–83.

Chen, S. J. and Chen, S. M. (2005). 'Aggregating fuzzy opinions in the heterogeneous group decision-making environment, *Cybernetics and Systems: An International Journal*, 36: 309–338.

Cheng, A. C., Yang, B. K. and Hwang, C. (1999). 'Evaluating attack helicopters by AHP based on linguistic variable weight', *European Journal of Operational Research*, 116: 423–435.

Corbett, J. J., Wang, H. and Winebrake, J. J. (2009). 'The effectiveness and costs of speed reductions on emissions from international shipping', *Transportation Research D: Transport and the Environment*, 14: 593–598.

Eide, M. S., Endresen, O., Skjong, R., Longva, T. and Alvik, S. (2009). 'Cost-effectiveness assessment of CO_2 reducing measures in shipping', *Maritime Policy & Management*, 36 (4): 367–384.

Eide, M. S., Longva, T., Hoffman, P., Endresen, O. and Dalsoren, S. B. (2011). 'Future cost scenarios for reduction of ship CO_2 emissions', *Maritime Policy & Management*, 38 (1): 11–37.

El-Raey, M. (1997). 'Vulnerability assessment of the coastal zone of the Nile delta of Egypt, to the impacts of sea level rise', *Ocean and Coastal Management*, 37 (1): 29–40.

Füssel, H. M. (2007). 'Adaptation planning for climate change: Concepts, assessment approaches, and key lessons', *Sustainability Science*, 2 (2): 265–275.

Geerlings, H. and van Duin, R. (2011). 'A new method for assessing CO_2 emissions from container terminals: A promising approach applied in Rotterdam', *Journal of Cleaner Production*, 19: 657–666.

Hanak, E. and Moreno, G. (2012). 'California coastal management with a changing climate', *Climatic Change*, 111: 45–73.

Hanson, S., Nicholls, R., Ranger, N., Hallegatte, S., Corfee-Morlot, J., Herweijer, C. and Chateau, J. (2011). 'A global ranking of port cities with high exposure to climate extremes', *Climatic Change*, 104: 89–111.

Jevrejeva, S., Moore, J. C. and Grinsted, A. (2012). 'Sea level projections to AD2500 with a new generation of climate change scenarios', *Global and Planetary Change*, 81: 14–20.

Keohane, R. O. and Victor, D. G. (2010). 'The regime complex for climate change. Discussion paper 2010-33', *The Harvard Project on International Climate Agreements*. Cambridge, MA, January 2010.

Klinke, A. and Renn, O. (2002). 'A new approach to risk evaluation and management: Risk-based, precaution-based, and discourse-based strategies', *Risk Analysis*, 22 (6): 1071–1094.

Kuusela, H., Spence, M. T. and Kanto, A. J. (1998). 'Expertise effects on pre-choice decision processes and final outcomes: A protocol analysis', *European Journal of Marketing*, 32: 559–576.

Kwong, C. K. and Bai, H. (2003). 'Determining the importance weights for the customer requirements in QFD using a fuzzy AHP with an extent analysis approach', *IIE Transactions*, 35: 619–626.

Liu, S. K. (1997). 'Using coastal models to estimate effects of sea level rise', *Ocean and Coastal Management*, 37 (1): 85–94.

Maunsell Australia Pty Ltd. (2008). *Impact of Climate Change on Infrastructure in Australia and CGE Model Inputs*. Report prepared for the Garnaut Climate Change Review (GCCR).

McGinnis, M. V. and McGinnis, C. E. (2011). 'Adapting to climate impacts in California: The importance of civic science in local coastal planning', *Coastal Management*, 39: 225–241.

Mizumoto, M. (1995). 'Improvement of fuzzy control methods.' In: Li, H. and Gupta, M. (Eds.), *Fuzzy Logic and Intelligent Systems*. Norwell, MA: Kluwer Academic Publishers, pp. 1–16.

Mullett, J. and McEvoy, D. (2011). 'Climate resilient seaports.' In: *Coasts and Ports 2011: Diverse and Developing: Proceedings of the 20th Australasian Coastal and Ocean Engineering Conference and the 13th Australasian Port and Harbour Conference*. Barton, A.C.T. Engineers Australia, 2011: 523–528.

Ng, A.K.Y. (2006). 'Assessing the attractiveness of ports in the North European container transhipment market: An agenda for future research in port competition', *Maritime Economics & Logistics*, 8 (3): 234–250.

Ng, A.K.Y. and Liu, J. J. (2014). *Port-Focal Logistics and Global Supply Chains*. Basingstoke: Palgrave Macmillan.

Ng, A.K.Y. and Song, S. (2010). 'The environmental impacts of pollutants generated by routine shipping operations on ports', *Ocean and Coastal Management,* 53(5–6): 301–311.

Ng, A.K.Y., Chen, S. L., Cahoon, S., Brooks, B. and Yang, Z. (2013). 'Climate change and the adaptation strategies of ports: The Australian experiences', *Research in Transportation Business and Management*, 8: 186–194.

Nicholls, R. J. and Hoozemans, F.M.J. (1996). 'The Mediterranean: Vulnerability to coastal implications of climate change', *Ocean and Coastal Management*, 31(2–3): 105–132.

NRC. (2010a). *Adapting to the Impacts of Climate Change*. Washington, DC: National Academies Press.

NRC. (2010b). *Advancing the Science of Climate Change*. Washington, DC: National Academies Press.

Osthorst, W. and Manz, C. (2012). 'Types of cluster adaptation to climate change: Lessons from the port and logistics sector of Northwest Germany', *Maritime Policy & Management*, 39 (2): 227–248.

Peters, G. P., Marland, G., Hertwich, E. G., Saikku, L., Rautiainen, A. and Kauppi, P. E. (2009). 'Trade, transport and sinks extend the carbon dioxide responsibility of countries: An editorial essay', *Climatic Change*, 97: 379–388.

Posas, P. J. (2011). 'Exploring climate change criteria for strategic environmental assessments', *Progress in Planning*, 75: 109–154.

Preston, B. L., Westaway, R. M. and Yuen, E. J. (2011). 'Climate adaptation planning in practice: An evaluation of adaptation plans from three developed nations', *Mitigation and Adaptation Strategies for Global Change*, 4: 407–438.

Psaraftis, H. and Kontovas, C. A. (2010). 'Balancing the economic and environmental performance of maritime transportation', *Transportation Research Part D: Transport and the Environment*, 15: 458–462.

Rosenzweig, C., Solecki, W. D., Blake, R., Bowman, M., Faris, C., Gornitz, V., Horton, R., Jacob, K., LeBlanc, A., Leichenko, R., Linkin, M., Major, D., O'Grady, M., Patrick, L., Sussman, E., Yohe, G. and Zimmerman, R. (2011). 'Developing coastal adaptation to climate change in the New York City infrastructure-shed: Process, approach, tools and strategies', *Climatic Change*, 106: 93–127.

Sanchez-Arcilla, A., Jimenez, J. A., Stive, M.J.F., Ibanez, C., Pratt, N., Day Jr, J. W. and Capobianco, M. (1996). 'Impacts of sea-level rise on the Ebro Delta: A first approach', *Ocean and Coastal Management*, 30(2–3): 197–216.

Schaeffer, M., Hare, W., Rahmstorf, S. and Vermeerm, M. (2012). 'Long-term sea-level rise implied by 1.5°C and 2°C warming levels', *Nature Climate Change*, 2: 867–870.

Scott, M. J., Edmonds, J. A., Mahasenan, N., Roop, J. M., Brunello, A. L. and Haites, E. F. (2004). 'International emission trading and the cost of greenhouse gas emissions mitigation and sequestration', *Climatic Change*, 64 (3): 257–287.

Shea, E. L. and Dyoulgerov, M. F. (1997). 'Responding to climate variability and change: Opportunities for integrated coastal management in the Pacific Rim', *Ocean and Coastal Management*, 37 (1): 109–121.

Stenek, V., Amado, J.-C., Connell, R., Palin, O., Wright, S., Pope, B., Hunter, J., McGregor, J., Morgan, W., Stanley, B., Washington, R., Liverman, D., Sherwin, H., Kapelus, P., Andrade, P. and Pabon, J. D. (2011). *Climate Risks and Business Ports: Terminal Martimo Muelle et Bosque Cartagena Colombia*. World Bank. Accessible at: http://www-wds.worldbank.org/external/default/WDSContentServer/WDSP/IB/2011/06/21/000386194_20110621014319/Rendered/PDF/626410PUB00Ports0Box0361488B0PUBLIC0.pdf (accessed 12 May 2015).

Tobey, J., Rubinoff, P., Robadue Jr, D., Ricci, G., Volk, R., Furlow, J. and Anderson, G. (2010). 'Practicing coastal adaptation to climate change: Lessons from integrated coastal management', *Coastal Management*, 38: 317–335.

UNCTAD. (2012). *Ad Hoc Expert Meeting on Climate Change Impacts and Adaptation: A Challenge for Global Ports: Geneva, Palais des Nations, 29–30 September 2011: Main Outcomes and Summary of Discussions*. Geneva, Switzerland: UNCTAD.

UNFCCC. (2012). *Glossary on Climate Change Acronyms*. Accessible at: http://unfccc.int/essential_background/glossary/items/3666.php (accessed 16 April 2015).

Verchick, R.R.M. (2010). 'Adaptation, economics and justice.' In: Driesen, D. M. (Ed.), *Economic Thought and US Climate Change Policy*. Cambridge, MA: MIT Press, pp. 277–295.

Villalba. G. and Gemechu, E. D. (2011). 'Estimating GHG emissions of marine ports – the case of Barcelona', *Energy Policy,* 39: 1363–1368.

Wang, L. X. (1997). *A Course in Fuzzy Systems and Control*. New Jersey, USA: Prentice-Hall.

Wheeler, S. M., Randolph, J. and London, J. B. (2009). 'Planning and climate change: An emerging research agenda', *Progress in Planning*, 72: 210–222.

Wilby, R. L., Troni, J., Biot, Y., Tedd, L., Hewitson, B. C., Smith, D .M. and Sutton, R. T. (2009). 'A review of climate risk information for adaptation and development planning', *International Journal of Climatology*, 29 (9): 1193–1215.

Willows, R. and Connell, R. (2003). 'Climate adaptation: Risk, uncertainty and decision making', *UKCIP Technical Report*, May 2003.

Yang, Z., Bonsall, S. and Wang, J. (2008). 'Fuzzy rule-based Bayesian reasoning approach for prioritization of failures in FMEA', *IEEE Transactions on Reliability*, 57: 517–528.

Yang, Z., Bonsall, S. and Wang, J. (2010). 'Facilitating uncertainty treatment in the risk assessment of container supply chains', *Journal of Marine Engineering and Technology*, A17: 23–36.

Yang, Z., Zhang, D., Caglayan, O., Jenkinson, I. D., Bonsall, S., Wang, J., Huang, M. and Yan, X. P. (2012). 'Selection of techniques for reducing shipping NOx and SOx emissions', *Transportation Research D: Transport and the Environment,* 17: 478–486.

Appendix: The Survey Questionnaire

Instructions: From Q2 to Q4, based on your opinions and past experiences, please put "X" into the relevant boxes as appropriate. Please answer in the following order: Q1 (a to b) (pg. 1), Q2 (a to c) (pg. 3), Q3 (a to c) (pg. 4–5) and, finally, Q4 (pg. 4–5).

Q1a Your current position(s): ______________________________

____

Q1b Organization(s) that you are currently affiliated: ______________

____

Q2 What are your opinions on the following variables WITHOUT imposing any adaptation measures?

Q2a Timeframe of the potential threats of the environmental drivers due to climate change (ED) on the studied port

Q2b Severity of consequence of the potential threats of ED on the studied port

Q2c Likelihood of the potential threats of ED on the studied port

Q3 What are your opinions on the following variables WITH the imposition of adaptation measures?

Q3a Timeframe of the potential threats of the environmental drivers due to climate change (ED) on the studied port

Q3b Severity of consequence of the potential threats of ED on the studied port

Q3c Likelihood of the potential threats of ED on the studied port

Q4 What is your opinion on the cost of adaptation so as to address the potential threats of the environmental drivers due to climate change on the studied port?

Legends

ED Environmental driver due to climate change

Description of variables (Q2–Q8)

Timeframe (Q2a & Q3a) *(Note: When will you expect to see a particular threat to pose a significant problem on the studied port for the FIRST time?)*

VS	Very Short	Less than 1 year
S	Short	Approximately 5 years
M	Medium	Approximately 10 years
L	Long	Approximately 15 years
VL	Very Long	More than 20 years

Severity of consequence (Q2b & Q3b) *(Note: What is the extent of seriousness that a particular threat will pose on the studied port at the indicated timeframe (Q2a & Q3a) that you have chosen?)*

CA	Catastrophic	Very severe economic loss and/or disruption on the facilities/systems/services requiring a very long period and very high cost of recovery
CR	Critical	Severe economic loss and/or disruption on the facilities/systems/services requiring a long period and long cost of recovery
MA	Major	Significant economic loss and/or disruption on the facilities/systems/services requiring certain length of time and cost of recovery
MI	Minor	Some economic loss and/or disruption on the facilities/systems/services requiring some time and cost of recovery
NE	Negligible	A bit of disruption on the facilities/systems/services, and possibly with some economic loss, but with no real impacts on the continuance of services, nor does it require significant time and cost of recovery

Likelihood (Q2c & Q3c) *(Note: What is the probability that you believe that a particular threat will affect the studied port at the indicated timeframe (Q2a & Q3a) and severity of consequence (Q3b & Q3b) that you have chosen?)*

VH	Very High	It is very highly likely that the stated effect will occur, with a probability around 90% of at least 1 such incident within the indicated timeframe
H	High	It is highly likely that the stated effect will occur, with a probability around 70% of at least 1 such incident within the indicated timeframe
A	Average	It is likely that the stated effect will occur, with a probability around 50% of at least 1 such incident within the indicated timeframe
L	Low	It is unlikely that the stated effect will occur, with a probability around 30% of at least 1 such incident within the indicated timeframe
VL	Very Low	It is very unlikely that the effects will occur, with a probability around 10% of at least 1 such incident within the indicated timeframe

Cost of adaptation (Q4) *(Note: What is the cost of adaptation that you expect so as to address a particular threat on the studied port?)*

VH	Very High	Involves a very high financial cost so as to comprehensively address the stated potential effect
H	High	Involves a high financial cost so as to comprehensively address the stated potential effect
A	Average	Involves a significant financial cost so as to comprehensively address the stated potential effect
L	Low	Involves a financial cost (though not that significant) so as to comprehensively address the stated potential effect
VL	Very Low	Involves a minimal financial cost so as to comprehensively address the stated potential effect

Q2 (a–c): Timeframe, severity of consequence and likelihood of the potential threats of environmental drivers due to climate change on the studied port WITHOUT any adaptation measures being undertaken:

Environmental driver due to climate change (ED)	*Potential threat of ED on the studied port*	*Timeframe (Q2a)*					*Severity of consequence Q2b)*					*Likelihood (Q2c)*				
		VS	*S*	*M*	*L*	*VL*	*CA*	*CR*	*MA*	*MI*	*NE*	*VH*	*H*	*A*	*L*	*VL*
Sea level rise	High waves that can damage the studied port facilities															
	Port installations (like cranes and warehouses) in the studied port get flooded															
	Transport infra- and superstructures in the studied port get flooded															
	Coastal erosion at or adjacent to the studied port															
	Deposition and sedimentation along the studied port's channels															
Storm surge intensity and/or frequency	Waves that can damage the studied port facilities															
	Flooding within the studied port due to storm surge															
	Downtime in the studied port operation due to high winds															
	High wind damage to port installations (like cranes and warehouses) in the studied port															
	Coastal erosion at or adjacent to the studied port															
	Deposition and sedimentation along the studied port's channels															
Changing quality and quantity of agricultural and seafood production	Reduce the competitiveness of the studied port dedicated to such products															
	Negatively affect the economic well-being of surrounding communities which largely depend on the studied port															

Q3 (a–c) & Q4: Timeframe, severity of consequence and likelihood of the potential threats of environmental drivers due to climate change on the studied port WITH adaptive measures despite having imposed the adaptation measures as indicated in the table below:

Environmental driver (ED) due to climate change	*Potential threat of ED on the studied port*	*Adaptation measure to address the potential threat of ED on the studied port*	*Timeframe (Q3a)*					*Severity of consequence (Q3b)*					*Likelihood (Q3c)*					*Cost of adaptation (Q4)*				
			VS	*S*	*M*	*L*	*VL*	*CA*	*CR*	*MA*	*MI*	*NE*	*VH*	*H*	*A*	*L*	*VL*	*VH*	*H*	*A*	*L*	VL
Sea level rise	High waves that can damage the studied port facilities	Move facilities																				
		Build new breakwaters																				
		Increase breakwater dimensions																				
	Port installations (like cranes and warehouses) in the studied port get flooded	Raise port elevation																				
	Transport infra- and superstructures in the studied port get flooded	Improve transport infra- and superstructures resilience to flooding																				
	Coastal erosion at or adjacent to the studied port	Protect coastline and increase and beach nourishment programs																				
	Deposition and sedimentation along the studied port's channels	Increase and/or expand dredging																				

Environmental driver (ED) due to climate change	*Potential threat of ED on the studied port*	*Adaptation measure to address the potential threat of ED on the studied port*	*Timeframe (Q3a)*					*Severity of consequence (Q3b)*					*Likelihood (Q3c)*					*Cost of adaptation (Q4)*				
			VS	*S*	*M*	*L*	*VL*	*CA*	*CR*	*MA*	*MI*	*NE*	*VH*	*H*	*A*	*L*	*VL*	*VH*	*H*	*A*	*L*	VL
Storm surge intensity and/or frequency	Waves that can damage the studied port facilities	Move facilities																				
		Build new breakwaters																				
		Increase breakwater dimensions																				
	Flooding within the studied port due to storm surge	Raise port levels, move facilities, build coastal defense																				
	Downtime in the studied port operation due to high winds	Increase port size to deal with bottlenecks																				
	High wind damage to port installations (like cranes and warehouses) in the studied port	Increase the future standards of the studied port's construction to deal with higher winds																				
	Coastal erosion at or adjacent to the studied port	Expand beach nourishment programs																				
	Deposition and sedimentation along the studied port's channels	Increase and/or dredging																				

Environmental driver (ED) due to climate change	*Potential threat of ED on the studied port*	*Adaptation measure to address the potential threat of ED on the studied port*	*Timeframe (Q3a)*					*Severity of consequence (Q3b)*					*Likelihood (Q3c)*					*Cost of adaptation (Q4)*				
			VS	*S*	*M*	*L*	*VL*	*CA*	*CR*	*MA*	*MI*	*NE*	*VH*	*H*	*A*	*L*	*VL*	*VH*	*H*	*A*	*L*	VL
Changing quality and quantity of agricultural and seafood production	Reduce the competitiveness of the studied port dedicated to such products	Enhance communication between the studied port and surrounding regions, and encourage more inputs from surrounding regions on climate adaptation																				
	Negatively affect the economic well-being of surrounding communities which largely depend on the studied port	The studied port acts as the 'network manager' to liaise with all related stakeholders and coordinate adaptation plans and strategies (both inside and outside port areas)																				

4 Climate change and adaptation strategies of Canadian ports and shipping

The case of the St. Lawrence-Great Lakes system

Brian Slack and Claude Comtois

Introduction

It is an incontrovertible fact that Canada is more exposed to climate change than most other countries in the world. Its sheer size ensures that its territory will be impacted by a wide range of future climate conditions. Furthermore, it is predicted that northern latitudes will be those most exposed to abrupt and severe change, and Canada, which extends from mid-latitudes to the high Arctic between longitudes 52W to 141W, falls within the global zone of maximum predicted outcomes. While generally perceived as a country of continental dimensions, Canada is also a maritime nation with 265,523km of coastline, the most extensive in the world.

Many of the climate change impacts affecting ports around the world (Ng et al., 2013), such as sea level rise or increased storminess, will inevitably affect Canadian ports too, and many of the adaptations that will have to be made will be similar to those discussed in other chapters. However, within Canada there is an extensive system of inland ports where sea level rise will not be felt, but where a range of other climate factors come into play. This chapter deals with the St. Lawrence-Great Lakes system (SLGLS), an extremely important navigation system that extends from tidewater at Quebec City and extends 2,400km westward to the head of Lake Superior. It is a system that has a dual character, transporting goods between ports within the inland network, and exchanging freight between inland ports and the outside world.

This chapter is divided into three sections. The first provides a brief overview of the physical characteristics of the system and examines its economic importance and trade flows. The second section considers two major climate change impacts: first, the challenge of low water for the ports and the system as a whole; and second, extending the navigation season due to global warming. Predicting the magnitude and scope of future changes due to climate change is shown to be extremely difficult. The third section discusses adaptation measures that are being considered to ensure the commercial viability of ports and shipping on the system. An assessment is made of the difficulties in measuring the cost and operational effectiveness of adaptation measures.

The St. Lawrence Great Lakes system (SLGLS)

Most references to the SLGLS define it extending from the open Atlantic Ocean to the head of Lake Superior. This includes the Gulf of St. Lawrence and the estuary of the river, which, while providing access to the interior of the continent, is tidal. The presence of salt water confers on it a different character than those sections to the west. The tidal range is felt as far inland as just above Quebec City, and it is for this reason that we have adopted a narrower definition of the system to include only those sections characterised by fresh water, and where, incidentally, channel restrictions are encountered.

Thus defined, the SLGLS comprises three sections. First is the non tidal section of the river up to Montreal which provides uninterrupted navigation with a ship channel that has been dredged to 11.3m. Above Montreal the river contains many rapids, and thus a six-lock system of the St. Lawrence Seaway raises vessels from the 7m level of the river at Montreal to the 75m level of Lake Ontario, and this comprises the second section. The locks enable vessels drawing 7.92m to reach Lake Ontario. Between Lake Ontario and Lake Erie, a seven-lock system in the Welland Canal raises ships 99.4m, by-passing Niagara Falls. These locks have the same dimensions as the lower set of locks. The third section of the system comprises the Upper Lakes, which includes Lakes Erie, Huron and Michigan that are more or less at the same level. It also includes Lake Superior whose 6.4m difference in elevation is connected by the Soo Locks, which are longer than those on the St. Lawrence River and Welland Canal, thus allowing larger ships, 309m long, to sail within the four Upper Great Lakes.

The region adjacent to the SLGLS system is one of the most important populated areas of North America. It includes Quebec and Ontario, the most densely populated provinces of Canada, as well as six of the most industrialised states in the US. The system generated 93,000 direct jobs in the port and shipping industries and contributed $33.6 billion in direct and indirect economic activity in 2010 (Mowat, 2014).

There are 39 Canadian ports in the SLGLS that in 2011 handled a total of 99.9m tons of freight. The top ten ports handled 80 per cent of the traffic (see Table 4.1). Twice that traffic total is handled by ports on the US side of the Lakes. There are two distinct characteristics of the traffic of Great Lakes' ports. First, bulk cargoes dominate shipments. The GLSLS is overwhelmingly oriented towards industrial raw materials, such as iron ore, coal, limestone, sand and gravel, as well as agricultural goods such as cereals and petroleum products. The second is that intra-system traffic dominates. Most traffic is between Great Lakes ports. The exceptions to both features are the ports on the St. Lawrence River, where general cargo, especially containers, and international shipments are more important.

Table 4.1 Top ten Canadian ports, SLGLS, 2011 (in millions of tons)

Port	*Traffic*	*Location*
Montreal	27.9	River
Hamilton	10.0	Lake Ontario
Thunder Bay	7.5	Upper Lakes
Sorel	6.4	River
Nanticoke	6.1	Upper Lakes
Sault Ste Marie	5.4	Upper Lakes
Windsor	4.9	Upper Lakes
Goderich	4.8	Upper Lakes
Sarnia	3.3	Upper Lakes
Trois Rivières	3.2	River

Source: Statistics Canada Shipping Reports

The physical characteristics of the system, along with the nature of the freight, explain the characteristics of the vessels on the SLGLS. Most important are Lakers, bulk carriers that are built to the specifications of the locks. In fact there are two variants: on the Upper Great Lakes the vessels are longer and therefore have higher carrying capacities, but since their length precludes their passage below Lake Erie they are confined to ports above the Welland Canal; the other Lakers are shorter but can pass through the entire system and even beyond to ports on the lower St. Lawrence River. Ocean-going vessels can penetrate the system, but locks limit the capacity of ships, and because the ship channel is deeper than the seaway most ocean-going vessels venture no further than Montreal.

The question of water depths

The Great Lakes represent the largest body of fresh water in the world. Combined, the lakes cover 244,106sq.km. The system is dependent on precipitation for its replenishment. There are no major rivers draining into the lakes, and the watershed of the basin extends rarely further than a few hundred kilometres inland from the shores. The St. Lawrence River drains the lakes, and thus its flow is largely dependent on the water level of Lake Ontario (Mowat, 2014).

Historically, the water levels in the Great Lakes have fluctuated by season and by year. Melting snow in the watershed usually provides the highest levels of water in the spring, and evaporation in the summer gives rise to the lowest water levels in the fall and winter. This seasonal cycle is incorporated within annual variations, with periodic high and low water years. The last three decades of the twentieth century were marked by high water levels in the Great Lakes, but from 1998 this situation has been transformed by a virtually unbroken series of years with lower than average monthly water levels in Lakes Superior, Huron and Michigan, which have seen levels falling close to historic minima. A similar pattern prevailed in Lake Erie until 2004, since when levels have returned closer to average values.

The situation in Lake Ontario is somewhat different, since there are means to control its level because of the Moses-Saunders Dam on the Seaway. This has produced a pattern of flow that is more balanced around levels established by the

International Joint Commission, but even here all the lowest monthly levels have been recorded since 1998. Since the St. Lawrence River derives most of its flow from Lake Ontario, it is not surprising that it exhibits less dramatic fluctuations of water levels than those of the Upper Great Lakes. Certainly, the general trend of water levels at the Port of Montreal has been downward since 1998, with 2007 and 2012 being years when low water was experienced for extended periods (see Figure 4.1).

The fluctuations of water levels in the SLGLS appear to suggest a cyclical pattern, with 20 to 30 years of generally lower water levels followed by periods of variable length of higher water. The oscillations have been measured for nearly 200 years of record. The causes of the fluctuations are not fully understood, so there is little chance of being able to accurately predict when conditions will revert to a former high or low water state. There is evidence that some of the recent historic low levels of the Upper Great Lakes are due to dredging and sediment mining in the St. Clair River between Lakes Huron and Erie, which lowered the bed of the river by 16 inches.

Climate change

Whatever the past and present states of water levels in the SLGLS it is assumed that future conditions will be shaped by climate change. The importance of the system for navigation, power generation, recreational boating, and the use of

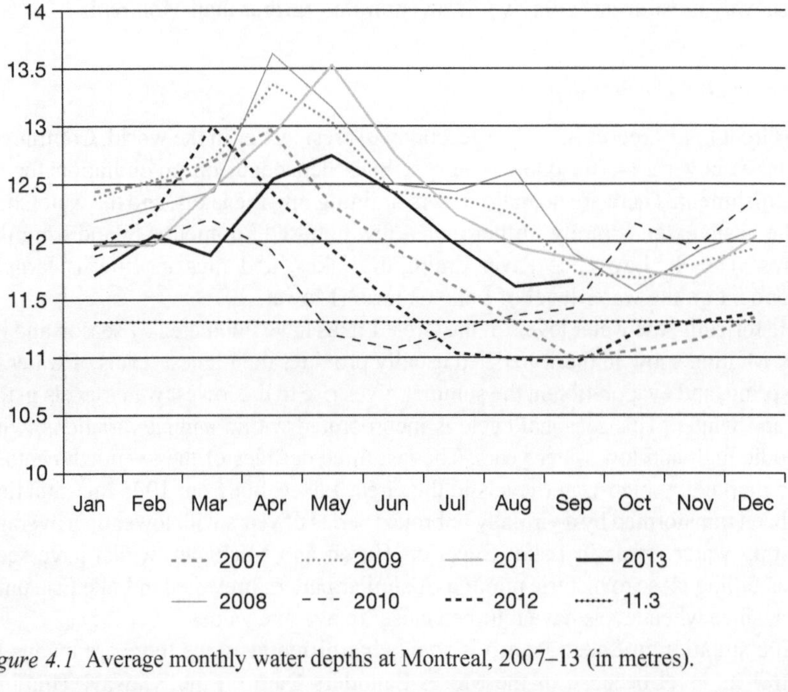

Figure 4.1 Average monthly water depths at Montreal, 2007–13 (in metres).

riparian environments and land has given rise to a large body of speculative research that has sought to predict the consequences of the changing climate (Angel and Kunkel, 2010; Lentz, 2006; Mortsch et al., 2000; Quinn, 2002).

Low water levels

Both American and Canadian climatologists have taken the existing global models of climate change and extrapolated the potential regional conditions. The models are based on estimates of future changes in precipitation patterns and higher evaporation rates, and most predict continued low water levels throughout the system. For example, Millerd (2005) used three model predictions from the Canadian Climate Centre, each based on different paths of CO_2 increases: CGM1 where global CO_2 doubles and then stabilises; CCMa 2030, where CO_2 increases continuously at a fixed rate to 2100, and where the estimate is based on the average level between 2021 and 2040; and CCMa 2050 where the average fixed rate growth between 2051 and 2060 is measured. Like other predictions made early in the new century, significant drops of water levels were predicted. In the case of Millerd's (2005) analysis the difference between the historic average in Lakes Huron, Michigan and Erie were estimated as –.72m for CCMa 2030, –1.01m for CCMa 2050 and –1.62m for CGM1. For the port of Montreal the difference from the historic average was predicted as –.45m for CCMa 2030, –.62m and –1.41m for CGM1. These predictions, if borne out, would represent a major challenge to shipping and ports in the region, and a number of studies have tried to quantify the impacts (Millerd, 2005).

For the St. Lawrence River, Lefaivre (2005) provided an estimate of the effects of climate change on river water levels. This study considered four models of climate change, two from the Canadian Centre for Climate Modelling and Analysis and two from the UK Hadley Centre for Climate Prediction and Research. The two Canadian models were based on two scenarios of warmer and drier conditions in the future; the two UK model scenarios were based on two variants of warm and more humid conditions. A base year of 1969 was chosen as a year when conditions on the river corresponded to historic average, and 2002 was included as a year when the river was lower than average. In extrapolating from these two base years to 2050, the results indicated that only one model prediction, that of the warmest and driest scenario, predicted that water levels would fall below chart datum, and this for the month of October only. When the extrapolations from 2002 to 2050 were calculated a much more serious impact was revealed, with between three to seven months of the year when the levels would be below chart datum.

More recent modelling has tended to indicate less striking climate change impacts (Angel and Kunkel, 2010; Hayhoe et al., 2010). They are based on the predictions of the most recent IPCC general climate models which are then downscaled to the fit the regional characteristics. These more recent predictions place water levels within the margins of the historic range (Mowat 2014), but still lower than former maxima.

This brief review of climate change predictions highlights the uncertainties in forecasting future water levels. The uncertainties begin with the imperfect understanding of the actual changes that have occurred in the past. Thus trying to predict future conditions is fraught with difficulties. The climate models themselves are imperfect and provide a wide range of scenarios that make it difficult to plan adaptation measures appropriate to future climate changes. While yearly means may fall, it is likely that monthly variations may still provide suitable conditions for navigation over many months, thus complicating impact assessments.

Longer navigation season

Climate change predictions indicate a warming throughout the region, with expected reductions in ice conditions. At present, navigation on the Great Lakes is closed for some 10 to 12 weeks per year. Ice conditions occur in most Lakes during the winter season, but the particular problem is that of ice formation in the locks. Opening and closing lock gates becomes hazardous. With expected warming there are predictions that the navigation season can be extended, with some even suggesting year-round open conditions. These most optimistic scenarios are unlikely. It is during winter that maintenance of the locks is undertaken, thus preventing interruptions during the regular navigation season. More realistic are predictions that the system as a whole could close for six to eight weeks a year, with certain sections being navigable with few interruptions.

Impacts on shipping and ports

A number of studies have provided measures of the impacts of low water on Canadian shipping and ports (Millerd, 2005; Mowat, 2014). All have been forward looking, in that the water levels obtained from future climate scenarios are compared with a base condition of a recent year or period, and the difference in levels is applied to losses in the carrying capacities of vessels, and the need for reconstruction of port infrastructures and dredging. Estimates from the most recent study (Mowat, 2014) provide a worst-case low-water level scenario which could cost shipping $1.18 billion over the period from 2012 through 2030 and $1.92 billion over the period through 2050 (converted to 2012 value and stated in USD). In the case of ports, the impacts would be $730 million (62 per cent of the total) through 2030 and $750 million (39 per cent of the total) through 2050. The primary impact on ports would be the costs of repairs and maintenance, with a secondary impact from dredging.

A weakness of these studies is that they base their predictions on theoretical vessel loadings, generalised shipping costs that ignore capital costs or underestimate bunker costs, and ignoring costs of returning empty vessels. They also do not differentiate between vessels that are chartered on the spot or whether they operate under long-term contracts. They ignore those ocean vessels that enter the system. For ports, the estimated costs do not separate normal dock repairs and maintenance as well as normal channel dredging that would have to be undertaken anyway from those made necessary by the climate-change-induced low water conditions.

Two recent studies have addressed some of these criticisms. In a study of the US-flagged fleet on the Great Lakes (MARAD, 2013), the costs of low water on iron ore shipments were calculated along with increases in bunker fuel, and included capital costs. The results indicate a potential increase in cost of 17 per cent for a 1000ft Laker and 19 per cent for a 800ft Laker. It may be assumed that these estimates would be similar for the Canadian flagged Great Lakes fleet.

A detailed analysis of actual vessel sailings in the port of Montreal during periods of low and high water between 2007 and 2013 revealed that nearly all ships sailing during high-water periods arrive or depart below their theoretical maximum capacities anyway, and that even in periods of lowest water, such as during parts of 2007 and 2013, ship loadings were impacted to a small degree (Slack and Comtois, 2014). The most sensitive trade is containers, where all the ships are ocean-going and many draw more water than the ship channel permits. In contrast, the bulk trades are dominated by Lakers which have no draft constraints in the river even under extreme conditions.

The longer shipping season on the Great Lakes that climate change may permit could help the shipping industry recover some of the losses incurred by low water. More revenue-generating voyages per shipping season would be possible thereby affording opportunities to offset lower loadings due to draft constraints.

Adaptations

The prospect of low water in the SLGLS has led several studies to suggest measures to offset draft limitations in ports and for shipping, both on the Great Lakes (Quinn, 2002; Millerd, 2011) and on the St. Lawrence River (D'Arcy et al., 2005). The list of potential responses remains relatively limited. Furthermore, there has been little effort to quantify the adaptations and to compare those expenditures to the predicted costs of low water on the shipping industry and ports.

Structural changes

Dredging

To the present day throughout the SLGLS, dredging has been the means by which ship channels and port berths have been deepened to assure safe navigation. Deep as the Great Lakes are by themselves, the connecting waterways and the access channels to ports could never have permitted access to larger ships and growing trade without significant removal of obstacles. Major deepening has not been undertaken in the system for the last 30 years, with a few exceptions. Rather, maintenance dredging is carried out each year to try to remove shoals and maintain the channels at current depths. Nevertheless, significant amounts of material have to be removed just to maintain levels. For example 5.5 million m^3 was dredged in the St. Lawrence River, mostly in the tidal sections between 1983 and 1994

(Environment Canada, 1994). On the Great Lakes between 1985 and 1989 15.8 million m^3 of sediment was dredged, mainly in Lakes Erie and Michigan (Environment Canada, 1994). More recent data for the US sections of the Great Lakes indicates that 5 million yds^3 are dredged annually.

Dredging most likely will be the measure of remediation undertaken in response to future low water. Unless bedrock is encountered, which would necessitate blasting, the technology of dredging sediment-laden channels would be relatively simple. There would be considerable constraints, however. Dredging is expensive. According to the amount of sediment that has to be removed, costs vary between $12 and $50 per cubic metre. The US Army Corps of Engineers (USACE) estimates that $200 million is required to bring the system in the US Great Lakes back to legislated channel depths. Further deepening would involve still higher costs. For Canadian ports, this will be a major challenge, since the federal government policy is one of cost recovery, and major expenditures for even the largest ports will be beyond their means without government support.

There exist even more serious issues than dredging costs in most cases. Many of the sediments in the SLGLS are contaminated due to more than a century of industrial activity. Most of the major ports are also centres of manufacturing, and disturbing river and lake sediments laced with heavy metals and other pollutants is hazardous to aquatic and riverine ecosystems, as well as threats to domestic water supplies. Regulations require that contaminated material be transferred to Confined Disposal Facilities, but most of those in existence are nearing capacity. Even if the materials are disposed in this way remediation of the aquatic ecosystems that have been subjected to dredging is extremely difficult.

Deepening of channels through dredging could have a negative effect on general water levels in parts of the SLGLS. The deepening of the St. Clair River and its effects on water levels in Lakes Huron and Michigan has already been noted. Dredging the St. Lawrence River Channel beyond 11.3m could increase the velocity of the river and have an impact on the shorelines and ecology. This would be most evident in the Lac St. Pierre section between Sorel and Trois Rivières (D'Arcy et al., 2005). Thus dredging in certain key sections could exacerbate any low water levels induced by climate change.

Other capital projects

The low water problem on the St. Lawrence River in theory could be managed by the construction of a barrage at the exit of Lac St. Pierre. Beyond this site the river becomes increasingly tidal, but a barrier could hold back river flow during low water periods. Such a barrier could be similar to those in Europe, such as Rotterdam, London, St. Petersburg and Venice, whose purpose is the opposite: to control flooding by high storm water surges from the sea. The gates are open during normal conditions, but closed when needed. On the St. Lawrence River the gates would close to reduce the river flow and raise water levels.

There are a number of difficulties with this solution. First is the cost. D'Arcy et al. (2005) estimated the cost of such a barrage on the St. Lawrence River at $469.6

million. Second, is the potential disruption to navigation when the gates are closed since a by-pass lock system would have to be provided. This would represent significant additional capital expenditure as well as delaying normal shipping activity. Third, is the impact on shorelines, since riparian ecosystems would be impacted by sudden fluctuations in levels.

The construction of lateral dikes along shorelines is a less direct but useful technique of enhancing water levels in relatively constrained water bodies, such as rivers or connecting channels. The dikes slow the currents and in this way raise water levels. Such dikes are already used on the Rhine where five dikes per kilometre of shore are located in certain sections (Krekt et al., 2011).

As simple as a dike system appears when compared with barrages and locks, costs would increase with the length of sections to be covered. For example, the Rhine dikes vary between €0.5 million and 1.5 million per kilometre. A particular problem for the St. Lawrence River is winter ice, which could damage the dikes and thus present additional maintenance costs.

Non-structural changes

There are several measures that could address some of the problems of low water that would not require the expenditure of capital in ports that will be most prone to seasonal and annual low water conditions. As detailed below, this does not mean that they are cost-free, since several would generate higher operating costs for the transport industry and customers, while another would require capital expenditures in one port and not others.

Adjusting seasonal cargo loadings

To a certain degree this 'solution' is being applied already. It involves lightening the ships below maximum capacity in order to reduce drafts. With the most severe climate change conditions this would result in most ship loadings being affected for longer periods, with the consequential increases in ships' operating costs. These extra costs are the factors considered in the model predictions of Millerd (2005 and 2011) and Mowat (2014). The 'lost' cargo would require extra sailings to fulfill cargo orders.

For ports this would not necessarily be a bad solution, since their incomes would rise by the additional ship calls, a fact already documented for ports on the Rhine River, and the shipping lines would enjoy more ships' revenue to offset the increases in cost. Although these extra costs would be likely passed onto the customers, on the Great Lakes the major customers are industries that ship or receive bulk commodities. In bulk supply chains, stockpiling is the norm, and the closure of the waterways in winter accentuates this feature on the Great Lakes. Thus, the iron, coal, limestone and steel industries would be in a position to insulate themselves somewhat from shipments stretched over longer periods.

Topping off

Another shipping adjustment already adopted for cargoes moving outside the Great Lakes is 'topping off', where a vessel loads at a Great Lakes port to whatever the draft limits are, and then sails to a port on the St. Lawrence River to load to capacity before going overseas. This practice could be expanded to provide a partial solution to future low water constraints on the Great Lakes.

Low water on the St. Lawrence River has the potential to disrupt container business at the port of Montreal. Should low water periods due to climate change become more frequent and accentuated, the lightening of vessels could have a significant impact on the port's competitive position. Its major competitor, the port of New York and New Jersey, is completing the dredging of its channels to 15m, while those of Montreal are 11.3m. The inability to provide even this 11.3m level could lead to further difficulties. Topping off at another port down river would not be feasible for several reasons. First, there are no ports equipped to handle containers elsewhere on the St. Lawrence. Second, containers would have to be transported by truck or rail to a topping-off port where specialised facilities would have to be installed, representing an additional cost and additional delays in transits for a time-sensitive supply chain. Third, the potential rail or truck traffic to the topping-off port could be captured by New York.

Port relocation

The difficulties in topping off container traffic and the threat of low water at Montreal has led some to suggest that the container business of the port of Montreal be relocated downriver where draft limits are avoided. There are several problems with this proposal. First is the question of market accessibility, a key determinant of container port location. There have already been feasibility studies for establishing a container terminal at the port of Sept Iles, which have concluded that despite its great depth of water, it is too isolated by road (one section of which involves a ferry) with no rail connections to markets. Similar constraints apply in varying degrees to all other potential sites on the lower St. Lawrence. The only city with a sizeable domestic market is Quebec City, but being 250km from Montreal, 90 per cent of the traffic would incur additional market distribution costs. Second is the question of the costs of building a new container facility. It has been estimated that relocation costs alone would exceed $1 billion, even if a suitably large site could be found. Third is the threat of loss of traffic to New York (see above).

Modal shift

If ports become unable to provide adequate access for ships it might be possible for freight to be transported by other modes. Given the bulk character of most of the waterborne trade, it is unlikely that trucks would be capable of handling the volumes at a rate that would be acceptable to shippers. Rail would hold some potential for some of the goods. Some grain shipped from the port of Thunder Bay on Lake Superior for transhipment at a St. Lawrence River port is already transported by rail during the winter closing of the seaway. However, a total transfer to rail would represent a 7 per cent higher transport cost than shipment by a Laker (Miller, 2011). Furthermore, there is a question as to whether the railroads would be able (or willing) to handle additional freight from the Prairies to the St. Lawrence Valley, given the 2013–14 problems of hauling grain by rail from the Prairies to the port of Vancouver when a bumper crop overwhelmed the capacity of the rail system.

Other commodities would be even more difficult to transfer. Iron ore from lower St. Lawrence River ports would have to be transhipped at Montreal to the railways for delivery to US steel plants, thereby incurring higher transfer costs. Road salt produced in mines in Ontario and shipped throughout the province from the ports of Windsor and Goderich could in theory be transferred to road or rail modes, but because of smaller scale economies, unit costs would be higher. Sand and gravel, used throughout the region for construction, would likely not be able to transfer easily to other modes because of the low value of the material. In the case of Canadian bulk cargoes shipped across the Lakes to US ports, routing by road or rail would be much longer and therefore more costly.

Any transfer of freight from ships would involve greater environmental costs. Not only are the capacities of other modes more limited, but their fuel consumption and emission levels are higher. A modal shift as a response to low waters would have unfortunate economic as well as environmental consequences.

Technological changes

Modern shipping is itself a product of technology. Vessel size, cargo handling, navigation, and improved environmental impact are all products of refinements in engineering and management. Several developments could help ports and shipping adapt to low water conditions in the SLGLS.

Vessel adjustments

Elsewhere in the shipping industry technology has facilitated the achievement of ever higher scale economies. In shallow water conditions there are opportunities for technological and regulatory changes to permit ships to maintain capacity while drawing less water. This is achieved by widening ships. Broader beams would permit vessel drafts to be reduced, while maintaining loading capacity. There is one physical difficulty with this solution: that of the effect of vessel squat (D’Arcy et al., 2005, p. 26). Widening ships paradoxically lowers their draft in the water column when they are moving. The effect of vessel squat can be compensated by reducing the ship’s velocity. In 2013 the Canadian Coast Guard raised the beam limit of vessels on the St. Lawrence River from 32.5m to 44m. Theoretically this enables the port of Montreal to receive ships of 6,000 TEU capacity. However, their speed is subject to more stringent regulations. If the same principle of increasing vessel beams is applied to ships built with lower drafts, present capacities could be maintained.

Unfortunately this solution would apply only to vessels serving ports on the St. Lawrence River, since the width of the seaway locks prevents such design parameters being applied to Great Lakes shipping. There is, nevertheless, a vessel redesign possibility for low water adjustment on the Great Lakes, by using barges. A survey of US Lakes carriers indicates a growing interest in the use of non-propelled vessels as a means of reducing costs, and as a way of adapting to low water conditions (MARAD, 2005). Most respondents indicated a preference for barge units in their future vessel investments. Barge traffic dominates cargo movements on the Mississippi River, where 1,500-ton shallow draft barges are used for all types of freight. A typical unit comprises 15 barges lashed together in a 3x5 combination and pushed by a powerful tug. Below St. Louis, barge combinations maybe extended to 30-barge units. Individual barges are dropped off en route, and new ones added. This provides for flexibility as well as reducing manpower in an environment of high labour costs.

On the Great Lakes, different types of barge units are appearing. The lock system on the Great Lakes prohibits the large barge combinations found on the Mississippi, but intra-Lake traffic could adopt a comparable configuration. Several old Lakers have been modified by stripping out engine rooms and crew quarters and replacing them with a steel notch into which a push barge can fit to direct the gearless vessels. Significant labour cost savings are claimed. In this way the threat of more low water may be the catalyst to renovate the Great Lakes fleet and make them more cost-effective.

Depth of water measurements and predictions

For many years the Canadian Hydrographic Service has maintained a monitoring service along the St. Lawrence River from Montreal and the estuary. There are 12 monitor points in the river between Montreal and Quebec City out of a total of 22. These points record water depths in real time. This allows ships' captains and the pilots to accurately determine the loading limits of the vessels. Unfortunately this real-time system is not particularly useful for ships that are leaving overseas ports, since they may take seven to ten days of sailing before entering the River, and by then conditions may have changed.

A system has been developed by the Canadian Hydrographic Service to predict water levels for 0 to 48 hours, and from 2 to 30 days. As useful as this tool has been, the predictions beyond a few days have been found to be off by as much as 22cm at Sorel after just two days, and therefore not very useful in loading preparation in overseas ports. An Australian company, OMC International, has recently been awarded a contract to try to improve the performance of predictions.

With the expected variations of water levels throughout the SLGLS, such hydrographic measurements and predictions become a tool that may help the shipping industry offset uncertainty.

Conclusions

The effects of climate change on ports and shipping in the SLGLS are difficult to predict. This is in part due to natural variability in water levels, as well as the uncertainties of predictions about future climate conditions. Had the early predictions of significant water level declines of over 1m been acted upon, both industry and governments would have been called upon to make decisions that might have ultimately been seen as an overreaction in light of more recent sets of forecasts. Equally, the recent and more moderate climate change predictions might be shown in the future to be exceeded by actual conditions. This uncertainty makes the selection of appropriate adaptation responses very difficult. This problem is accentuated by the fact that the climate modellers are predicting conditions 20–50 years into the future. Ports and shipping industries deal with much shorter timeframes. Who can predict with accuracy what market and technological changes are likely to take place even ten years ahead? Making these industries plan today for possible very long-term changes, changes which may be unpredictable, will be an exceptional challenge.

The review of possible adaptation measures indicates that there are several options available for ports and shipping to adapt to climate change in the SLGLS. Many, such as barrages and port relocation, are very costly and potentially disruptive, but there are sufficient alternatives to be considered. What is clear is that many options will not only have local and regional impacts within the SLGLS, but that they may impact the competitiveness of the system, diverting traffic to other modes on the Great Lakes, and shifting the competitive balance from the St. Lawrence River ports to the US East Coast.

While many possible adaptations involve engineering and technological solutions, their implementation is not only influenced by cost and effectiveness concerns. There are also significant environmental and safety considerations at play. Dredging, modal shift and ship design all involve regulatory issues. There are strict environmental regulations concerning physical changes to river and lake systems, and, as mentioned above, changes in vessel design require new measures to adjust ships' speed and operating conditions. Regulations are likely to be an extremely important factor shaping climate change responses. In the case of the SLGLS regulatory factors are likely to be particularly pertinent, since both provincial and federal governments share jurisdiction. Also, as the waters are shared with the US, the International Joint Commission, as the bi-national body overseeing the system, would have a major role to play.

References

Angel, J. R. and Kunkel, K. E. (2010). 'The response of Great Lakes water levels to future climate scenarios with an emphasis on Lake Michigan-Huron', *Journal of Great Lakes Research*, 36: 51–58.

D'Arcy, P., Bibeault, J.-F. and Raffa, R. (2005). *Changements climatiques et transport maritime sur le Saint-Laurent. Étude exploratoire d'options d'adaptation*. Ottawa et Québec: Comité de concertation navigation du Plan d'action Saint-Laurent.

Environment Canada. (1994). *Répercussions environnementales du dragage et de la mise en dépôt des sédiments.* Document préparé par Les Consultants Jacques Bérubé inc. pour la Section du développement technologique. Direction de la protection de l'environnement, régions du Québec et de l'Ontario. No de catalogue En 153-39/1994F.

Hayhoe, K., Vandorn, J., Croley II, T., Schlegal, N. and Wuebbles, D. (2010). 'Regional climate change projections for Chicago and the US Great Lakes', *Journal of Great Lakes Research*, 36: 7–21.

Krekt, A. H., van der Laan, T. J., van der Meer, R.A.E., Turpijn, B., Jonkeren, O. E., van der Toorn, A., Mosselman, E., van Meijeren, J. and Groen, T. (2011). *Climate Change and Inland Waterway Transport: Impacts on the Sector, the Port Of Rotterdam and Potential Solutions*. Utrecht: Knowledge for Climate.

Lefaivre, D. (2005). *Effets des changements climatiques sur les niveaux d'eau du fleuve Saint-Laurent entre Montréal et Québec: Projections pour les années 2050.* Rapport préparé pour le Comité de concertation navigation.

Lentz, B. (2006). *Great Lakes, Grand Problem: The Impacts of Significantly Lowered Lake Levels on the Industries, Ecosystems, and Individuals in the Great Lakes Region.* The Coastal Institute White Paper, May 12.

MARAD (US Maritime Administration). (2005). *Great Lakes Operators.* Industry Survey Series. US Department of Transportation.

MARAD (US Maritime Administration). (2013). *Status of the US-Flagged Great Lakes Water Transport Industry.* US Department of Transportation.

Millerd, F. (2005). 'The economic impact of climate change on Canadian commercial navigation on the Great Lakes', *Canadian Water Resources Journal*, 30: 269–280.

Millerd, F. (2011). 'The potential impact of climate change on Great Lakes international shipping', *Climatic Change*, 104: 629–652

Mortsch, L., Hengeveld, H., Lister, M., Lofgren, B., Quinn, F., Slivitzky, M. and Wenger, L. (2000). 'Climate change impacts on the hydrology of the Great Lakes-St. Lawrence system', *Canadian Water Resources Journal*, 25: 153–177.

Mowat Centre. (2014). *Low Water Blues.* Council of the Great Lakes Region.

Ng, A. K., Chen, A. K., Cahoon. S., Brooks, B. and Yang, Z. (2013). 'Climate change and the adaptation strategies of ports: The Australian experiences', *Research in Transportation Business & Management*, 8: 186–194.

Quinn, F. H. (2002). 'The potential impacts of climate change on Great Lakes transportation'. In: *The Potential Impacts of Climate Change on Transportation: A Federal Research Partnership Workshop*, pp. 115–124. US Department of Transportation Center for Climate Change and Environmental Forecasting.

Slack, B. and Comtois, C. (2014). *Economic Analysis for Adaptation to Climate Change: St-Lawrence Great Lakes System and the Port of Montreal.* Paper presented at the Annual Conference of the International Association of Transport Economists. Norfolk, VA, July 18.

5 Climate change and the adaptation planning of inland port and rail infrastructures in the province of Manitoba in Canada

Tianni Wang, Spencer Samsom, Adolf K. Y. Ng and Paul Earl

1. Introduction

The Canadian province of Manitoba (hereinafter called 'Manitoba'), due to its central location in North America and natural resources, is a well-known agricultural province. One of the most important pillar industries is transportation, whose multiple modes (road, air, rail and ship) not only support the national economy but also catalyze US–Canada as well as global trade. Meanwhile, since ports (including inland ports) are always regarded as a crucial node and the railway as a vital link connecting ports and supply chains (e.g. the Hudson Bay Railway in Canada), the impacts of climate change in the transportation system could significantly affect the efficiency of regional supply chains.

As Manitoba is located in a relatively interior area and northerly latitude, the impacts of climate change in Manitoba are disproportionate and diverse. Some impacts of climate change on Manitoba's transportation system have been identified. Generally, with the tendency of global warming, Manitoba is expected to suffer from warmer and wetter winters, together with longer, drier and hotter summers: as the most dramatic climate change in the past 100,000 years, the temperature could rise by 3–4°C in the summer and 5–8°C in the winter by 2080 (Government of Manitoba, 2002), meanwhile, the precipitation could rise by 5–10 percent in spring and decrease by 10–20 percent in summer (Manitoban Government website, 2015). However, this prediction is uncertain because there have been more dramatic and frequent climate change events (e.g. flooding and melting permafrost) and extreme weather events (e.g. storms, tornadoes and hurricanes), impacting or indirectly affecting, for example, the communities' economy, transportation infrastructure and patterns, as well as the integrity of transportation and supply chain infrastructures, such as CentrePort Canada (hereinafter called 'CentrePort') and the Hudson Bay Railway.

Although a considerable amount of assistance has been given to CentrePort and the Hudson Bay Railway (e.g. the political action plan and funding from the federal and provincial governments, the infrastructure investments of Highway

75, CentrePort Canada Way, Canadian National Railway (CN), Manitoba Infrastructure and Transportation (MIT), Bison Transport, etc., and the relevant academic research and workshops together with forecasting and risk-analysis tools for climate change), there is a lack of specific adaptation plans for climate change at both the corporate (CentrePort and port of Churchill) and governmental (both provincial and city) levels. However, given the potential risks as stated above, strategic adaption planning is very important, in which it would play a pivotal role in minimizing risks and uncertainties posed by climate change, thus maximizing the benefits of every stakeholder in this supply chain.

Understanding such, this chapter investigates the potential risks posed by climate change, and the adaptation strategies and planning by transportation and supply chain infrastructure provider in Manitoba. We will focus on CentrePort and the Hudson Bay Railway (which connects the Canadian National Railway to the port of Churchill located along the Hudson Bay's coastline), and analyze the current plans for climate change in this area. In the first section, on the basis of several interviews undertaken with the relevant stakeholders in the supply chain, corporate documents and governmental policies, the authors identify the climate change risks (floods) and uncertainties (the extreme weather events on supply chain interruption and transportation patterns and infrastructure) on CentrePort by positioning it as a transportation hub for North America. Afterwards, we will focus on adaptation planning and analyze the necessity for, and the resources and barriers involved in, the planning process. We will then reiterate the significance of adaptation planning for climate change in this supply chain and attempt to encourage more attention to, and discussion of, detailed issues (such as public participation and insurance) in adaptation planning.

We strongly believe that adaptation planning is necessary, and should form a significant part of regional and national development in this part of Canada. On the basis of analyzing the existing preparations for climate change at CentrePort, it is suggested that adaptation planning for climate change calls for attention from all of the relevant stakeholders. Similarly, the next section examines climate change on the Hudson Bay Railway as a two-edged sword creating threats (e.g. melting permafrost) and opportunities (e.g. a longer season for commercial shipping) along with new challenges for OmniTRAX, the owner and operator of the Hudson Bay Railway and port of Churchill in Northern Manitoba, Canada. Also, this requires a specific adaptation plan for further supply chain management. By examining climate change and adaptation of the supply chain in Manitoba, as well as the adaption planning process, we attempt to offer insightful, workable recommendations for this emerging issue.

2. Research process and methodology

The research process involves in-depth case studies on CentrePort (a dry port/ inland terminal) and the Hudson Bay Railway. As the studies focus on the planning process, we have undertaken a qualitative research approach and have accessed a large amount of qualitative information through documental reviews and a number of in-depth interviews (see below), so as to enable researchers to analyze relationships and social process, of which it would be difficult to do so using only quantitative methods, especially when there are lacking of abundant data (Miles and Huberman, 1984).

In the fall of 2014, we completed seven semi-structured, in-depth interviews with relevant personnel affiliated with CentrePort, Hudson Bay Railway and OmniTRAX Canada. They were visited via one to two key informants at each port from the targeted population that includes CEOs/senior port directors, planners, environmental managers, development managers, policymakers, environmental academics and other relevant port stakeholders. During the interviews, we investigated with the interviewees on several important issues, as follows:

1 What are the current climate-change risks and uncertainties, and what specific climate-change events have impacted the interviewees' ports in the past decades? Meanwhile, what are the perceptions and considerations of the interviewees in implementing adaptation and mitigation strategies for climate-change impacts?
2 What are the main priorities and principles for adaptation planning for climate change in terms of the short-term and long-term planning at the interviewees' port? How does the decision maker assess the risks or uncertainties posed by climate change?
3 How do the interviewees conceive the unique conditions of their ports' adaptation planning? What kind of resources do they utilize when conducting adaptation planning for climate change?
4 What are the interviewees' perceptions of adaption planning for climate change? Who are the internal and external participants supporting and conducting adaptation planning for climate change?
5 What is the planning horizon of adaptation planning for climate change? What are the primary aspects and key elements in a successful adaptation plan for climate change?

An interview framework had been designed with a general question list being provided to interviewees before the conduction of each interview. Corresponding to the above five issues, the interview questions were broken into four categories: Part A, identifying the vulnerabilities of ports posed by climate change; Part B, assessing risk and planning priorities; Part C, recognizing the characteristics and differences of ports' condition; Part D, determining the environment of stakeholders for an adaptation plan; Part E, implementing an adaptation plan and developing adaptation strategies. However, there was no fixed wording for interviewers to ask questions, and the respondents were not restricted to answering the questions according to the categories. The process was expected to reflect the real situation of specific climate-change impacts on targeted ports, as well as the real thinking of interviewees' perception on the port planning process for climate change, adaptation strategies and planning for climate change, and climate-related policies, etc.

Overall, all the interviewees were asked the same basic questions in the same order. At the same time, based on the previous research about the port's history and planning background for climate change (e.g. Has the port suffered from the climate change in the past decades? Has the port implemented an adaptation plan already?), the specific wording and sequence of the questions varied among different interviewees. This strategy allowed us to increase the possibility that all the prepared questions in the checklist would be adequately addressed within the limited timeframe of the interview.

In complying with the guidelines on case studies proposed by Miles and Huberman (1984), a within-case analysis and cross-case analysis were utilized to address the above interviews' questions and to fulfill the design purposes. First, for each case, a within-case analysis is very important to investigate the existing climate-change risks and potential uncertainties, as well as the implementation of adaptation strategies and planning for climate change on a typical inland port in Canada. Furthermore, a cross-case analysis provided for comparing and contrasting the specific situational context of climate change impacts, adaptation planning for climate change and other associated issues of CentrePort and Hudson Bay Railway in Manitoba under the background of a port supply chain. Meanwhile, to guarantee the construct validity, a diverse data source of evidence including interview data and archival data in every case study was identified and analyzed for triangulation. The reliability and traceability was ensured by a series of consistent procedures of case study from the case design and data collection to the analysis, as well as a standardized form of interview from preparation, personal conversation, audio-recording to finalized transcription (McCutcheon and Meredith, 1993).

3. CentrePort

3.1 Flooding in Manitoba

CentrePort is a 20,000-acre inland port in Winnipeg, Manitoba. Located along the longitudinal centre of Canada, it aims to become a hub for trade and transportation corridors, connecting Canada to major markets, such as the USA, Mexico, Latin America, Europe and Asia (CentrePort Canada website, 2015). While enjoying strong transportation and economic and infrastructural support from the provincial government in the city of Winnipeg, CentrePort as a part of the community also shares the risks and uncertainties posed by climate change involving the city and its transportation network. One of the most significant risks is flooding along the Red River Valley. As a tri-modal (rail, road and air) inland port, although there is no confirmed evidence that CentrePort has been severely affected by flooding, it is recognized that CentrePort can be vulnerable to the risks of flooding in highway, railway and air transportation.

Indeed, the flooding history of Manitoba can be traced back as early as the 1800s. Four exceptional events took place in 1950, 1997, 2009 and 2011 in the areas along the Red River, Assiniboine River and Saskatchewan River, respectively

(Manitoban government website, 2015). The impacts of flooding in the last century were wide-ranging and far-reaching in ground water, surface water, livestock, crops and farm property, as well as the transportation system. For instance, the recent flooding on the Assiniboine River and Lake Manitoba in 2011 was expected to create the highest water level in the past 330 years. This one-in-2,000-year event (e.g. Owen, 2012) resulted in "the closure of 850 roads, including part of the Trans-Canada Highway" (Manitoban government website, 2015). Meanwhile, Highway 75 in major floods of the Red River was cut off for a month, so the supply chain was disrupted for a month, which interrupted the flow of cargo between the USA and CentrePort and impacted further demand and supply between southern and northern Manitoba. The reasons for these floods are complex and varying from year to year. The flooding could be attributed to a heavy rainstorm in spring or melting heavy snow water during the previous winter. Also, a heavy storm or tornado in summer would increase the possibilities and intensities of floods (City of Winnipeg website, 2014).

Unsurprisingly, since the 1950s, the Manitoba government has adopted a series of policies and action plans to fight and compensate floods. The report entitled *Flood Fighting in Manitoba* (Government of Manitoba, 2013) illustrated the background and reviewed the history of Manitoba's flood protection (Manitoban government website, 2015). The main flood control structures for protecting the city of Winnipeg include the Red River Floodway, the Portage Diversion and the Shellmouth Reservoir. One of the most significant measures for CentrePort is the Red River Floodway constructed in 1968 (and expanded in 1997). It contains the Assiniboine River Primary Dikes, the West Dike, and the Red River Primary and Ring Dikes. Starting in 1997, more than \$1 billion has been invested in flood mitigation and control, which effectively prevented over \$7 billion in damages throughout the province of Manitoba (Manitoban government website, 2015). Since 2005, a \$665 million investment in expanding and excavating the floodway channels in the Red River Floodway has greatly helped minimizing damages in communities, agriculture lands and the transportation network. Simultaneously, a relatively complete forecasting (by Hydrological Forecast Centre) and strong emergency management system (by a partnership led by Manitoba Emergency Measures Organization (EMO) and MIT) with top-down mitigation acts and enhanced training, infrastructure and equipment has been contributing to protecting Winnipeg and CentrePort from the risks posed by the unprecedented flood in 2011.

Nevertheless, Manitoba is still facing dilemmas caused by flooding. As an emergency management summit held in Alberta in 2013, entitled *Flood Mitigation through Innovation: A Manitoba Perspective*, has summarized, there are difficulties in predicting flooding in particularly severe climate-change conditions. Additionally, since investment in flooding mitigation always has long-term returns, it is necessary for investors and decision makers to consider a variety of factors including return on investment, protection standards, and the anticipated frequency and intensity of floods. Understanding such situations, we might suspect that CentrePort is still confronted with similar flooding challenges posed by climate change as the city and province. Thus, a specific adaptation plan for climate change in CentrePort is considerable from the flooding perspective.

3.2 Uncertainties and current measures

It is worth noting that the climate-change impacts for the central area of Manitoba include more than flooding (Figure 5.1). They include *more extreme weather events* (e.g. tornadoes, heavy storms, extreme cold events, longer and frequent droughts might effect port operation), *variable ecological conditions* (e.g. grassland would move to northern areas so as to edge out boreal forest which might increase the severity of flooding risks), as well as *significantly less snow cover* (which could lead to less moisture for agriculture which in turn might indirectly impact the port's business), and all introduce uncertain elements for

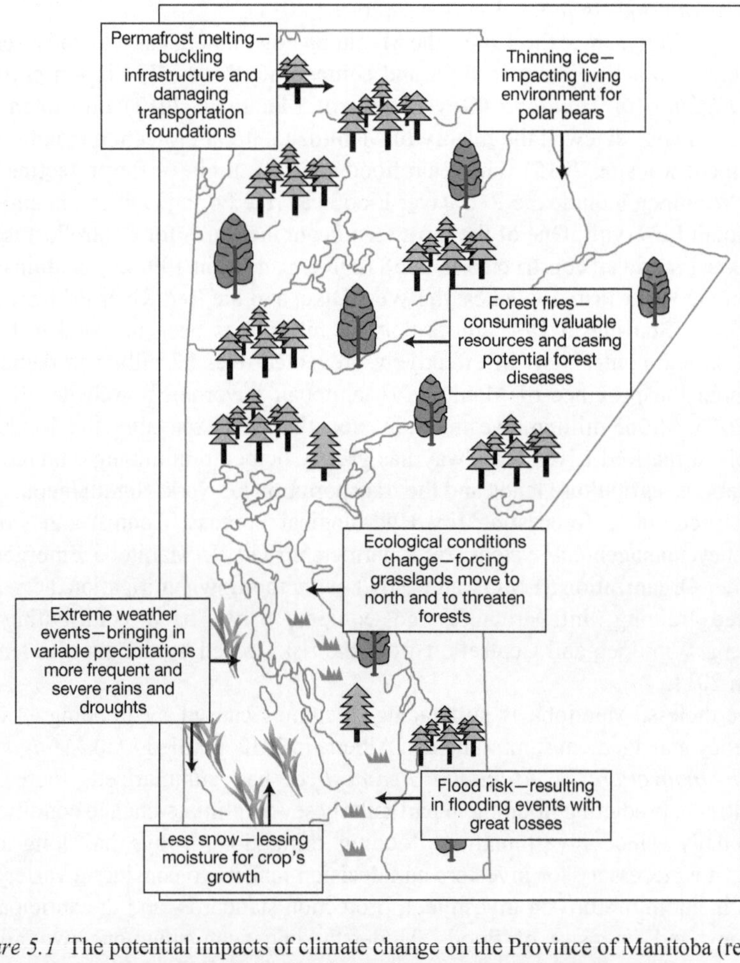

Figure 5.1 The potential impacts of climate change on the Province of Manitoba (redrawn from the Government of Manitoba, 2014; blank map source: d-maps.com).

CentrePort (Manitoban Government's website, 2015). Moreover, the climate-change events happening in other jurisdictions might trigger variations and uncertainties in the traffic and business of CentrePort due to the domino effect on the supply chain. For example, floods, tornadoes and hurricanes in other Canadian seaports and the USA might impact freight movement by rail, road and air networks, as well as delay the transport schedule and damage the infrastructure in all the supply chains with which CentrePort is involved.

Fortunately, the Manitoba government has been pursuing relatively holistic policies and initiatives to handle climate-change impact in the past years. *Tomorrow Now: Manitoba's Green Plan* is a recent provincial government document released in 2012 to guide green economic activities until 2020. One of the highlighted themes is climate change, which contains both mitigation and adaptation initiatives to prepare and prevent climate-change impacts in Manitoba. This important guide provides for a wide range of public participation from municipalities, environmental organizations, industry associations and academics as well as households. For the new mitigation plan, Manitoba attempts to utilize a comprehensive strategy in reducing greenhouse gas (GHG) emission by encouraging stakeholders to get involved in the new plan, enhancing the mandatory reporting of GHG emission, establishing a new biomass economy network, activating a transportation strategy etc. Simultaneously, the adaptation plan stresses the collaboration, attention and innovation from the whole of society, and the pathway is divided into three phases from governmental-wide risk assessment and province-wide risk assessment to adaption strategies and action plans. Considering that climate change impacts on transportation arise from various sources, it might be suggested to focus on multiple areas in this guideline. For example, for flooding issues in Manitoba, it might need to consider disaster prevention, prediction and monitoring (e.g. by 'Light Detection and Ranging (LIDAR) data acquisition' mapping and analyzing techniques and water retention and storage measurements), as well as infrastructure upgrading (e.g. sustainable drainage and green infrastructure) in water management.

Although CentrePort has considered climate-change factors in its business plan (including anti-flooding measures in developing the common-use rail facility and modernizing the highway network and infrastructure, and green sustainable philosophy in reducing GHG and encouraging rail usage), this inland port was established in 2011 and is mainly focused on facilitating investment, regional economics and employment, and promoting and marketing itself in Canada, North America and around the world (CentrePort Canada website, 2015).

3.3 Adaptation planning

It is obvious that there is a lack of attention and specific planning on how to adapt to uncertainties posed by climate change at CentrePort and in the port supply chains overall from both the corporate and government perspective. The factors that require attention include financial budgets and lower-priority awareness among other factors. The chapter argues that an adaptation plan for climate change considering the flooding risks and other uncertainties is very necessary for CentrePort, and will greatly contribute to realizing the objectives of CentrePort's forthcoming five-year plan (2015–19) and developing a strategic inland port. The authors will analyze the necessity, as well as the strengths and weaknesses of adaptation planning for climate change at CentrePort in the following context.

First of all, from the political perspective, adaption planning accommodates the demand for sustainable and green development of the provincial, federal government and international society. Adapting to the impacts of climate change requires open information exchange and communication between internal (CentrePort) and external audiences (municipal politics, international institution, public, etc.). The governmental initiatives for climate change impact environmental policies ("Tomorrow now") and investment in transportation infrastructure (e.g. Highway 75, CentrePort Canada Way) plays a vital role in guaranteeing the traffic safety of CentrePort. Meanwhile, the feedback from CentrePort and its customers in operations and utilization are equally important to modifying and improving the higher-level plans. Therefore, an adaptation plan for climate change at CentrePort will greatly help to bring higher-level knowledge to CentrePort and also provide information to governments regarding its environmental plans.

Most importantly, an adaption plan for climate change would be related to the strategic priorities in CentrePort's five-year plan. For instance, the adaptive flood-proofing measures will minimize the risks in the common-use rail facility in order to facilitate cargo security and efficiency and inland port management. Furthermore, because any chains are only as strong as their weakest link and the uncertainties posed by climate change on CentrePort will impact the efficiency of the supply chain in either a direct or indirect way, an adaption plan which attempts to minimize the risks and uncertainties posed by climate change and maximize the benefits to the relevant stakeholders (the customers, partners, suppliers) will boost marketing and investment promotion in the long run.

There are resources available to support the development and implementation of an adaptation plan for climate change at CentrePort. First, the corporation itself possesses advantages from provincial and municipal governments in terms of funding and infrastructure assistances. In 2013–14, CentrePort Canada Inc., based on the funding agreements, has received $517,880 from the provincial government and $1,284,800 from Western Economic Diversification, and those numbers are expected to continually increase in 2014–15 (CentrePort Canada website, 2015). The relatively rich budget will allow the corporation to become financially self-sufficient by 2017 and make an adaptation plan for climate change in the future. In addition, government investment in main transportation infrastructure including Highway 75 and the newly constructed CentrePort Canada Way have already taken some climate-change factors into account (e.g. flood-proof designing), and have laid a solid foundation for creating an adaptation plan at CentrePort.

As an inland port largely connected by trucks, CentrePort is less vulnerable to climate-change risks than seaports that mainly depend on shipping and rail. Given that the tri-modal transportation network is attributed to multiple partnerships with Bison Transport, Manitoba Infrastructure and Transportation, Canadian

Pacific Railway (CPR), Canadian National Railway (CNR), Boeing and Winnipeg James Armstrong Richardson International Airport etc., an adaptation plan for climate change at CentrePort will benefit from the experiences and achievements of these cooperating entities in adapting to the impacts of climate change. For example, technical and engineering expertise could be provided by different entities such as MIT, and the Manitoba Trucking Association (MTA).

Although the importance of adapting to the impacts of climate change on the transportation sector has only begun to be appreciated by companies and governments in the past few decades, there have been a considerable number of studies and discussions on the relevant topics in Manitoba. For instance, during a workshop entitled *Transportation and Climate Change in Manitoba* (held in Winnipeg in February 2013), participants summarized some significant impacts of climate change on Manitoba's transportation industry (e.g. the risks of melting permafrost on remote communities and the opportunities of winter roads on the port of Churchill), the mitigation measures on reducing GHG emissions in transportation-related areas, as well as adaptation initiatives (e.g. climate-change-damage-tolerant infrastructure) at industry as well as provincial and federal government levels (University of Manitoba Transport Institute website, 2015). Also, a recent initiative project entitled *Climate Risk Assessment of Transportation Requirements for the Manitoba-Nunavut (MB-NU) Supply Chain* provides useful information on the establishment of adaptation plans for climate change along the CentrePort and Hudson Bay Railway supply chains. For instance, the risk-assessment tool PIEVC, in examining climate vulnerabilities for major forms of transportation infrastructure used to connect Northern Manitoba and the Kivalliq Region, could be considered in the case of CentrePort.

4. Hudson Bay Railway

4.1 Introduction

Climate change is a two-edged sword, creating opportunities along with new challenges for OmniTRAX Canada, owner and operator of the Hudson Bay Railway and port of Churchill in Northern Manitoba, Canada. The Hudson Bay Railway connects the Canadian National Railway to the arctic deep-water port of Churchill, transporting commodities such as grain, lumber and petroleum as well as other essential goods to northern communities. Although climate change is responsible for longer shipping seasons and access to previously impassable waterways such as the Northwest Passage, it is also creating unique challenges for the railway that is operated by this company.

4.2 The problem

The Hudson Bay Railway was laid over permanently frozen peatland, which has high water content and becomes very unstable when thawed, unlike bedrock, which provides a stable base whether it is frozen or not. The frozen peatland actually serves as an insulating layer, keeping the ground permanently frozen and relatively stable. During construction, portions of this insulating layer were excavated and removed to even the surface while the track was laid. Actions such as this allow summer heat to penetrate into the permafrost and allow the peatland to melt and become active, reducing stability dramatically. Also, adding to the stability issue faced by the Hudson Bay Railway are the effects of climate change. Permafrost, the type of ground that the railway sits on, remains frozen for at least two consecutive years (Weather Underground, 2014). The top layer of permafrost is known as the active layer, which has been known to thaw in summer. Increased temperatures due to climate change have caused previously inactive areas of permafrost to become increasingly active, resulting in larger areas of permafrost that are susceptible to thawing in the summer months (Weather Underground, 2014). This phenomenon is partially responsible for the increasing instability of the Hudson Bay Railway.

4.3 The impacts

The resulting impact of the unstable railway has resulted in an overall lack of reliability. Recent derailments have caused supply-chain interruptions, resulting in delayed shipments to the port of Churchill that proved costly due to the lack of alternate routes. These interruptions extend beyond delayed movements of goods to ports and essential supplies to communities in northern Canada. Aside from costly air travel, the railway is the only way for passengers to move between Churchill and the more populated areas of southern Manitoba. The Hudson Bay Railway provides a crucial connection for people residing in northern Manitoba. Recently, a derailment of thirteen-grain cars resulted in cancelled passenger services to and from the town of Churchill (Canadian Broadcasting Corporation, 2014).

Even when the railroad is open, slow travel speeds are often required due to uneven and heaving track. This can result in a journey that is relatively rough for passengers, and makes it difficult to transport much of anything other than bulk commodity goods. The lack of stability and reliability of the rail line has recently been cited as a reason for the disapproval by government and interest groups of a proposal by OmniTRAX to move large quantities of petroleum to the port of Churchill via the Hudson Bay Railway (Canadian Broadcasting Corporation, 2014).

4.4 The current plan

Rail upgrades have recently been undertaken to remedy the stability problem. OmniTRAX claims to have invested nearly $110 million in recent track upgrades, with an additional $40 million contributed by the provincial and federal governments (Winnipeg Free Press, 2014). A significant amount of money was spent to enhance the stability of the bed itself, particularly in the area stretching between the towns of Churchill and Gillam in Manitoba.

Although the Hudson Bay Railway is privately owned and operated, it serves as the main connection between the remote areas of northern Manitoba and the more densely populated areas of southern Manitoba for passengers and goods. Despite private ownership, this railway has certain aspects of a public good. The railway was built and operated by the Canadian National Railway until 1997, when it was privatized and portions of rail were sold. Although the railway is now owned and operated by the private sector, the provincial and federal governments still have interests in keeping the track operational due to the number of citizens that rely on the railway.

4.5 The need for an adaptation plan

An adaptation plan should be developed to cope with the negative impacts of climate change while simultaneously providing an outcome that also makes use of the opportunities that climate change can provide. In this case, an appropriate plan would allow OmniTRAX to operate the Hudson Bay Railway despite the increasing destabilization of the ground beneath the tracks as well as provide the company with an opportunity to make use of longer shipping seasons and access to new routes through the port of Churchill. In this case, two approaches come to mind, the first of which is excavation. Removing the peatland that makes up the layer of permafrost that consists of high water content and replacing it with a course material such as crushed rock could increase stability temporarily; however, it would also allow warm summer temperatures to have an access route to penetrate the frozen ground by removing the insulating layer of earth, ultimately resulting in reduced stability. This would likely impact the areas surrounding the tracks negatively, and would be an expensive undertaking regardless. This is not, therefore, a practicable solution. The second approach would be to lift the track and lay an insulating bed of gravel that might reduce the way in which heat can penetrate the top layer of permafrost (according to an anonymous interviewee). Although this is not a permanent solution and would require regular maintenance of (and addition to) the gravel bed, this solution might reduce the negative impact on the surrounding environment and allow the railroad to operate more reliably.

Currently, a formal adaptation plan has not been officially developed. OmniTRAX and the Churchill Gateway Development Corp are working with universities to study climate change and the impact on the port and the railway (Interview with OmniTRAX Canada). The majority of the research is focused on determining predictability of the changes. Once predictability can be adequately determined, a comprehensive adaptation plan will be put in place. The plan will be focused on dealing with the negative impact that climate change has forced onto the railway. The primary focus will involve improving the rail bed to increase stability and reliability to allow the owner to shift attention to the secondary focus, which is to determine how to utilize the opportunities provided by climate change to their advantage. Currently, advantages provided by the longer shipping season are difficult to achieve due to insurance restrictions for vessels travelling to and from the port of Churchill outside of the regular shipping season (Interview with OmniTRAX Canada).

5. Discussions and recommendations

Even with the above resources in both cases, barriers and dilemmas stand in the way of making adaptation plans for climate change. The most concerning issue is the financial constraint: there is a constant negotiation between the public and private sector on who pays for reducing risks. Ultimately governments, public organizations, private entities (e.g. CentrePort itself), and co-investors are responsible, but for what percentage? For CentrePort, within a rather tight budget, how much and at what exact time point should it spend for the future on resiliency, on preparing for 1-every-100-year or 1-every-300-year events? At the same time, macrocosmically, there is little doubt that adaptation planning is necessary and becoming increasingly important with more attention and help being given. However, when institutional guidelines are lacking, how can the high-level knowledge be downscaled into corporate plans and businesses? Since this adaptation plan at CentrePort will involve almost every stakeholder in the supply chain (i.e. Manitoba governments, OmniTRAX, port and town of Churchill, the First Nation communities along the rail line from Thompson to Churchill, Manitoba Hydro and academic institutions, such as the University of Manitoba), the next challenge is how to analyze the trade-offs among different stakeholders, and prioritize and maximize the benefits of the whole supply chain.

In the case of the Hudson Bay Railway, does a privately owned transportation mode that is relied upon by an entire community warrant government intervention and funding to ensure proper maintenance, safety and reliability? Although climate change is not just a government or private-sector issue, at what point does maintenance of a privately owned good become the responsibility of the public sector? On the other hand, in the case of CentrePort, it is noticeable that because adaptation planning is in the conceptual stage, it needs to draw advanced experiences and suggestions from similar inland ports when adapting to climate change risks and uncertainties. However, as CentrePort has its own unique geographic and vulnerable conditions for climate change, accommodating the planning of others into its own pattern is another question that needs to be considered.

A successful adaptation plan is always in line with high public and community participation, and is universally recognized in policy responses to climate change (Few et al., 2007). Recent research on adapting to climate change on a coastal area in the UK describes the significance of effective public participation, including, in particular, local participation, as an important consideration in the decision-making of adaptation planning (IISD et al., 2003; Willows and Connell, 2003). Nevertheless, since climate change is not always the priority in the public mind and political agenda, it would be difficult to collect information in the initialization stages of the plan, as well as to maintain and upgrade information after the plan has been implemented. Thus, how to increase the public participation rate and diversify the sources of participation should be taken into account in long-term adaptation planning.

When conducting adaptation planning, it must be noted that climate change is not simply a public- or private-sector issue; it is an issue that impacts all parties, and therefore all parties must work together to adapt. Our recommendation consists of a five-step approach to climate adaptation planning (see Table 5.1).

Table 5.1 The five-step approach to achieving effective climate adaptation planning for transportation infrastructures

Step	*Description*
1	To increase awareness and understanding of the challenges and risks imposed by climate change. Specific guidelines must be set by governments and interest groups to guide adaptation planning.
2	To maintain transparency and openness when sharing information regarding the magnitude of the risks (a lack of cooperation can result in expensive redundancies and reduce the overall effectiveness of planning).
3	To balance and prioritize the trade-offs between all of the relevant stakeholders. Trade-offs should be balanced in a way that maximizes the outcome and interests for the parties involved.
4	To utilize information from the experiences of other ports in similar situations. Although these experiences are not directly applicable to the port in question, they can often provide valuable information when conducting adaptation planning. It is important to adjust the plan to the local situation. Applying the experiences of other similar ports will help to create a plan that is relevant and useful.
5	To get all relevant stakeholders to participate in the decision-making process. All relevant stakeholders, whether they are governments, industry groups, private-sector firms, or environmental groups, must be involved in the decision-making process to ensure that the interests of each party are represented and all parties buy into the process.

Source: Authors

It is our recommendation to use this five-step approach to determine an adaptation plan that accurately represents the issues addressed in this study as well as the interests of all parties involved. Although we have listed specific suggestions, further research is urgently needed before an effective adaptation plan can be implemented to deal with the negative and positive aspects of climate change.

Conclusion

With the strengthening of adaptation strategies for climate change, the development of ports and port supply chains requires information from worldwide advanced experiences in further adapting to the impacts of climate change. This chapter focuses on climate change and adaptation of transportation and supply chain infrastructures in Manitoba, Canada. By focusing on CentrePort and the Hudson Bay Railway, it examines the way climate change is influencing the port and port supply chains in both positive and negative ways. The provincial and federal governments have undertaken relevant action plans and initiatives and considerable infrastructure investments in both cases. Given favorable flooding prevention and protection from governments, CentrePort has started with a high standard in adapting to flooding as a primary risk posed by climate change in Manitoba. Simultaneously, rail upgrades have recently been undertaken to remedy the stability problems of frozen peatland and permafrost posed by climate change. However, both of the cases failed to realize the potential risks (e.g. tornadoes, heavy storms, extreme cold events, etc.) posed by climate change on the inland port or railway itself and on stakeholders in the supply chain. At the time this study was undertaken, neither CentrePort nor Hudson Bay Railway had a specific adaptation plan for climate change, which is mainly attributed to the deficiencies of top-down policies, prevention awareness as well as advanced experiences in adapting to climate-change uncertainties.

Understanding such, we urge planners, policymakers and other relevant stakeholders to establish relevant climate adaptation plans for CentrePort and the Hudson Bay Railway, and undertake detailed analysis on related needs and resources. It is especially important to focus on the common dilemmas in the trade-offs between stakeholders, public participation and other topics in adaptation planning. Through this process, it should trigger more discussions, and provide workable recommendations for other ports and port supply chain development in adapting to the risks and opportunities posed by climate change.

Acknowledgments

The study is partially funded by the Social Science and Humanities Research Council of Canada (SSHRC)'s Connection Grant Program (43620). The usual disclaimers apply.

References

Canadian Broadcasting Corporation. (2014). 'Via Rail cancels trains to Churchill', June 25,2014,accessibleat:http://www.cbc.ca/news/canada/manitoba/via-rail-cancels-trains-to-churchill-after-derailment-1.2663265

CentrePort Canada website: http://www.centreportcanada.ca (last accessed January 15, 2015).

City of Winnipeg's website: http://www.winnipeg.ca (last accessed November 30, 2014).

d-maps.com (last accessed December 10, 2014).

Few, R., Brown, K. and Tompkins, E. L. (2007). 'Public participation and climate change adaptation: Avoiding the illusion of inclusion', *Climate Policy*, 7 (1): 46–59.

Government of Manitoba. (2002). *Kyoto and Beyond: A Plan of Action to Meet and Exceed Manitoba's Kyoto Targets*. Winnipeg, MB: Province of Manitoba Climate Change Adaptation Plan.

Government of Manitoba. (2013). *Floodfighting in Manitoba.* Retrieved August 20, 2014, at: http://www.gov.mb.ca/asset_library/en/spring_outlook/flood_fighting_2013.pdf

Government of Manitoba. (2014). 'How will climate change affect Manitoba?' accessible at: http://www.gov.mb.ca/conservation/climate/climate_effect.html

Government of Winnipeg. (2014). 'EmergWed: Action Plan, August 20, 2014, accessible at: http://www.winnipeg.ca/emergweb/ActionPlan/SevereStorms.stm

IISD. (2003). *Natural Resource Management and Climate Change Adaptation in a New Approach to the Reduction of Vulnerability and Poverty.* Winnipeg, Canada: International Institute for Sustainable Development.

Manitoban Government's website: http://www.gov.mb.ca/conservation/climate/climate_effect.htm, last accessed on January 28, 2015.

McCutcheon, D. M. and Meredith, J. R. (1993). 'Conducting case study research in operations management', *Journal of Operations Management*, 11 (3): 239–256.

Miles, M. B. and Huberman, A. M. (1984). *Qualitative Data Analysis: A Sourcebook of New Methods*. Thousand Oaks, CA: Sage.

Owen, A. C. (2012). *Industrial and Nutritional Hemp in Manitoba: A Case Study Exploring Stakeholder Strategies and Legitimacy*. Doctoral dissertation, Humboldt State University, Arcata, CA.

University of Manitoba Transport Institute's website: http://www.umti.ca, last accessed on January 29, 2015.

Weather Underground. (2014). 'Permafrost in a warming world', June 17, 2014, accessible at: http://www.wunderground.com/resources/climate/melting_permafrost.asp.

Willows, R., Reynard, N., Meadowcroft, I. and Connell, R. (2003). *Climate Adaptation: Risk, Uncertainty and Decision-Making*. UKCIP Technical Report, UK Climate Impacts Programme, Oxford.

Winnipeg Free Press. (2014). 'Northbound freight service resumes after derailment', June 25, 2014, accessible at: http://www.winnipegfreepress.com/breakingnews/Northbound-rail-service-resumes-after-derailment-262801151.html

6 The impacts of Hurricane Sandy on the Port of New York and New Jersey

Lessons learned for port recovery and resilience

Tiffany C. Smythe

1. Introduction

Improving the resilience of ports to the effects of coastal storms and climate change is an emerging area of research and policy discussion (for example, Becker et al., 2013, Sturgis et al., 2014). Ports and maritime facilities regularly face direct threats from coastal storm events and are especially vulnerable to the impacts of sea-level rise, coastal flooding, and storm intensification that are expected as a result of climate change. Port resilience considerations must include both designing port facilities and systems for these long-term effects, and enabling existing port facility operators to efficiently respond to and recover from the short-term effects of extreme coastal storm and flooding events.

Hurricane Sandy made landfall on the New Jersey coast of the eastern United States on October 29, 2012 as a post-tropical cyclone. The storm resulted in enormous storm surges throughout the greater New York City metropolitan region and caused the Port of New York and New Jersey (see Figure 6.1), which is the largest port on the US east coast, to close to most vessel traffic for nearly a week. The port recovery effort that was undertaken in response was the largest such storm-related effort in the US since Hurricane Katrina in New Orleans in 2005. As post-storm recovery is often considered a "window of opportunity" for improved mitigation and planning for future storms (see, for example, Platt, 1998), the researcher conducted a study during the early stages of the Sandy port recovery process to assess lessons learned from this storm event. The research objective was to gain input from port and maritime professionals on the storm's impacts and the response and recovery process, in order to address the overarching research question: *What lessons learned from Hurricane Sandy can inform long-range planning for future storms and other anticipated climate-change impacts?* This chapter discusses key findings and future research needs that emerged through this study. Given the extraordinary size of this storm and the resultant port recovery effort, emphasis is placed on synthesizing port and maritime professionals' experiences and perceptions as reported during research interviews with the goal of sharing these first-hand perspectives with climate scientists and policymakers.

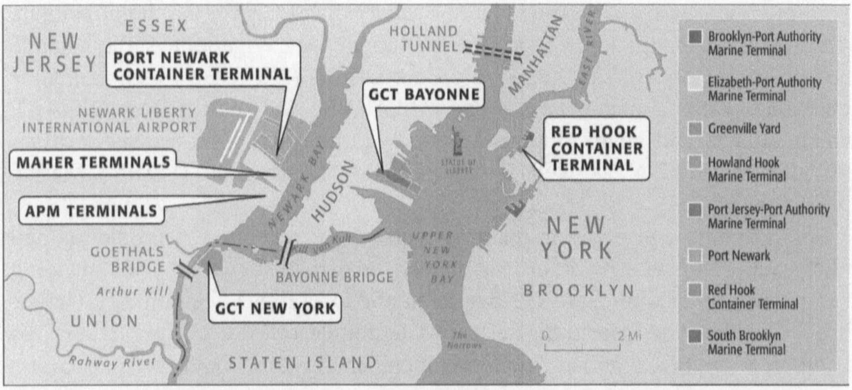

Figure 6.1 Map of the Port of New York and New Jersey, featuring locations of Port Authority marine terminals (Courtesy of the Port Authority of New York and New Jersey).

2. Background and case study context

The Intergovernmental Panel on Climate Change (IPCC) predicts an average global sea level rise of between 0.9 and 2.7 feet by 2100 (IPCC, 2013). Scientists also predict an increase in high-intensity Atlantic tropical storms and associated storm surges (Bender et al., 2010, Grinsted et al., 2013). Due to their coastal location, port facilities and infrastructure are especially vulnerable to the impacts of these trends. Storms and associated flooding can cause major disruptions by interrupting operations, damaging infrastructure, and causing the release of hazardous contaminants into the marine environment. For example, Hurricane Katrina caused $1.7 billion in damages to southern Louisiana ports (Santella et al., 2010). Future storms that may be associated with climate change are expected to have a disproportionate impact on the port/maritime sector due to its exposure and close links to climate.

The Port of New York and New Jersey ("Port") is the third largest port in the US and the largest port on the Atlantic coast (Port Authority of NY and NJ, n.d.). The Port comprises a wide range of maritime operations, facilities, and activities, situated in various locations on the New York and New Jersey waterfronts (Figure 6.1), which are vulnerable to extreme storms and climate change. Cargo operations in the Port are supported by a range of facilities including container ports, oil terminals, vehicle terminals, and other bulk and break bulk cargo facilities. These are not managed comprehensively; the bi-state, public Port Authority of New York and New Jersey (PANYNJ), as a landlord port, owns some of the land hosting these diverse facilities, while other facilities, including oil terminals, are privately owned. Passenger ferries and terminals such as the New York City Department of Transportation's (NYCDOT) Staten Island Ferry are also important to the Port, and provide important transportation redundancy in a region of islands, bridges, and tunnels. Port support services include docking and harbor pilots, such as the private Sandy Hook Pilots, and government vessels and services such as those provided by the US Coast Guard Sector NY (SECNY), that help facilitate safe and secure navigation. Other agencies supporting navigational safety include the US Army Corps of Engineers New York District (USACE), which maintains federal navigational channels, and the NOAA Office of Coast Survey (NOAA), which surveys and charts marine waters. Maritime first response services are provided by SECNY as well as the NY Fire Department (FDNY) Marine Operations unit.

Port planning and decision-making, with regard to both coastal storms and other safety and security issues, is shaped in part by two port-wide standing committees: the Harbor Safety, Navigation and Operations Committee ("Harbor Ops") and the Area Maritime Security Committee (AMSC). The Harbor Ops committee comprises broad, inclusive membership from both the public and private sectors who convene regularly to discuss a wide variety of Port and MTS safety and security matters. Many study participants were members of the Harbor Ops committee, and had been attending meetings and working together for years. The Port's AMSC was first established in 2004 pursuant to the Maritime Transportation Security Act of 2002 (P.L. 107-295), which called for the establishment of maritime security committees in the nation's ports to implement area maritime security plans. Membership is limited to those public and private actors with specific interest in maritime security issues; all study participants represented organizations included in the AMSC.

The Port's Marine Transportation System Recovery Unit (MTSRU) is based out of a subcommittee of the AMSC. The MTSRU is a specialized inter-organizational unit, led by the Coast Guard, and is stood up only when MTS recovery is needed or anticipated. The Coast Guard first established the MTSRU in 2006, pursuant to the Maritime Transportation Security Act, in part in response to lessons learned from Hurricane Katrina about the potential impacts that MTS disruption may have on the economy and community well-being (Torres, 2012). The function of the MTSRU is to facilitate the reopening of the Port and the resumption of maritime commerce in the aftermath of a disaster. The Port's MTSRU, as established during Sandy, included representatives from SECNY; agencies including the federal Maritime Administration (MARAD), NOAA, and USACE; the PANYNJ; the Sandy Hook Pilots; and private-sector representatives from container terminals, oil terminals, and other port businesses (Morrissey, pers. comm., 2013). In the case of Hurricane Sandy, the MTSRU was stood up on October 27, two days before the storm made landfall. Although the MTSRU was only stood up immediately before the storm, members had been pre-identified and had participated in routine organizational meetings and phone calls, as well as annual exercises, in preparation for actual port recovery events.

Several pre-existing storm preparation and response plans provide the foundation for much of the Port's response and recovery work. These included, but are not limited to, the Coast Guard *Hurricane and Severe Weather Plan* (US Coast Guard SECNY, 2012b), and the *Captain of the Port New York Hurricane and Severe Weather Plan for the Port of NY and NJ* (US Coast Guard SECNY, 2012a), which lay out the procedures by which the Coast Guard works with Port partners to prepare for and respond to a major storm event. Other key Coast Guard plans include the MTS Recovery section of SECNY's *Area Maritime Security Plan* (US Coast Guard SECNY, 2009), which lays the groundwork for the MTSRU. Additionally, many organizations have their own in-house preparedness and response plans.

3. Methods

Research was conducted in spring 2013 and constituted a series of key informant interviews. Study participants were identified in coordination with SECNY using a purposive sampling method (Bernard, 2011) in order to identify individuals and organizations who had been directly involved in the port recovery effort. Thirteen interviews were conducted with sixteen individuals representing nine different organizations: SECNY; FDNY; NYCDOT Staten Island Ferry; PANYNJ; NOAA; USACE NY District; Sandy Hook Pilots; the federal Maritime Administration (MARAD); and the NJ Office of Homeland Security and Emergency Preparedness. Interviews employed a semi-structured approach (Bernard, 2011) and were recorded, transcribed, and coded using NVivo 9 qualitative data analysis software (QSR International, 2011). Interview transcripts were coded using a "broad-brush or 'bucket'" coding method (Bazeley, 2007) in order to identify storm events and impacts and key lessons learned.

4. The impacts of Hurricane Sandy

Hurricane Sandy made landfall in the northeastern US near Brigantine, NJ. At that point the storm was a post-tropical cyclone with sustained winds of 70 knots, and was notable because of its size—the storm's diameter extended to 870 nautical miles prior to landfall. As a result it drove an enormous storm surge into the New York, New Jersey, and Connecticut coastlines. The National Hurricane Center reports the highest storm surge as 12.65ft in western Long Island Sound, whereas record storm tides reached 14.06ft above Mean Lower Low Water at the Battery (the southern tip of Manhattan) and 14.58ft at Bergen Point West Reach (the north shore of Staten Island). Inundations on normally dry land were 4 to 9 feet in Staten Island and Manhattan, 3 to 6 feet in Brooklyn and Queens, and 2 to 9 feet in areas of New Jersey containing port infrastructure (Blake et al., 2013).

As a result of these conditions, and because of the storm duration and extent of flooding, there were extensive and prolonged disruptions to many port activities and facilities. Much port activity was shut down for nearly a week. The Coast Guard Captain of the Port closed the entire port before the storm on October 28 and the Port was not fully reopened to all traffic until November 4. The PANYNJ followed a similar schedule in reopening cargo terminals and related facilities. The Staten Island Ferry suspended service for five days, resuming service on November 2. However, numerous port facilities, including container and oil terminals, did not resume full operations once waterways were open due to facility damage and loss of power.

Areas of the Port were closed to traffic for days because SECNY, NOAA, and the USACE first needed to conduct waterways surveys to ensure navigational aids were on station, marine debris such as floating shipping containers was cleared, and shoaling had not created navigational hazards. The USACE and NOAA share responsibility for conducting harbor surveys, but the USACE NY District's waterfront facility, which contained scientific equipment that supported such work, was seriously damaged during the storm, and as a result NOAA performed a greater share of the survey work.

Local response organizations including SECNY and FDNY did not fully shut down during the storm but took numerous steps to reduce exposure while remaining on mission. While both agencies endeavored to perform essential response functions immediately after the storm, extensive flooding and damage at a few key facilities disrupted their work. SECNY suspended all operations at its three small boat stations and its two Aids to Navigation stations, evacuated personnel and families, and moved vessels to safe haven locations according to established hurricane evacuation plans so as to ensure the safety of people and equipment. As soon as it was safe to resume operations after the storm's passage, each SECNY unit began restoration activities to mitigate the damage their infrastructure received. While some units were able to return to full service within hours, others continued operating at a degraded capability for more than six months after the storm while awaiting repairs to waterfront facilities and infrastructure (Pierro, pers. comm., 2013). Additionally, FDNY Marine Unit 9 ran on generator power for over four months after the storm and continued running on temporary utility power for several more months after that (Schug, pers. comm., 2013).

Most participants were still assessing losses and rebuilding costs when interviewed for this study. The PANYNJ reported that it was estimating \$170 million in costs—\$130 million of which was capital (Rooney, pers. comm., 2013). SECNY reported that repair costs for the sector and sub-units were estimated at \$76 million (US Coast Guard SECNY, 2013). These are direct costs only, and do not include indirect costs such as business losses incurred during the port closure.

5. Lessons learned

The following sections summarize key lessons learned as identified by study participants. Overall it is notable that despite the storm's extraordinary impacts, nearly all participants described this response and recovery effort as a success given storm size and surge extent. There was no loss of life within the Port community and no real damage to vessels, and most participants agreed that the Port was reopened, and basic operations restored, within a very short period of time given the extent of damage and disruption. However all study participants identified numerous challenges, many of which were beyond their control, which require consideration for future port resilience planning.

A. Successes

Many of the lessons learned identified herein highlight the success of the response and recovery effort and the strengths that exist in the Port community. These successes represent strengths that the Port community can leverage to deal with future disasters and other problems, and that other port communities can consider within their future planning.

Coordination within the Port. Most participants described their planning, response, and recovery activities as a series of coordination efforts, both within their respective organizations and with external partners. Inter-organizational coordination was especially important with regard to the Port closing and the post-storm recovery of the marine transportation system. Coordination activities included communication between SECNY and Port partners about weather conditions and the closure of the Port; coordination of multiple agencies to support post-storm harbor survey and cleanup activities; and coordination between the public and private sector to facilitate the resumption of port commerce. There was a resounding consensus among participants that port-wide coordination was not only efficient and effective, but evidence of the Port community's strength and resilience.

The extent and efficiency of port-wide coordination may be due to a number of factors. Most important among these was the MTSRU, which had only been implemented in the Port twice before during relatively insignificant storms (Morrissey, pers. comm., 2013). All participants who had participated in the MTSRU emphasized its effectiveness, which they attributed to its well-established structure and plans as well as pre-existing relationships between MTSRU members. A chief example of the MTSRU's effectiveness was the post-storm work between SECNY, Sandy Hook Pilots, NOAA, and the USACE to survey the Port's waterways for navigational hazards. Before the storm, SECNY and Sandy Hook Pilots had arranged to use the pilots' vessels to conduct an initial post-storm waterways assessment. Because the pilot vessels, unlike SECNY and NOAA vessels, had remained in the Harbor throughout the storm, the Pilots were able to get underway immediately after the storm with SECNY and NOAA officials aboard to conduct this assessment. This public–private effort, involving two federal agencies and a business, was coordinated through the MTSRU. In a second example coordinated through the MTSRU, NOAA expanded their survey operations and conducted roughly 68 percent of all waterways surveys (Pounds, pers. comm., 2013) because the USACE's facilities and surveying equipment were damaged during the storm.

Some participants also attributed effective coordination to the Harbor Ops committee and the AMSC, discussed above. All study participants represented organizations involved in at least one of the two committees, and many commented on how their longstanding involvement with these committees made it easier to know who to contact and how best to work together. For example, one participant stated that Harbor Ops helped to build this coordination because it "brought everyone together on a regular basis … People are used to working together for the betterment of the port" (McGovern, pers. comm., 2013).

Relationships and trust. Effective coordination was also facilitated by a network of relationships and trust between Port partners, seemingly built through the committees and other prior experiences of working together. One participant described the Port as "a pretty tight community—everybody knows everybody else and everybody's got everybody else on their speed dial" (Tavolaro, pers. comm., 2013). Another noted "we've been working all these issues for so many years, and developing that trust, so when something happens it just naturally flows. People … are used to working with each other and that trust is already built up" (McGovern, pers. comm., 2013). Some described the resilience of the Port as grounded in these relationships. SECNY representatives emphasized this; one stated "You don't want to meet them in a crisis. You want to meet them when things are quiet, a day like today, and establish those relationships" (Fiumano, pers. comm., 2013). Notably, many of these relationships span jurisdictional or public–private sector boundaries or the competition inherent in the private sector. One participant recounted, "in the height of the fuel crisis, Coast Guard and NOAA were having difficulty finding gasoline to fuel NOAA harbor survey vessels; a MTSRU participant who is a representative of one cargo terminal, a private business, overheard this conversation and offered gasoline to fuel the vessels at no charge" (Sturgis, pers. comm., 2013).

Prior experience. Effective coordination, response, and recovery was also facilitated by Port partners' prior experience dealing with storms and other disasters. Most participants referenced the Port community's experiences during 9/11, and in some cases, individuals recounted detailed stories of how the maritime community coordinated and improvised during the 9/11 response. One noted that the relationships described above are founded in that event: "You saw this in 9/11 and other events … people who would normally compete when it comes to the port, they would work together" (McGovern, pers. comm., 2013). Another participant even described Sandy in those terms: "in the maritime environment, this was our 9/11" (Rooney, pers. comm., 2013).

Others described how the Sandy response and recovery was informed by the Port's experience during Hurricane Irene in 2011. Hurricane Irene had been predicted to have extreme impacts on the greater NYC region, and was the first time that SECNY had fully implemented the MTSRU as an initiative involving both the Coast Guard and Port partners (Morrissey, pers. comm., 2013). A SECNY representative commented, "we were really able to practice in a real-world situation without there being catastrophic damage, and we were really grateful for that. It got the port community ready" (Morrissey, pers. comm., 2013). Given the importance of these prior experiences, as disaster response and recovery simulations are an established practice, future research should investigate how such simulations might be enhanced to better replicate some of the complex response and recovery challenges and lessons learned by Port partners from these real-world disasters.

Expertise and improvisation. Another key to success was Port partners' ability to improvise before, during, and after the storm, drawing upon the relationships described above, as well as prior experiences and professional expertise. This was necessary due to the extraordinary size of the storm and associated surge, which flooded areas and caused damage that for some was unanticipated. One example of improvisation during and after Sandy was recovery and continuity of operations at Coast Guard Station NY. The Station was so severely damaged that it could not resume full operations on site right away, and so firemen at FDNY Marine Unit 9, located nearby on Staten Island, invited Station NY's crew to live and work at their firehouse. As a result, the two units, representing one federal agency and one city agency, cohabitated for nearly three months. Another example is SECNY's work to reopen port facilities despite prolonged power outages. One participant described working to develop alternative compliance measures for security requirements, and working with staff in the field to evaluate conditions: "I said, 'if a proposal seems right while you are visiting the facilities, go ahead and approve it, and we'll work through documenting their plan amendments later.' There was a lot of risk-based operational decision-making physically in the field" (Sturgis, pers. comm., 2013).

B. Challenges

Other lessons learned identified in this study highlight challenges that the Port community, and other ports, may choose to prioritize for future planning and capacity building.

Storm surge. Participants described Sandy as a "surge" event, and extensive flooding damage to waterfront infrastructure—in areas that participants had never known to be flooded or at risk of flooding—was ubiquitous. Storm surge had a significant impact on commercial port operations, due in part to the concentration of infrastructure and resources in low waterfront areas adjacent to commercial berths. Flooding affected these operations by damaging equipment, such as electric motors powering cranes used for offloading cargo; truck chassis used for transporting cargo beyond the Port; docks and piers that support first responder, passenger ferry, and other vital operations; and scientific equipment used for conducting harbor surveys. This suggests a need for storm-surge planning. As several participants pointed out, hurricane plans typically address wind, not surge, events, and no participant indicated having seen flooding like this before in the Port. A PANYNJ representative acknowledged that storm surge had not been part of their hurricane plan; she indicated that this was in part because the "history" of flooding was simply not there, and because the floodplain maps they had been using suggested a risk of minimal flooding in areas that were severely damaged (Rooney, pers. comm., 2013). This illustrates how port managers need access to the best available data on current and projected future storm-surge risks, in a form that is usable by non-scientists, for use in both short- and long-range planning.

Power. Many participants identified the widespread, prolonged power outages following the storm as among the worst of the problems resulting from Sandy. Loss of power, and perceived challenges communicating with electrical companies, impacted the Port in numerous ways. Loss of power meant a loss of communications—landlines, cell phone towers, and Internet—such that many vital recovery operations were conducted using personal devices. Loss of power also meant an inability for terminals to handle product. In the case of PANYNJ facilities, loss of power also resulted in safety and security concerns. Security fences—in some cases surrounding lots full of recently imported cars with the keys inside them—created the risk of theft, and the absence of traffic lights and firefighting equipment created potentially dangerous conditions for staff. Some also emphasized how even after the waterways had been reopened for commerce, oil terminals were still unable to handle and distribute petroleum products because they had no power. This suggests a need for improved coordination between the Port sector and power companies in preparation for future storm events.

Fuel. The power problem was closely related to the fuel problem. Fuel shortages in the immediate aftermath of the storm were widely publicized, with media highlighting lines at gas stations and the implementation of rationing policies. Petroleum products' movement through the Port was delayed not only because of the Port's temporary closure while waterways were being surveyed, but because oil terminals did not have power and therefore could not move product, brought in by tankers, within and beyond their facilities. This problem had a ripple effect through the region and the Port community. Critical-response work such as harbor surveys was temporarily inhibited due to the fuel shortage, and some Port personnel could not get to work. In addition, some participants reported that this also became a public relations problem when politicians and other leaders began criticizing the Coast Guard for ostensibly holding up the flow of fuel and other cargo. Many Port partners were able to improvise to deal with this problem; for example, the PANYNJ brought in a fuel truck to provide gasoline for staff members (Rooney, pers. comm., 2013). To address this issue, organizations may decide to maintain reserve fuel tanks with their mobile response equipment; for example, NOAA indicated that this might be recommended (Pounds, pers. comm., 2013). Perhaps more importantly, however, oil terminals and other facilities may need to improve the resiliency of their energy infrastructure or build in emergency power sources that are designed to withstand storm events.

Waterfront buildings and structures. While effective preparation resulted in no loss of life and nearly no damage to vessels, the storm surge resulted in severe damage to some waterfront buildings and infrastructure that support vital waterfront services. These included the USACE NY District's waterfront facility, two Coast Guard SECNY small boat stations, and a Coast Guard Aids to Navigation station. These also include the Sandy Hook Pilots' building. Damage included extensive flooding inside buildings that destroyed building structure and materials, and damage to adjacent piers and floating docks. Some buildings were rendered virtually uninhabitable. A Sandy Hook Pilot representative recounted having about 4 to 5 feet of standing water in their building, with waves on top of that, and described anecdotes in which a steel door and frame were completely torn out and wood decking on their pier was rolling up and down with the water "like you see on a piano" (McGovern, pers. comm., 2013). While all of these organizations continued to perform their missions during and immediately after the storm, they were all nonetheless faced with extreme and costly damage to their facilities.

In the densely developed metropolitan area surrounding the Port, there are few options for relocating waterfront facilities to reduce vulnerability to storms; one participant commented, “You can’t say we’re gonna move everything … there’s no alternative, just because of where we’re located” (DeSimone, pers. comm., 2013). Instead, waterfront buildings and infrastructure that support water-dependent services will likely need to be redesigned to accommodate future storm events. For most participants, it was too early in the recovery process to have clear visions of how they will redesign or rebuild. However, participants introduced initial ideas about either elevating or redesigning their structures. The Sandy Hook Pilots stated unequivocally that they intend to fully elevate the entire building, and referenced examples from US Gulf Coast pilot stations to illustrate their point. Other participants indicated that they might consider elevating parts of their building, or consider design features and materials that would allow their building to be effectively floodproof. For example, one participant suggested that “wood and sheetrock need to be removed from the building and you need to go in with glass blocks, stainless steel, and appropriate venting. And just assume that the space is going to flood, and design it so that it can flood. And then you can just hose it out” (DeSimone, pers. comm., 2013).

Waterfront electrical infrastructure. In addition to building and structural damage, much waterfront electrical infrastructure was wiped out because of saltwater intrusion or other types of damage. This meant that a large amount of infrastructure on the first floors and in basements of buildings was completely destroyed, and in some cases generators intended to provide alternate power also did not work. A notable example of this was the failure of electrical engines used to power large cargo cranes. Participants described how these cranes once ran on saltwater-tolerant diesel engines, but were switched to electrical power as part of the Port’s efforts to improve air quality. This problem suggests that much waterfront electrical infrastructure should be elevated and perhaps redesigned given the likelihood of future saltwater flooding. Participants described many ideas for elevating, floodproofing, or fundamentally rethinking electrical infrastructure; for example one described installing “almost like a ship’s watertight door” to a space containing many of his facility’s electrical panels (DeSimone, pers. comm., 2013).

Coordination: outside of the Port. While inter-organizational coordination was very strong within the Port community, study results suggest that coordination with entities outside of the Port community—but that the Port relies upon—may be less robust. This issue was evident with regard to power, as discussed above – many participants emphasized that improved communications with the electric companies was vital for future response and recovery efforts—as well as other sectors ranging from inland transportation to water. One participant described this within a broader context of “interdependencies,” commenting: “The Coast Guard coordinates really well … they’re sharing lots of information in the universe of the Port and keep all the major Port players prepared. [But…] There needs to be further coordination between sectors. The port is a sector, let’s say. But then you have the water and wastewater sector, the power sector ... We need to be planning more this way" (Picciano, pers. comm., 2013).

Data and information. Many participants agreed that the sharing of information is nearly always an area for improvement. While many of the information needs are not surprising—timely information on weather conditions, port closures, and areas of the harbor surveyed—two items merit specific attention. One participant indicated the need for improved floodplain maps as well an improved understanding of the Port's vulnerability to climate change and sea-level rise (Rooney, pers. comm., 2013). Another participant identified a similar need, noting that improving the availability of site-specific data on projected storm surge would help both port facilities and adjacent businesses prepare for surge events (Picciano, pers. comm., 2013). This speaks both to the importance of sharing information between organizations, and to the ways in which the scientific community can support the maritime community.

"Messaging" the Port. A somewhat unexpected issue was the problem many participants had communicating about port recovery with politicians, public officials, and other senior leaders in government. In the midst of response and recovery activities, SECNY representatives were fielding inquiries about why the Port was closed and the distribution of petroleum products delayed. Several participants even mentioned incidents in which SECNY's closure of the Port was blamed for causing the regional fuel shortage. Because of this, Coast Guard officials who were overseeing the port recovery work were simultaneously engaged in "messaging"—conducting their own public relations by explaining why harbor surveys must be conducted before reopening the Port and that even once the Port was fully reopened, oil terminals and refineries still must have power and functioning infrastructure in order to offload and distribute fuel. As such, Port partners were effectively educating the broader community during the port recovery effort about how the Port, critical infrastructure, and the supply chain work (Sturgis, pers. comm., 2013). This issue suggests that future port resilience planning must include educating political leaders and the broader public about the Port and the recovery process.

Personnel management. A final challenge was that of personnel management. One problem was transportation: the region's mass transportation infrastructure and roadways were all compromised, and then fuel shortages became evident, so personnel who had evacuated Port facilities during the storm had trouble getting back to work to support recovery efforts. As a PANYNJ official noted, "It's difficult to send them out and bring them back" (Larrabee, pers. comm., 2013). Further, many personnel had experienced storm disruption in their personal lives; many had damage, no power, and in some cases total losses, at their homes, and were required to temporarily relocate their families. SECNY personnel and families were required to evacuate housing at sites like Sandy Hook. Another problem was safety and security at port facilities. A PANYNJ participant noted how power outages meant that there were no traffic lights and life support equipment (i.e. firefighting), rendering conditions potentially unsafe, such that facilities were open only to essential staff, during daylight hours, for days following the storm (Rooney, pers. comm., 2013). These problems suggest that personnel management may need to be more explicitly and comprehensively integrated into the Port's future storm planning.

6. Future research questions

This study also identified numerous questions—many which were explicitly raised by participants—which merit future research. Investigation of these questions will help this and other port communities plan for future storm events and the expected impacts of climate change.

How can events like Sandy generate political will in support of change?

Some participants noted that Sandy may mark a significant change or shift in thinking about and planning for storm events. One remarked that Sandy "made believers out of some cynics who said this climate change stuff is a bunch of nonsense … it's made believers out of all of us" (Larrabee, pers. comm., 2013). However many of the same participants expressed concern that it might not result in real change, especially when it comes to spending money on costly infrastructure improvements and resilience measures. "The real question is not so much what do we do, but how do we find the will to fix some of these things?" commented one participant (Larrabee, pers. comm., 2013). Research is needed to determine how events like Sandy can be most effectively leveraged, within the real constraints of politics and budgets, to generate political will in support of change over both the short and the long term.

What storm and flood protection ideas are practical, cost effective, and feasible?

While many ideas, proposals, and reports have been introduced following the storm, participants emphasized the importance of finding solutions that are practical, feasible, and affordable within the context of current and projected future storm-related hazards. In particular, many spoke with skepticism about a highly publicized proposal to install a storm-surge barrier across the mouth of the harbor under the Verrazano Narrows Bridge. Many others noted with skepticism the ideas of relocating facilities or elevating infrastructure within a system as old and complex as the Port of NY and NJ. One participant commented that her organization really needed "feasibility studies" to help it make decisions about "reasonable mitigation measures" (Rooney, pers. comm., 2013). Port operators need applied research and practical decision support tools that they can use to make practical decisions about realistic and cost-effective protection measures given the reality of future storm and flooding risks.

How can the Port community's social capital be expanded and replicated elsewhere?

The powerful relationships between port partners that are discussed above are a form of social capital, which has been identified as a key to achieving resiliency to coastal disasters and climate change (see, for example, Adger, 2003). This research suggests there is a great deal of social capital in the Port of NY and NJ. While this study has not compared this port with other ports or other storm events, it is likely that some other ports could benefit from augmenting their social capital. Further research is needed to better understand the Port of NY and NJ's social capital—both how it can be expanded to improve coordination with non-port entities, and how social capital can be built in other port communities.

What are the attributes of a resilient port?

Nearly every participant used the term "resiliency," even though the researcher did not initially introduce this term. Participants offered widely varying conceptualizations of resilience. Some described it in economic terms, emphasizing the resiliency of the supply chain and maritime commerce (see, for example, Trepte and Rice, 2014). Others described it in physical terms, describing it as the ability of buildings, transportation and electrical systems, and other infrastructure to withstand or bounce back quickly from a disaster, whether a natural disaster (see, for example, National Academies, 2012) or an act of terrorism (see, for example, Walklate et al., 2013). Still others described it in entirely social terms, emphasizing the interconnectedness between sectors, or the relationships and trust that made working with the port partners so effective (see, for example, Adger, 2003). It may be that a resilient port is all of those things. However, given the widespread use of this language, including by government agencies spearheading resiliency initiatives, achieving a common, practical, applied vision of port resiliency—one that can easily be operationalized by port managers—may help guide and improve port resilience planning efforts.

How can ports move forward in long-range climate change adaptation planning?

Hurricane Sandy illustrates the Port of NY and NJ's extreme vulnerability to extreme storms and flooding events, which are expected to become more common due to the effects of climate change. This study revealed that the Port community has many strengths, but that the community does not appear to be leveraging these strengths to address the need for climate change adaptation planning. Research is needed to examine how this and other ports can utilize existing assets—port governance mechanisms, leaders, experienced professionals, and social capital—to facilitate climate-change adaptation planning.

Conclusion

This study has identified a series of lessons learned, based on the Port of NY and NJ's experience with Hurricane Sandy, that can be used to inform planning for port recovery and resilience. Strengths and successes include coordination within the Port community; relationships and trust; prior experiences; and professional expertise and the ability to improvise. Challenges and areas for improvement include storm-surge planning; prolonged power outages and fuel shortages; impacts to waterfront buildings, structures, and electrical infrastructure; coordination beyond the Port community; sharing data and information; "messaging the Port"; and personnel management. In addition, this study has identified a series of research questions: How can storm events like Sandy generate political will for change? What storm protection measures are practical and appropriate? How can the social capital of the port be leveraged within the port and replicated elsewhere? What is a resilient port? And how can the strengths of the port community be leveraged in support of climate change adaptation planning?

Future research is necessary to help port and maritime professionals improve the practice of port resilience planning, which includes responding to and recovering from extreme storms and flooding events. The lessons learned and research questions identified herein require further investigation and merit priority attention because they were identified by port professionals who responded to the biggest storm-related port recovery event in the US since Hurricane Katrina. Future port resilience and climate-change adaptation research should seek to leverage port communities' existing strengths and prioritize the development of practical decision support and planning tools that port communities can use to enhance their port resilience planning efforts.

Acknowledgment

This research was funded in part through the University of Colorado Natural Hazards Center's Quick Response Grant Program, funded by National Science Foundation grant CMMI 1030670. Research was supported by the US Coast Guard Academy Center for Maritime Policy and Strategy.

Disclaimer

The views expressed in this report are those of the author and do not represent the official policy or position of the US Coast Guard, Department of Homeland Security, or the US government.

References

Adger, W. N. (2003). 'Social capital, collective action and adaptation to climate change', *Economic Geography*, 79 (4): 387–404.

Bazeley, P. (2007). *Qualitative Data Analysis with NVivo*. London: Sage Publications.

Becker, A., Acciaro, M., Asariotis, R., Cabrera, E., Cretegny, L., Crist, P., Esteban, M., Mather, A., Messner, S., Naruse, S., Ng, A., Rahmstorf, S., Savonis, M., Song, D., Stenek, V. and Velegrakis, A. (2013). 'A note on climate change adaptation for seaports: A challenge for global ports, a challenge for global society', *Climatic Change*, 120 (4): 683–695.

Bender, M., Knutson, T., Tuleya, R., Sirutis, J., Vecchi, G., Garner, S. and Held, I. (2010). 'Modeled impact of anthropogenic warming on the frequency of intense Atlantic hurricanes', *Science*, 327 (5964): 454–458.

Bernard, H. R. (2011). *Research Methods in Anthropology*, 5th ed. Lanham, MD: Alta Mira Press.

Blake, E. S., Kimberlain, T. B., Berg, R. J., Cangialosi, J. P. and Beven II, J. L. (2013). *Tropical Cyclone Report: Hurricane Sandy (AL182012), 22–29 October 2012*. National Hurricane Center.

Grinsted, A., Moore, J. and Jevrejeva, S. (2013). 'Projected Atlantic hurricane surge threat from rising temperatures', *Proceedings of the National Academy of Sciences*, 110 (14): 5369–5373.

Intergovernmental Panel on Climate Change. (2013). *Climate Change 2013: The Physical Science Basis*. Online at: https://www.ipcc.ch/report/ar5/wg1/.

National Academies Committee on Increasing National Resilience to Hazards and Disasters and Committee on Science, Engineering, and Public Policy. (2012). *Disaster Resilience: A National Imperative*. Washington, DC: The National Academies Press.

Platt, R. H. (1998). *Disasters and Democracy*. Washington, DC: Island Press.

Port Authority of NY and NJ. (n.d.). "About the Port." Online at: http://www.panynj.gov/port/about-port.html (last accessed November 17, 2014).

QSR International. (2011). *NVivo 9 Qualitative Data Analysis Software*.

Santella, N., Steinberg, L. and Sengul, H. (2010). 'Petroleum and hazardous material releases from industrial facilities associated with Hurricane Katrina', *Risk Analysis*, 30 (4):635–649.

Sturgis, L., Smythe, T. and Tucci, A. (2014). 'Port recovery in the aftermath of Hurricane Sandy: Improving port resiliency in the era of climate change', *Voices From the Field* Series. Center for a New American Security.

Torres, C. (2012). 'Protecting the supply chain: The marine transportation system recovery unit', *Coast Guard Proceedings*, Winter 2011–2012: 67–68.

Trepte, K. and Rice, J. (2014). 'An initial exploration of port capacity bottlenecks in the USA port system and the implications on resilience', *International Journal of Shipping and Transport Logistics*, 6 (3): 339–355.

US Coast Guard Sector NY. (2009). *Area Maritime Security Plan*.

US Coast Guard Sector NY. (2012a). *Captain of the Port New York Hurricane and Severe Weather Plan for the Port of NY and NJ*.

US Coast Guard Sector NY. (2012b). *Hurricane and Severe Weather Plan*.

US Coast Guard Sector NY. (2013). *Hurricane Sandy 101* (Sector NY Briefing).

Walklate, S., McGarry, R. and Mythen, G. (2014). 'Searching for resilience: A conceptual excavation', *Armed Forces and Society*, 40 (3): 408–427.

7 Climate adaptation of German North Sea ports

The example of Bremerhaven

Winfried Osthorst

1. Introduction

During the last years, adaptation to climate change has been recognized as a serious challenge for practitioners from various sectors that requires responses from private as well as public actors. The relevant scientific discourse reviewed a first generation of existing policies and strategic approaches (for example, Bauer et al., 2012; Swart et al., 2009, for the European context). Here, the governance dimension of adaptation has been highlighted and core principles of successful adaptation have been identified for different contexts, underlining most notably the demand for policy integration as *"the heart of adaptation"* (Mickwitz et al., 2009: 37).

The discourses on understanding adaptation governance, however, reflect insights into environmental policy and climate policy research, and have to be tied to the literature on port management and port governance. The latter has predominantly been occupied with the liberalization of the sector and its effects, with competition among locations or elements of the value chains within specific markets, and with regional actor constellations in regard to developmental demands. Linking the two debates allows for an understanding of individual sectoral actors' roles and rationales within adaptation processes. This yields insights into the factors supporting the development of integrative port adaptation strategies.

Obviously, the rationales of European ports, including the major German ones, are governed by aspects that touch directly upon the preconditions for successful climate adaptation strategies: the demand for sectoral development within a growing market, at environmentally sensitive locations necessarily vulnerable to climate change, makes the *degree of integration* of the ports' adaptation efforts the crucial issue.

This article investigates a recent climate adaptation project of the port of Bremerhaven, Germany, as an empirical case. It describes how the port seeks to combine market orientation with sustainability. Further, varying responses and vulnerabilities of different parts of the value-chains are identified. The case study reveals that existing ambitious development efforts need to be re-examined, if climate adaptation is taken seriously.

The next section relates climate adaptation research to the literature on port governance and identifies preconditions and prerequisites for integrative adaptation approaches within the port sector. The third section provides an overview about German adaptation policy and its relevance for the empirical case. The forth section presents the climate adaptation project of Bremerhaven as an inter-organizational dialogue, favoring the development of a problem perception shared by decisive local stakeholders over planning technical measures. The brief discussion focuses on the port authority's role in structuring regional processes and mediating environmental conflicts as factors shaping port development.

2. Climate adaptation as an issue of policy integration—climate adaptation and port governance

As a consequence of both country-specific new public management strategies (Baird, 2000) and globalization in broader terms (Juhel, 2001), port management research has focused on the various forms of organizing port functions along the public–private divide, discussing their effects on efficiency, the extent of marketization, and the organization of regulatory tasks (for an overview, see Brooks and Cullinane, 2007). Research highlights the broad range of mixed constellations, reflecting political constraints, path dependencies resulting from state-specific task distribution (among levels, for instance), and differences between market segments (Ng and Pallis, 2010). A generally established result, however, is that ports have increasingly become elements of value chains (Robinson, 2002). In some world regions, including in Europe, competition among ports for market shares serving wider economic hinterlands, and for dominant positions within spatially extended port systems has become a reality (Notteboom, 2010). Hence, maintaining locational quality has become a major issue for the ports, including the demand for continually increasing hinterland infrastructure capacities, and good access conditions for ever larger ships. The integration of the cluster into the regional institutional landscape has thus become a decisive prerequisite for ensuring political and financial support, as well as administrative and training capacities. At the same time, conflict management has become crucial, as environmental impacts of the port operations and competition for space need to be addressed (de Langen, 2007).

Such considerations are also relevant for climate adaptation. First, the specific delimitation of regulatory and operational functions between public and private actors shape their capacities and rationales. Here, the notion of distinct state traditions being a relevant factor in distributing regulatory roles, calls for an understanding of the specific regulatory framework governing ports, as well as climate adaptation. Moreover, a differentiated analysis of roles or market niches disaggregating abstract functions (for example, Baltazar and Brooks, 2001) allows for a better understanding of potential constraints when organizing adaptation within a particular setting. Second, the port literature's emphasis on market pressure, growth orientation, and locational qualities raises the question of balancing sectoral demands with competing interests which are also affected by climate change.

Regarding the requirements for adaptation governance, policy integration refers to instruments which strengthen the prioritized cross-cutting issues in decision-making processes by internal procedures (for example, impact assessments or cross-sectoral consent in decision making), monitoring mechanisms, and sectoral regulations, supporting, for example, environmental assets (Jänicke and Jörgens, 2004). When it comes to climate change, climate policy integration (CPI) has emerged as an independent requirement for the development of adaptation strategies, emphasizing that sectoral adaptation efforts must neither undermine climate protection nor other sectors' sustainability targets (for example, biodiversity) (Mickwitz et al., 2009).

Hierarchic regulation and network-based voluntary approaches are both considered to be important for adaptation policies. While the former is supposed to reduce target conflicts between sectors and to define reliable frame conditions for *all* affected sectoral developments, the latter is expected to support the development of trust, to foster learning and to deal with uncertainty. Empirically, however, comparisons of European adaptation policies prove the dominance of voluntary approaches and a shortfall of regulative measures (Swart et al., 2009). A further function of hierarchic regulation should be the provision of coherence, pragmatically understood as the unambiguous definition of frame conditions for different sectors. In the light of persistent contradictions between crucial instruments supporting, for instance, agriculture, energy provision, or nature protection, climate adaptation is presently being seen as taking place under incoherent conditions (Beck et al., 2009).

However, the literature on adaptation governance also underlines the indispensability of sector-based processes which ensure the relevant actors' commitment, the recognition of sectoral conditions, and internal knowledge. Inevitably, these approaches are partly in conflict with integrative principles, since sectoral actors are bound to specific normative frames and problem perceptions. Hence, sector-based approaches bear the risk of enforcing underlying target conflicts between different sectors. A minimum requirement for avoiding maladaptation is seen in the voluntary *"commitment to minimize contradictions"* (Mickwitz et al., 2009: 22, OECD, 2002).

Under present conditions, regional initiatives are considered to be crucial for ambitious adaptation concepts, since local governments are supposed to have access to actors who are relevant for pursuing projects, including private and public organizations operating at different levels. As a consequence, a network-based approach is virtually inevitable when it comes to project implementation. A further condition for success is the existence of an active promoter that functions as lead agency and manages the interfaces with the various partners. In addition, inclusiveness is considered to be a procedural precondition (Swart et al., 2009, referring to criteria developed in OECD, 2002 and EEA, 2003).

These requirements pinpoint the preconditions for successful climate adaptation of the northwest European ports. Here, the distinction between a "Latin" centralized port management tradition, a Hanseatic tradition of municipally owned ports, and an Anglo-Saxon tradition of independent public or private port management seems to matter, despite strong similarities in their recent developments (Verhoeven, 2011). With their strong regional ties, Hanseatic-type port authorities seem to have the organizational preconditions for fulfilling the role of a lead agency within sectoral or regional projects with basically integrative aspirations.

The developmental demands of Europe's most important ports, however, may affect other sectors that are particularly challenged by the consequences of climate change. Relevant examples are water management, agriculture as well as nature-, coastal- and flood-protection in coastal regions (European Commission, 2013: 3–5). The foreseeable increase in traffic, at sea and in the hinterlands, will result in increasing emissions, contradicting politically defined CO_2 reduction targets (Lloyd's Register et al., 2013).

Another aspect is the governance challenges currently present in European port systems which influence their strategic orientation. One field is European environmental regulation, including the Fauna, Flora and Habitats Directive (FFH; 92/43/EEC) and the Water Framework Directive (WFD, 2000/60/EC). While the FFH-Directive aims at the conservation of ecologic functions, and requires the protection of species and habitats, including estuaries and river systems, the WFD focuses, among others, on the improvement of environmental statuses of river systems and estuaries. If taken seriously, both regulations establish rigid requirements for impact assessments, ecosystem management, approval procedures, and compensation standards. Moreover, the regulations define competencies of the EU Commission and the European jurisdiction within the respective project-related procedures, thereby increasing their uncertainty and complexity. Despite the port authorities' increasing capacities in handling such challenging requirements, the sector considers the consequences of these regulations to be examples of policy incoherence (Verhoeven, 2009). Growing demands for operational efficiency across value chains also enforce *logistics integration*. The emphasis here is on restructuring the ports' internal processes, hinterland operations, and cooperation with inland terminals, in order to meet the foreseen demand for increasing port capacities. The strategies of dominant market players challenge ports, especially through decisions about major shipping lines and terminal capacities that shape the ports' stance within the competitive market.

In light of these frame conditions, the *degree of integration* of the ports' adaptation efforts becomes the crucial issue.

Taking into account the industries' engineering competencies and financial capacities, the ports' capacities to implement technical and operational measures that reduce their vulnerability to the consequences of climate change may be considered high in comparison to other sectors or parts of society. However, along all elements of the value chain (ranging from the waterways to the ports themselves to their hinterland infrastructures and inland terminals), the orientation and design of adaptation measures may take on different forms. Structured according to the CPI requirements, types range from *growth-oriented adaptation*, focusing solely on the ports' functionality and port-specific climate change consequences, to *green-efficiency-oriented adaptation*, contributing to a broader range of regional adaptation demands, to *climate-proof-oriented adaptation*, which seek to address port climate adaptation and market-oriented development within the context of overarching societal developmental demands (Osthorst and Mänz, 2012).

3. German climate adaptation policy as context for the northwest as a port region

In order to fulfill international obligations, the German government developed a National Adaptation Strategy (NAS) and specified measures (Federal Government, 2008), together with an action plan to ensure their implementation (Bundesregierung, 2011). Main concerns apply to the rise of mean global temperatures and the ensuing consequences, such as an increase in the number and intensity of extreme weather or climate events. The NAS refers to the Intergovernmental Panel on Climate Change's (IPCC) concept of vulnerability, seeing it as a function of exposition, sensitivity, and adaptive capacity of affected systems (IPCC, 2007). It identified 15 sectors and areas in Germany as being vulnerable, including human health, the building sector, transport and transport infrastructure. Since possible vulnerabilities vary across regions and sectors, the NAS seeks to improve the knowledge base, create transparency and participation through a broad process of communication and dialogue, support awareness and information through comprehensive public relations and, finally, develop strategies to deal with uncertainty. Its emphasis lies on regions (Federal Government, 2008). Thereby, the responsible Federal Ministry for the Environment, Nature Conservation, and Nuclear Safety (BMU)[1] reacted not only to the conceptual particularities of climate adaptation, but also to the realities of the German federal system, since a number of legislative and administrative tasks fall into the purview of the *Länder*. While the ports themselves are an issue of the *Länder*, all major traffic infrastructures, including waterways, motorways, and railways, are regulated and financed by the federal government. At present, all major ports demand infrastructure extensions, including the deepening of the rivers *Elbe* (the waterway of Hamburg) and *Weser* (the waterway of Bremerhaven, Brake, and Bremen).

As is routine in the German political system, the development of the NAS and the action plan took place in close cooperation with the respective ministries of the *Länder* as well as other stakeholders (Swart et al., 2009: 218). As far as the NAS addressed federal tasks, it led to modifications of the legislative and norm-related framework, for example the amendment of the Spatial Planning Act in 2008, the amendment of the Water Resources Act in 2009, and the amendment of the Building Codex in 2011. Further actions supported activities of the lower levels, including the *Länder* and municipalities.

In general, the federal administration is constrained by the principle of subsidiarity, while simultaneously it is obliged to provide guidance and coordination and to organize consent among different polities. One established instrument applied under such conditions, understood as 'golden rein', is competitive programs, established for funding autonomous activities of the lower levels which fulfill the federal administration's (or European) criteria.

A pivotal activity is the research and development program 'KLIMZUG—Managing climate change in the regions for the future' of the Federal Ministry of Education and Research (BMBF) (2008–14; funding: €83 million). Following a call for proposals, networks of German regions could apply for funding in a two-stage process. Of the seven projects selected for funding, two included subprojects that addressed port issues: RADOST, focused on the port of Lübeck on the Baltic Sea (see Schröder and Hirschfeld, 2014), and NORDWEST 2050, focused on the ports neighboring the Weser Estuary.[2] The KLIMZUG program aimed at the implementation of innovative adaptation strategies to climate change and related weather extremes. This was to be accomplished in a region-specific manner through the creation of research networks of political and administrative actors, the economy, society, and science. Noteworthy, the funded activities were meant to foster the integration of the anticipated climate-change effects into regional planning processes, and intended to increase competitiveness by advancing new technologies, procedures, and strategies (Federal Ministry of Education and Research, 2012). Thereby, KLIMZUG shows strong similarities to other programs addressing climate change in the fields of spatial planning and urban development, originally designed by the BMVBS,[3] such as *KlimaMoro* and *KlimaExWoSt*.

NORDWEST 2050 was characterized by a cooperation of the involved sectors', academics', and administrative functionaries' networks. The "Metropolregion Bremen-Oldenburg im Nordwesten e.V." (Bremen-Oldenburg Metropolitan Region Association), a regional cooperation body embracing Bremen and its surrounding Lower-Saxon neighborhood, became the leading consortium partner. With an area of 13.750km^2 and 2.7 million inhabitants, the "Metropolregion" covers a considerable part of the northwestern part of the *Land* of Niedersachsen (Lower-Saxony), as well as the cities of Bremen and Bremerhaven that constitute the *Land* of Bremen. Additional major economic centers are Oldenburg and Wilhelmshaven. Methodically, the project applied two approaches. First, it assessed the vulnerability of crucial regional sectors, including the food, energy, and port sectors, as well as port-related logistics, and it addressed them by sector-specific innovations and sectoral road-maps. Second, an integrative long-term roadmap for climate adaptation in the entire Metropolitan Region defined demands for structural changes until 2050.

With regard to the port sector, the project included formal cooperation with the port authorities of the two states, *bremenports GmbH & Co. KG* and *niedersachsenports*, who are responsible for infrastructure management and development, and were reshaped as private enterprises in public ownership. While the city-state Bremen is clearly of the Hanseatic governance type, the governance structures in Niedersachsen are inconsistent, combining centralized elements, privatized constellations, and port-specific arrangements between different partners, including municipalities. Responding to agreements at the federal level, both states started to prepare a state adaptation strategy. In this context, both *bremenports* and *niedersachsenports* intended to use project resources for enhancing sector-specific competencies relevant for their work on port adaptation concepts. Capacity building and support of the communication with sectoral stakeholders, including superior departments, were particularly required. The cooperation with *bremenports*, however, aimed additionally and explicitly at supporting port development aspects and integrative solutions for a number of aspects.

The regional port industry, situated at the Weser estuary and the neighboring Jade bay, claims that about 85,000 of the 750,000 jobs within the region depend directly on the ports and related logistics (ISL and BAW, 2010: 9). The cluster, particularly its container segment, benefits from its role within the German export economy. Important ports are Bremen, Bremerhaven, and Wilhelmshaven. The port of Bremerhaven is considered to be one of the largest European container ports and to be the largest trans-shipment place for the automotive. Wilhelmshaven, on the other hand, is the German crude oil port and the location of the new Jade-Port container terminal. Other middle-sized ports with terminals specialized in particular goods or industries are located in the cities of Nordenham, Brake, and Cuxhaven. Additional inland-ports exist in Oldenburg and Cloppenburg.

4. Developing a climate adaptation approach for the port of Bremerhaven

An important frame condition for any attempt to design a sectoral climate adaptation strategy is the status of port development within the region. The container port of Bremerhaven, for example, has been extended through a sequence of large-scale terminal projects. Further, draught conditions of the Weser were improved through projects deepening the waterway. However, the most recent Weser deepening project is currently blocked by legal controversies. These projects, demanding substantial public investment at state level, were prioritised by economic associations, by the port sector itself, and by the state's political decision makers. Environmental actors, on the other side, criticized their negative impacts on the estuary and coastal ecosystems, and the region's growing vulnerability to flood events. Governed by a SPD–Green coalition since 2007, the relation between economic and environmental actors and positions remained a sensible issue. The respective sectoral administrations and agencies, including *bremenports*, struggle to reduce the sectoral projects' environmental impacts to a legally and politically acceptable degree. A number of crucial economic actors, however, still regard concepts representing or reflecting environmental problems, including climate protections and climate adaptation, as strategically threatening.

Under the condition of trust-based networking processes, the acceptance of ongoing port development activities as sectoral status quo, including the contested Weser deepening, thus became a precondition for the work on climate adaptation.

Starting in spring 2009, the project included a number of elements dealing with the regional port sector's climate adaptation: First, a vulnerability assessment applied existing climate data on regional and sectoral specifics. Second, a number of network-based approaches addressed different sectoral communities, namely the *port-related transport and logistics industry,* and a workshop series including experts of *niedersachsenports.* The enterprises located in the *Intermodal Logistic*

Zone Bremen (GVZ) were addressed by an interactive study, authored by ISL (2011), to identify location-specific adaptation demands. Most ambitious, however, was the workshop-based project *"Resilient Port Infrastructures,"* bringing together different branches of *bremenports* with ministerial actors, major port enterprises, and commercial associations. Its participants engaged in a strategic dialogue on port development and climate adaptation, focusing on the port of Bremerhaven as an exemplary case. Thus, after a brief overview of some of the core results of the other project elements, the following section will focus on the central features of this project component.

The *vulnerability assessment* for the port sector particularly emphasized *adaptation capacities*, taking into account sectoral market dynamics and actor relations. It was based on an analysis of the region's *exposition* to climate change that reinterpreted existing climate models (for details, see Schuchardt et al., 2010). All findings are characterized by high uncertainties and considerable value ranges for the time spans investigated. To say the least, compared to other parts of the world the overall exposition to the consequences of climate change can be considered moderate. In detail, the mean temperature is expected to increase between 1 and 2°C until 2050; whereas the sea level is anticipated to rise between 9 and 70cm due to the combined effects of climate change, tectonic processes, and changing tides. Extreme weather or climate events are estimated to occur more often: days with torrential rain are expected to increase by 1, storm days by 0.4 days per year. Particularly significant are increases of storm flood levels, estimated to reach between 19 and 111cm. However, changes are expected to advance progressively in the long term after 2050, dependent on the development of global greenhouse gas emissions.

Such climate effects may raise the sector's *sensitivity*. An increase and intensification of storms, strong rain, hail, or heat waves may damage buildings, installations, vehicles, and goods, resulting in possibly higher economic losses to the affected enterprises. The reliability of major transport infrastructures, constituting the backbone of port-related logistic processes, may be reduced.

For the affected *elements of port-related value chains*, however, optimized procedures, risk management, and the use of improved equipment are realistic and cost-effective reactions for enterprises, since they can be implemented in the long run, using normal replacement rates. Higher standards for equipment and building can be ensured by tightening regulation, particularly on the federal level. As a consequence, the *adaptation capacity* is considered high for this part of the sector, leading to *low vulnerability*. More problematic are *transport infrastructures*: given underfinanced budgets, complex planning and decision procedures of major traffic projects across political levels, and simultaneously growing demands for substantial capacity increases of motorways and railways, especially in port regions, ensuring transports infrastructures' reliability is already considered a political and financial challenge. Due to these institutional factors, the *adaptation capacities* of transport infrastructures are estimated *medium*, resulting in *heightened vulnerability* (Osthorst et al., 2012).

Considering these findings, the project started a number of workshops for the logistics sector, jointly organized by the *Chamber of Commerce* and *VIA Bremen*, a sectoral marketing and cooperation platform, with participation of sectoral business associations, individual enterprises, and training agencies. Obviously, the industry only recently started to consider adaptation to be an issue of importance. By contrast, energy preservation and enhanced energy efficiency have long been accepted as responses to climate protection demands, not least because of market pressures and legal obligations. Protecting equipment and buildings proved to be a challenge for the sector due to the short-term character of its contracts. Thus, improvements are supposed to take place within the context of usual replacements, while fulfilling legal requirements. Only larger enterprises acted strategically in reducing locational sensitivity, improving transport routing and protection of goods during the transport (for example, with regard to heat), and addressing climate protection as a market issue, with initiatives establishing CO_2-free transport chains. Interestingly, the industry's strong reliance on the states' regulatory competencies and on standards of publicly provided infrastructures is obvious.

Subsequently, the subproject *"Resilient Port Infrastructures"*, jointly designed by *bremenports* and the *University of Applied Sciences Bremen*, approached climate adaptation within the context of port development in a narrow sense, and with the focus on Bremerhaven, due to resource and capacity restrictions. Its explicit aim was to contribute to Bremen's state adaptation strategy and recognize the ports' developmental demands while contributing to the ports' sustainability. Hence, an improved congruence of problem perceptions and coordination between decisive actors was considered crucial. Thus, improving communication between regional actors was prioritized over identifying specific technical or financial adaptation demands. As a consequence, the subproject conducted a sequence of two workshop-based dialogues, accentuating their professional preparation and realization, including synthesizing intermediary and final results. The first series of workshops in 2011 served as a stocktaking of knowledge and of ongoing planning activities, involving mainly the different organizations' operational level, including the port authority's various departments, departments of the State Ministry for Ports and Economy, as well as the State Ministry of the Environment's representatives, and a terminal operator. The working group worked on identifying adaptation demands in a narrow sense, scoping potentials for innovation, and reassessing existing sustainability efforts. Based on its results, the network participants decided to align adaptation and sustainability demands more strongly with strategic port development. Thus, the second series of workshops (2012–14) was methodologically organized as a process of assessing the drivers and frame conditions for the ports' development, identifying options, narrowing down subsequent developmental goals, and prioritizing projects. Reflecting the specified focus, representatives of the Chambers of Commerce of Bremen and Bremerhaven and of another terminal operator joined the working process.

As a result, the first workshop series reassessed a number of ongoing projects from different perspectives (see Meincke, 2011).

With regard to explicit adaptation demands, existing and scheduled coastal defense measures were evaluated to be sufficient until 2050, whereas reactions to possible changes in sedimentation would require additional knowledge. As a consequence of hotter summers, a demand for extended air conditioning in various buildings, including warehouses, was identified. Further, a number of measures for handling extreme events were considered necessary, for example the waste-water system of a container terminal's storage had to be adjusted to increased precipitation. However, in a number of issues the underlying problem was unclear responsibilities among actors—in the aforementioned case, for example, the specific roles of the user (terminal operator), the landlord (port authority), and the infrastructure provider (municipality) had to clarified. Regarding sustainability, the so-called "greenports" strategy[4] of *bremenports* was appraised as a solid basis for joint initiatives of the enterprises and organizations located in Bremerhaven. Particularly, the standardization of indicators was identified as being essential. Here, *bremenports'* participation in the PERS-certification process and in designing the sustainability report according to GRI-standards was regarded as a yardstick. In terms of social aspects, again, a demand for joint efforts of all port users, including small employers, was identified. With respect to innovation potentials, responses to major market processes and infrastructure bottlenecks were considered to be most urgent, including reactions to the use of larger ships, to regulations requiring cleaner fuels, to the demand for an environmentally sound modal split, to the challenge of block train handling within the ports and to bottlenecks of hinterland infrastructures.

The second workshop series applied a rigid methodology, often used in scenario development, to derive a prioritization of measurable development targets and projects from the drivers. Steps included the reassessment of the drivers of port development, a reflection of developmental options, and the definition and prioritization of developmental aims. Reflecting crucial dimensions of port development as well as the participants' different professional perspectives, the working process focused on strategy development, (port internal) infrastructures, and sustainability. Partially, the topics were discussed separately, requiring a joint follow-up synthesis (details in Kupczyk and Osthorst, 2014).

The appraisal of *central drivers* for Bremerhaven's development identified both general and location-specific factors. The first group included global economic development, energy costs and environmental regulation, customers' demand for CO_2 reduction and constraints of hinterland infrastructure development due to political conflicts. The second group focused on the severe budgetary constraints of the *Land* of Bremen and on limits to spatial extension, resulting from the regional society's competing interests.

Development options as possible reactions were differentiated according to the dimensions described above. Regarding *strategy*, maintaining good cooperation between political actors, administration, infrastructure providers, and operators was seen as essential for realizing the anticipated growth targets, forecasting 12 million TEU until 2025. Other measures were improved conflict management with regard to traffic projects, extended cooperation between northern German ports in a number of fields, and projects aiming at a higher loco rate (for example, offshore wind energy). In terms of *infrastructure options*, innovations increasing spatial productivity (for example, for handling containers and cars) were ranked highest, followed by maintaining waterway quality, and developing regional dry ports. *Sustainability options* included the extension of the greenport strategy, fostering technical innovation in the realm of ship fuels, the ambitious implementation of the port enterprises' sustainability management systems, including the implementation of existing environmental standards, and the application of differentiated port tariffs, applying the Environmental Ship Index (ESI[5]). Further options included extended port cooperation in the realm of risk management and the pooling of compensation areas.

Based on SMART principles ("Specific, Measurable, Accepted, Realistic, Timely"), more specific *development targets* were prioritized with regard to relevance and resource demands. Out of 50 targets, seven were singled out for each dimension as the most crucial ones. Although the process involved different professional perspectives, a high congruence in problem perceptions was observable. The measures agreed upon as most important were those fostering spatial productivity, removing port-internal bottlenecks (pre-gates, truck management), improving rail systems (realization of a state cargo rail concept), as well as the realization of federal traffic infrastructure projects, including the A22 motorway, the planned "Y"-railway, and the Weser waterway deepening. Further, a number of initiatives focusing on renewable energy use and production as well as developing or implementing new fuels (LNG), were ranked high.

The character and outcome of the project activities for the port of Bremerhaven illustrate how the ports' adaptation to climate change is inherently linked to strategic choices in handling the challenges of port development in general.

In this case, the participating actors used the resources provided by the federal government to reconsider the port authority's collaboration with the most relevant enterprises and regional actors, their relations to higher political levels in ensuring competitive frame conditions, and their reactions to increasingly rigid constraints resulting from the challenge of sustainability.

Obviously, considering all activities re-evaluated during the project, climate adaptation is only one challenge among others, but adds additional risks to the underlying assumptions of measures or investment. The project allowed the operational level of different organizations below the decision-making level to maintain political support for its work on concepts and projects in light of anticipated challenges. Thus, it is not surprising that most of the measures considered during the project process reflected already existing concepts and did not generate true innovations. These project results thus document an incremental and efficiency-oriented approach to both developmental and integrative challenges. The working process, however, also allowed the port infrastructure agency *bremenports* to explore its role as a pro-active agency responsible for the location's sustainability and environmental management across organizational boundaries.

At the same time, the project was unable to address some crucial target conflicts between port development goals and the adaptation demands of other interested parties. The most relevant example was the planning of the Weser deepening, considered vital for Bremerhaven's locational quality, but being contested in court by environmental associations in response to an alleged violation of European regulation. However, the obviously stiffer application of WFD and FFH requirements in courts might serve as a hierarchic frame condition, enforcing the future development of more integrative concepts and more flexible responses to market demands, despite strong regional support for the port cluster.

As an immediate result, a number of considerations of the working process, including the relevance of sustainability and climate policy integration, influenced the discussion of Bremen's state port development plan 2025 during the year 2014.

Conclusion

Climate adaptation governance is characterized by contradictory requirements. In the absence of coherent regulatory frame conditions, adaptation governance often has to rely on voluntary, network-based processes with regional or sectoral participants. At the same time, addressing the interactions between different sectors holistically and interactively is considered crucial if maladaptation is to be prevented. In contrast, from the perspective of port governance, adaptation challenges are closely linked to the sectors' developmental perspectives. The article argued that in light of this constellation, different types of adaptation with various degrees of integration are to be expected.

The case study for the port of Bremerhaven confirmed that, at least for European North Sea ports, climate risks alone are not a threat, despite a number of measurable effects. It is their interaction with additional challenges, such as market pressures, costly infrastructure maintenance and extensions, budgetary squeezes at different political levels, and increasingly tangible ecological boundaries at sensible locations which could lead to serious problems. As a consequence, the project identified the lack of political coordination in the area of infrastructure development as the main reason for reduced adaptation capacities. Regarding the various segments of the value chains, larger enterprises, including terminal operators and logistic providers, react more ambitiously and with a stronger focus on integrative responses. A clear locational and long-term perspective is applied by the Hanseatic-type public infrastructure provider who is trying to take a pro-active stance in addressing sustainability issues, not least due to strong political pressure resulting from European environmental regulation.

The adaptation project prioritized the communication about core questions of port development between different sectoral stakeholders over the identification of technical adaptation demands. This highlights that climate adaptation is a matter of strategic choices. Underlining the limits of voluntary processes, controversial issues, such as waterway extensions, which are likewise relevant for the specific locational quality, the level of sustainability, and the coasts' vulnerability to climate change were not addressed.

Notes

1 In December 2013, the responsibility for urban development and spatial planning was transferred to the BMU which then became the Federal Ministry for the Environment, Nature Conservation, Building and Nuclear Safety (BMUB). The former Federal Ministry for Transport, Building and Urban Development (BMVBS) was renamed Federal Ministry of Transport and Digital Infrastructure (BMVI). The article refers to the former structures, relevant during the process of shaping the German adaptation policy.
2 For all projects funded, see http://www.klimzug.de/de/94.php.
3 See Note 1.
4 http:/www.bremenports.de/2048_2
5 http://www.environmentalshipindex.org

References

Baird, A. J. (2000). 'Port privatisation: Objectives, extent, process and the U.K. experience', *International Journal of Maritime Economics*, 2 (3): 177–194.

Baltazar, R. and Brooks, M. R. (2001). *The Governance of Port Devolution: A Tale of Two Countries.* Seoul, Korea: World Conference on Transport Research, July.

Bauer, A., Feichtinger, J. and Steuer, R. (2012). 'The governance of climate change adaptation in 10 OECD countries: Challenges and approaches', *Journal of Environmental Policy and Planning*, 14: 279–304.

Beck, S., Kuhlicke, C. and Görg, C. (2009). 'Climate policy integration, coherence, and governance: Germany'. UFZ-Bericht 1/2009, Leipzig: Helmholtz Centre for Environmental Research—UFZ.

Brooks, M. R. and Cullinane, K. (2007). 'Governance models defined'. In: Brooks, M. R. and Cullinane, K. (Eds.), *Devolution, Port Governance and Port Performance*. Amsterdam: Elsevier, pp. 405–436.

Bundesregierung. (2011). *Aktionsplan Anpassung der Deutschen Anpassungsstrategie an den Klimawandel vom Bundeskabinett am 31.08.2011 beschlossen.* Berlin.

De Langen, P. W. (2007). 'Stakeholders, conflicting interests and governance in port clusters'. In: Brooks, M. R. and Cullinane, K. (Eds.), *Devolution, Port Governance and Port Performance*. Amsterdam: Elsevier, pp. 457–478.

EEA (European Environment Agency). (2003). *Europe's Environment: The Third Assessment.* Luxemburg: Office for Official Publications of the European Communities.

European Commission. (2013). *An EU Strategy on Adaptation to Climate Change. Communication from the Commission to the European Parliament, the Council, the European Economic and Social Committee and the Committee of the Regions.* COM(2013) 216 final. Brussels.

Federal Government. (2008). *German Strategy for Adaptation to Climate Change*. Berlin.

Federal Ministry of Education and Research. (2012). *Climate Change in Regions. Adaptation Strategies for Seven Regions.* Bonn.

IPCC. (2007). *Climate Change: Impacts, Adaptation and Vulnerability. Contribution of Working Group II to the Fourth Assessment Report of the Intergovernmental Panel on Climate Change.* Cambridge.

ISL. (2011). *Prognose und Prozessbegleitung im Cluster Logistik in „nordwest 2050", Metropolregion Bremen-Oldenburg im Nordwesten – Schlussbericht.* Bremen: Institut für Seeverkehrswirtschaft und Logistik.

ISL and BAW. (2010). *Europa-Hub für Deutschland: Die gesamtwirtschaftliche Rolle der Logistikregion Nordwestdeutschland.* Bremen: Institut für Seeverkehrswirtschaft und Logistik (ISL); BAW Institut für regionale Wirtschaftsforschung GmbH.

Jänicke, M. and Jörgens, H. (2004). 'Neue Steuerungskonzepte in der Umweltpolitik', *Zeitschrift für Umweltpolitik & Umweltrecht*, 27 (3): 297–348.

Juhel, M. H. (2001). 'Globalisation, privatisation and restructuring of ports', *International Journal of Maritime Economics*, 3 (2): 139–174.

Kupczyk, M. and Osthorst, W. (2014). *Resiliente Hafenentwicklung: Von Rahmenbedingungen zu Aktionsansätzen.* nordwest2050-Werkstattbericht Nr. 28. Bremen.

Lloyd's Register, Qinetiq and Strathclyde University. (2013). *Global Marine Trend 2030.* Singapore and London.

Meincke, A. (2011). *Projekt „Resiliente Hafenstrukturen".* nordwest2050-Werkstattbericht Nr. 14. Bremen.

Mickwitz, P., Aix, F., Beck, S., Carss, D., Ferrand, N., Görg, C., Jensen, A., Kivimaa, P., Kuhlicke, C., Kuindersma, W., Máñez, M., Melanen, M., Monni, S., Branth, P. A., Reinert, H. and van Bommel, S. (2009). *Climate Policy Integration, Coherence and Governance.* PEER Report No 2. Helsinki.

Ng, A. K. Y. and Pallis, A. A. (2010). 'Port governance reforms in diversified institutional frameworks: Generic solutions, implementation asymmetries', *Environment and Planning (A)*, 42 (9): 2147–2167.

Notteboom, T. E. (2010). 'Concentration and the formation of multi-port gateway regions in the European container port system: an update', *Journal of Transport Geography*, 18 (4): 567–583.

OECD. (2002). *Improving Policy Coherence and Integration for Sustainable Development – A Checklist.* Organisation for Economic Cooperation and Development (OECD) Observer.

Osthorst, W. and Mänz, C. (2012). 'Types of cluster adaptation to climate change. Lessons from the port and logistics sector of northwest Germany', *Maritime Policy and Management*, 39 (2): 227–248.

Osthorst, W., Meincke, A., Nibbe, J. and Mänz, C. (2012). 'Hafenwirtschaft und Logistik'. In: Schuchardt, B. and Wittig, S. (Eds.) *Vulnerabilität der Metropolregion Bremen-Oldenburg gegenüber dem Klimawandel (Synthesebericht).* nordwest2050-Berichte Heft 2. Bremen / Oldenburg: Projektkonsortium, nordwest2050, pp. 111–122.

Robinson, R. (2002). 'Ports as elements in value-driven chain systems: The new paradigm', *Maritime Policy and Management*, 29: 241–255.

Schröder, A. and Hirschfeld, J. (2014). 'Anpassungsbedarfe und Strategien in der Hafenwirtschaft an der deutschen Ostseeküste'. In: Mahammadzadeh, M., Bardt, H., Biebeler, H. Chrischilles, E. and Striebeck, J. (Eds.) (2014). *Unternehmensstrategien zur Anpassung an den Klimawandel. Theoretische Zugänge und empirische Befunde.* München: Oekom Verlag, pp. 129–140.

Schuchardt, B., Wittig, S. and Spiekermann, J. (2010). *Klimaszenarien für, nordwest2050.* Teil 1/2, BioConsult 2/3. Werkstattbericht. Bremen.

Swart, R. J., Biesbroek, G. R., Carter, T. R., Cowan, C., Henrichs, T., Mela, H., Morecroft, M. D. and Rey, D. (2009). *Europe Adapts to Climate Change. Comparing National Adaptation Strategies.* Helsinki: PEER Report No. 1.

Verhoeven, P. (2009). 'European ports policy: Meeting contemporary governance challenges', *Maritime Policy & Management*, 36 (1): 79–101.

Verhoeven, P. (2011). *European Port Governance. Report of an Enquiry into the Current Governance of European Seaports.* Bruxelles: ESPO.

8 Port planning and climate change

Evidence from Italy

Maria Inés Cusano, Claudio Ferrari and Alessio Tei

1. Introduction

Climate change effects are modifying several characteristics of the Mediterranean basin, visibly increasing the annual average temperature and the risk for high-impact natural events (Medri et al., 2013). While these consequences of climate change are mainly spreading their effect in the common social life and only in some particular economic sectors (for example, agriculture and tourism), one of the most unperceivable effects will deeply impact waterborne trade: sea-level rise (SLR). According to some studies carried out by the Italian National Institute of Geophysics (Antonioli and Silenzi, 2007) more than 25 percent of the Italian coastline is suffering from widespread erosion and about 60 percent of the coastline is at high risk of sea flooding due to SLR. Fifteen out of twenty Italian regions are currently registering problems connected to the effects of climate change on their coastal areas. According to the best-case scenario presented in the Intergovernmental Panel on Climate Change (IPCC) 2013 the experts think sea-level rise will likely be about 0.4–0.6 meters by 2100 and 0.6–1.0 meters by 2300. While according to the more likely higher emission scenario, the results are 0.7–1.2 meters by 2100 and 2.0–3.0 meters by 2300. This picture seems even worse in some particular regions, such as the Northern Adriatic and the Venice lagoon, where SLR is affecting lands that are already below the natural sea-level (i.e. Venice) and that may be irreversibly damaged. In fact, as reported in Figure 8.1, according to the recent assessment on the Global and European sea-level rise published by the European Environment Agency (EEA, 2014), SLR has deeply affected the Mediterranean basin and will remain a big environmental concern in the coming years.

Despite this scenario, Italy currently lacks specific policy tools aiming at mitigating or adapting to the possible effects of climate change on economic and social activities. One step in that direction was taken in 2001 when the national government appointed the main responsibilities to protect the local environment and plan territorial development to the regional governments and authorities in charge of elaborating regional plans and promoting proper interventions among their provinces and municipalities (PRC, 2009). In this framework, the Ministry of the Environment maintains a role of coordination among different policies and

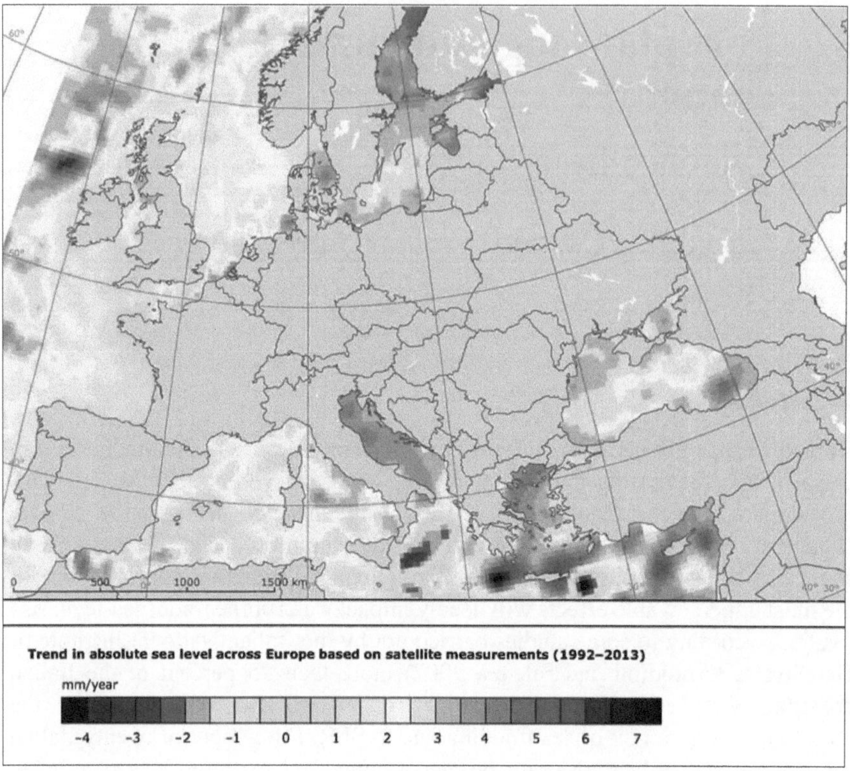

Figure 8.1 Sea-level rise. Source: European Environmental Agency (2014).

the possibility of promoting specific investments in accordance with the local governments. One of these integrated approaches comes directly from the European Union and is represented by the introduction of Integrated Coastal Zone Management (European Commission, 2002: COM 2002/413/CE) with a special approach to the Mediterranean partnership (European Commission, 2009: Barcelona Protocol, 2008). Under the integrated management, the ministry, its international partners, and local authorities, plan interventions aiming at reaching better adaptation level and preventing potential damages caused by environmental and climatic changes in the coastal zones.

Currently, each region has introduced its own plan to prevent or reduce the effects of climate change on its own territory, normally by monitoring differences among several timeframes (e.g. waves intensity and height), implementing physical barriers that may limit the effects of sea-level rise or promoting dedicated policies aiming at changing critical behaviours (e.g. agricultural practices, facility location). On this regards, the greatest Italian project currently under development is MOSE, a dynamic and adaptive system located in the Venice lagoon that may protect the area from a sea-level rise of about 60cm that will be mainly used during the high tide periods (Medri et al., 2013).

Within the abovementioned scenario, climate change in Italy will potentially affect seaport activities and hinder future goods distribution and new investments. Despite this, different approaches are currently in place by different actors involved: these differences lie in the local autonomy at regional, municipal and port level that affect the integrated interventions. Moreover, some physical characteristics of the Italian ports also affect the different approaches of local authorities to the climate-change problem. In the Tyrrhenian coast of Italy, while few ports are characterised by low banks that may be strongly affected by sea-level rise, other atmospheric changes (e.g. wind speed, storms) are already starting to heavily affect port operations during some periods of the year (e.g. Autorità Portuale di Genova, 2012). On the other hand, the Adriatic ports are normally characterised by an opposite situation in which the main threats are linked to SLR effects (e.g. Autorità Portuale di Venezia, 2013). Moreover, the port physical structure may also influence the magnitude of climate-change effects: for instance the few ports located closed to an estuary (e.g. Ravenna) are currently facing difficulties due to changes in the wave patterns and the marine stream (that also affects the needed dredging investments), the lagoon port of Venice is experiencing severe threats from sea-level rise while the north Tyrrhenian ports (e.g. Genoa) are affected by an increasing wind power.

In this context it is clear that local circumstances and regional planning differences have a strong influence on the real interventions to the climate-change problem. The current chapter addresses the existing Italian port approach to climate change, highlighting the critical issues related to the required interventions and planning efforts for future port development. After this short introduction, Section 2 introduces a literature review on the effects of climate change on seaports and on the main coastal economic activities. Section 3 is dedicated to the analysis of the current policies and investments already planned by the Italian ports in order to adapt or mitigate the effects of climate change on their own activities, while Section 4 underlines the difficulties of proper policies under the Italian framework. Finally, Section 5 is dedicated to the conclusions.

2. Literature and local framework

Coastal cities will also be heavily affected by climate change as a consequence of the continued growth of population on and around coastlines (reaching 3.1 billion people by 2025 according to the UN population division), therefore port cities will be put under severe stress to deal with the upcoming changes (Cusano et al., 2013). Academic literature (e.g. Acciaro et al., 2014; Daamen and Vries, 2013) as well as policy papers (e.g. OECD, 2011; Lim and Spanger-Siegfried, 2004) have deeply studied the consequences of port-related activities on the environment and the different ways and methods that may allow ports to reduce their environmental effects. In general, these interventions are grouped together in the so-called *green port policies*. Less common and accepted seems to be the debate related to the effects of the environmental changes on port operations. Therefore, while the discussion on the impact of human behaviour on its own living environment is well-known, the unpredictable counter-effects (i.e. climate change) have been underlined and accepted only relatively recently (e.g. Bosello et al., 2012). A proof can be found in the long period of attendance for the ratification of the Kyoto protocol (it was elaborated in 1997 but its entry into force only dates back to 2005) or some recent European policies devoted to this specific aspect (e.g. only in 2002, the EU released the communication on climate change in coastal zones). The main problem in studying climate-change effects is linked with the reliability of data and forecasts that strongly affect the possibility to analyse the consequences or to prepare specific strategies. Several hypotheses on these consequences can be easily found in the main literature on this topic: analyses are normally limited to specific regions (e.g. Esteban et al., 2009; USEPA, 2008) or use surveys in order to illustrate the perception of operators and/or collect ideas for possible alternatives in case of extreme scenarios (e.g. Becker et al., 2012). Moreover, due to the difficulties in collecting reliable data and information, studies dealing with the evaluation of costs and benefits normally consider not just the port industry, but more in general the effects on the port region, limiting the possibility to evaluate specific projects. An example can be found in the analysis made by Hanson et al. (2011) and Nicholls et al. (2007), in which climate change is studied in relation to the potential extreme effects on port cities.

As indirectly introduced above, most of the literature (e.g. Acciaro et al., 2014; Becker et al., 2012) separates mitigation from adaptation measures. The former concept is represented by all the efforts made in order to reduce the environmental impact of each economic unit (e.g. green ports policies) while the latter is linked to the efforts invested in adapting the specific economic unit to the changing environment (e.g. to the increasing storm frequency). Ideally, adaptation measures should aim at achieving resilience of the analysed system (Nelson et al., 2007). Even if these two aspects are strictly linked, they are normally analysed separately.

Considering the Italian case study, it is important to notice how the main interventions are focused mainly on mitigating the effects of port operations on the local environment while only general future plans foresee an adaptation approach that would lead to the creation of resilient cities and ports.

According to the current management model of Italian seaports ruled by the law no. 84 issued in 1994, Italian Port Authorities (PAs) are in charge of the planning and control functions, while the port operations are made by private companies. Therefore the Port Regulatory Plan (PRP) is the only strategic tool at the disposal of PAs to model new interventions in order to adapt seaports to the changing environments while the Triennial Operative Plan (TOP) is the short-term programme to accomplish the interventions planned in the PRP (Ferrari et al., 2015). Only a few PAs are currently investing in research projects that can forecast natural disasters and/or reduce the related risks. Moreover, projects are also part of international collaborations or European projects, such as the "Vento e Porti" [*Wind and Ports*] international project aiming at modelling the wind patterns in the north Tyrrhenian region to better adapt port operations to wind conditions (Autorità Portuale di Genova, 2012). Apart from the PRP handled by the PAs, there are only few official analyses on the effects of climate change at Italian level applied to the port sector (e.g. APICE, 2012) or data collection aiming at providing a specific database that can be used for reliable estimations. This lack of information and studies is strongly limiting the development of coordinated policies and interventions among ports and different regions: each PA is investing in planning its own adaptation path without a general national framework. In addition, the national port law (no. 84/1994) allows the PA to plan and coordinate activities and investments only within the port boundaries, thus reducing its power to influence decisions along the logistics supply chains: this fact, deeply discussed by Italian scholars (Ferrari and Musso, 2011; Carbone and Munari, 2006) for its implications in terms of port efficiency, separates the interventions within the ports from those occurring outside the port boundaries. This issue also limits potential coordinated policies between municipalities and PAs to elaborate an integrated port city plan.

From a private operator's viewpoint, the current legislation and the port concession practice limit the interest of stevedoring and other port operators to invest in adaptation measures. This low interest is mainly linked to the limitation of possible financial returns on investments and the belief that they will be able to avoid problems related to climate change during their current concession agreement duration (considering they cannot obtain an extension of the concession as a consequence of their new investments in environmental issues). On the other hand, according to the new legislation, from 1994 Italian PAs have awarded the majority of the port areas to private operators, around 60 per cent to 70 per cent of the main port's surface for an average period of more than 20 years in the main ports reaching peaks of 60 years for some great new terminals (Parola et al., 2012). The concessions establish compulsory investments and operational rules and thus limit potential changes in the agreement due to the legal rigidity characterising the Italian concessions' system (Brignardello, 2013). Due to the long duration of the agreements and the abovementioned rigidity in changing parts of it, PAs cannot force stevedoring operators to make investments that were not foreseen in the concession contract, limiting actions to those that are convenient for all the parties involved. The result is again a stress in reducing the pollution and the carbon footprint of port operators – partially foreseen in the initial agreement and because it gives a better image in terms of firm communication – while only a few new concessions also include some adaptive actions on behalf of terminal operators. Eventually, many concession agreements will come to an end around 2020 and by then it would be possible to include stringent environmental clauses in the renewal process – or in the new awarding process.

3. Climate change and port policies in Italy

As mentioned above, Italian PAs currently have a limited power to intervene in climate-change matters, controlling all the different aspects that may be important in order to adapt the port structure to future environmental challenges. Together with these difficulties, national and regional governments seem to be more interested in promoting studies that focus on the general characteristics of the coastal areas than in specific aspects of the port environment.

In order to analyse current and foreseen interventions on this matter we look at the propositions included in the PRP – or its planned implementations in TOP – that may adapt ports to climate change and the participation of the PAs in projects aimed at studying (and eventually investing) in port climate adaptation measures. Nevertheless, the most effective actions tackling the climate change issue derive from specific policies fostered by international and national institutions, such as the adoption of Integrated Coastal Zone Management. Figure 8.2 summarises the current structure of measures and the actors involved at different levels in Italy.

Currently there are 24 PAs in Italy among which less than 50 per cent of the ports exceeds 10 million tonnes of annual throughput (Assoporti, 2014) thus showing a general small size in the cargo handling that is reflected in the small financial and physical size of these harbours. PAs are located in all the main Italian coastal regions with few exemptions (i.e. the central regions of Abruzzi and Molise and the south region of Basilicata) with the majority of the big ports (Genoa, Gioia Tauro, La Spezia, Leghorn, Civitavecchia, Naples, Palermo, Cagliari) located on the Tyrrhenian bank and only a few big ports on the Adriatic (Trieste and Venice) and Ionic (Taranto) coasts. Figure 8.3 shows the Italian port system, including not only the major ports, but all the Italian PAs.

The presence of a great share of small ports limits the number of great investments connected to the environment and this is due to their restricted budget and personnel, thus the main projects are all promoted at central government level (and then with only general guidelines). Moreover, the same can be stated about

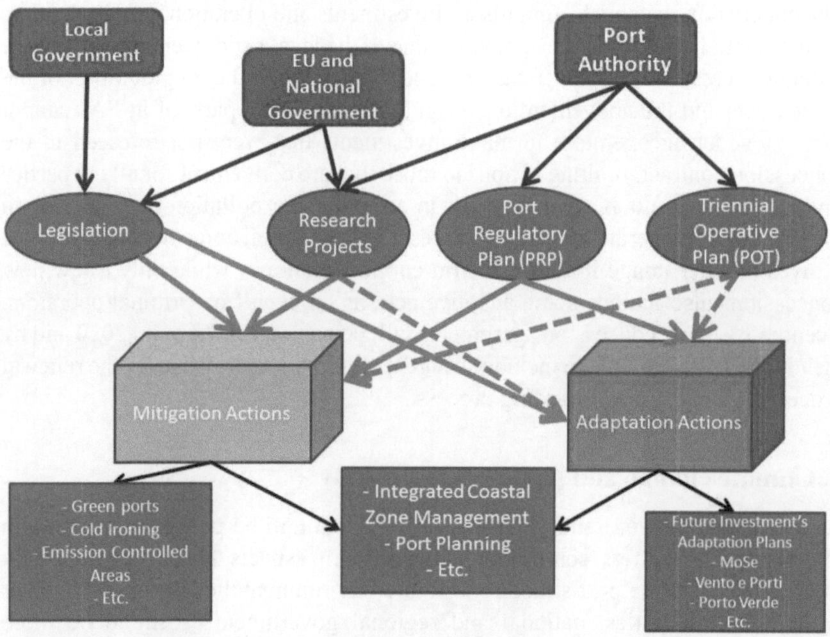

Figure 8.2 Actors and processes involved in mitigation and adaptation measures.

Figure 8.3 The Italian port system.

the publicity of systematic data aiming at sharing the risk with the population: looking at the websites of the Italian PAs, only 30 per cent of them show an updated PRP adopted after 2008 and this is reflected in their environmental plans or in the presence of effective green port policies. In general, common environmental prescriptions among the PAs' planning reports appear to be the development of an energy plan that considers alternative ways to supply port activities (e.g. cold ironing, green energy production) and guidelines to promote the reduction of water pollution and air emissions.

In the main Adriatic ports, the different interventions and studies promoted by the PAs focus on SLR and land erosion, aiming at reducing those effects that may hinder the daily operability of the terminals. In this regard, the best practice is Venice with the project 'Porto Verde' [*Green port*], a plan included in the PRP aiming at reducing the environmental effects of the port's activities. The plan also considers the abovementioned MOSE project (that should protect the lagoon from SLR) and the construction of a new off-shore container terminal in the coming years. Many of the guidelines included in the 'Porto Verde' project focus on energy issues more than on adapting strategies to the changing environment, moreover, some of the ideas to be applied appear somewhat incoherent with other expansion plans included in the same PRP. For instance, the efficiency of the off-shore terminal container may be at high risk in case of a relevant increase in SLR. In general, the project gives valuable advice without a proper contextualisation or a compulsory action to be confirmed by the actors involved or by other sections of the PRP. Similar comments may be raised about the PRP of Trieste: here a section is dedicated to environmental critical issues even if there are no proper projects on this topic. In general all the mentioned mitigation actions deal with emission reduction plans without a proper consideration of climate change (Autorità Portuale di Trieste, 2013; Autorità Portuale di Venezia, 2013).

Focusing on the Tyrrhenian ports, the main harbours on this shore are more affected by land erosion and increasing storm frequency than SLR. For this reason, the only ports that are currently investing in climate-change-related issues are located in the north Tyrrhenian region and they promote projects such as the 'Venti e porto' which monitors the increasing wind and proposes interventions in order to reduce impact on the operability of the ports and prevent possible damage to port facilities. The sensitiveness to wind issues can also be assessed in the investments in cranes that are less affected by wind during operations (e.g. Genoa terminal containers) that are being carried out by stevedores. Another important topic seems to be the increasing strength of the marine stream, deeply studied in accordance with some new investments foreseen in Naples and Leghorn and also in specific research projects, such as Climeport. Currently, the only infrastructural project specifically approved and studied in order to adapt the new structure to the foreseen environmental conditions is the planned seawall that should use the increasing wind and maritime stream in order to produce clean energy in favour of the port operators in the port of Genoa (Autorità Portuale di Genova, 2012). In accordance with the public information on the website and on the available PRPs, Table 8.1 provides some examples of the main adaptation actions carried out by the Italian ports, grouping them into four categories: current adaptation actions (i.e. actual interventions aiming at adapting the port basin to the changing environment); research projects (i.e. participation in research projects aiming at evaluating the impact of climate change); inclusion of environmental evaluations in the new port investments (in which climate change effects are included); and the provision of actual adaptation measures in the PRP.

Despite the recent scientific elaborations which show that the Italian coastal areas are at high risk regarding SLR and other effects of climate change, the analysis of both the planning tools and the current port plans highlight a low perception of the risks related to these issues. An interesting confirmation comes from a series of interviews with terminal operators and a nautical service provider carried out in the port of Genoa that confirm the low interest and concern regarding these issues, preferring investments in pollution reduction rather than in climate-change adaptation projects. The main reasons lie in the timeframe period in which effects may come: apparently, operators do not justify investments to mitigate the effects of climate change due to their uncertainty regarding their presence within the next 20 years or so.

Table 8.1 Résumé of the main actions taken by the Italian PAs

Current adaptation actions	*Development of research projects on climate change*	*Inclusion of adaptation studies in the investment plans*	*Inclusion of adaptation measures in the PRP*	*Mitigation actions*
Venice (MOSE)	Genoa, La Spezia, Leghorn, Savona	Genoa Leghorn Naples	Genoa	All the ports
	Venice	Trieste Venice	Venice	

Source: Authors' elaboration

4. Adaptation and future implications

The previous sections show how the different effects of climate change might affect the Italian ports and strengthen the vulnerability of the Mediterranean sea ports to SLR and increased extreme weather events such as wind speed.

Keeping in mind that some of the effects of climate change are not a consequence of human activities, for example some important changes are due to geological movements that are to be considered as natural, there is unequivocal evidence from *in situ* observations and ice core records showing that the atmospheric concentrations of important greenhouse gases such as carbon dioxide, methane and nitrous oxides have increased over the last few centuries (IPCC, 2013). The overall effects of climate change are so important as to completely modify the distribution of human activities in the next decades if no efforts will be done in order to adapt – and prevent if possible – those changes.

Seaports and marine navigation will be deeply impacted by some of those effects, namely SLR, as a consequence of ice thinning in the polar regions, higher exposure to strong winds and extreme precipitations. This increases the risks associated with seaport and navigational operations.

Undoubtedly, we are discussing changes that will happen in the long term, and that are currently not clearly perceived by the community nor by all the actors directly involved in the seaport industry. However, a rough idea of their relevance is given by the increasing process of ice melting in the Arctic waters that has encouraged the exploitation of the northern sea route: in 2013 the maritime traffic on that route doubled from the volume registered in 2012 and expectations for the coming years are very positive (Furuichi and Otsuka, 2015).

According to our research, the climate change topic is not being dealt with in the long-term policy reporting of port authorities and maritime authorities in Italy. To the best of our knowledge, there are no adaptation plans to climate change specifically carried out by seaports. However, climate change is taken into consideration in a few general regional plans; for example, the Adaptation Action Plan compiled by the Province of Genoa that defines strategies to increase the natural defences of the environment that could be further stressed by climate change.

The effects of climate change behave as economic externalities and they are not confined to a specific region but are spread, albeit unequally, all over the world. Therefore, it is a common problem with consequences that will differ among ports located in developed and developing countries, as well as among ports located near a city or in more isolated areas.

Even if, in general, climate change is not an ignored issue by port managers, as already recently surveyed by Becker et al. (2012), it is completely missing in the official planning reporting in Italy. Moreover, climate change is never explicitly mentioned in the socio-economic scenario that informs the long-term port planning tools. This is quite striking since the PRPs draw the evolutionary path of the port environment and its infrastructure, i.e. those artefacts that impact on the port structure for some decades that are directly affected by the possible consequences of climate change. Probably, there is also a sort of myopic effect due to the fact that SLR is perceived as dramatically impacting port operations only on particular occasions, forgetting that the main effect derives from land erosion and its pressure on port cities.

In contrast with the ports' lack of adaptation strategy to climate-change effects, is the numerousness of mitigation actions carried out by the Italian PAs: from the installing of on-shore power supply for ships (the so-called cold ironing) to the electrification of port vehicles and cranes, to the use of port infrastructure as green energy power supply (through wind turbines, for instance). These actions face distinct effects of climate change, but in large part try (and contribute) to reduce the carbon footprint of seaports.

The reasons behind this preference should be further investigated, however, it may be argued that it is due to the immediate perception of port operators, stakeholders and seaport-city citizens. Furthermore, the size of port authorities may push towards mitigation instead of adaptation actions; in fact, they are institutions that are too small in respect to the paramount effects of climate change and their power is limited to the port area; therefore since there is a lack of effective coordination of action on a large scale, the behaviour of Italian PAs is understandable.

In other words, the climate change topic and its effect on ports represent another proof of the need for a more effective coordination of the long-term planning function of seaports in Italy (Ferrari and Musso, 2011). Concerning this issue, the port reform law, currently under discussion, may increase the possibility of tackling climate-change problems. In fact, according to Ferrari et al. (2015) the introduction of regional logistics authorities – coordinating several ports currently administrated by different PAs – might facilitate the introduction of wider port policies that could also embrace environmental issues and may also establish green policies all along the logistics corridors.

In addition, the new port regulation might also affect the relations between terminal operators and regulatory authorities: as introduced before, one of the current obstacles to new investments in adaptation measures is the fact that port concessionaires are not interested in long-term interventions because a great many of the concession contracts are close to expiration. Therefore at this moment of

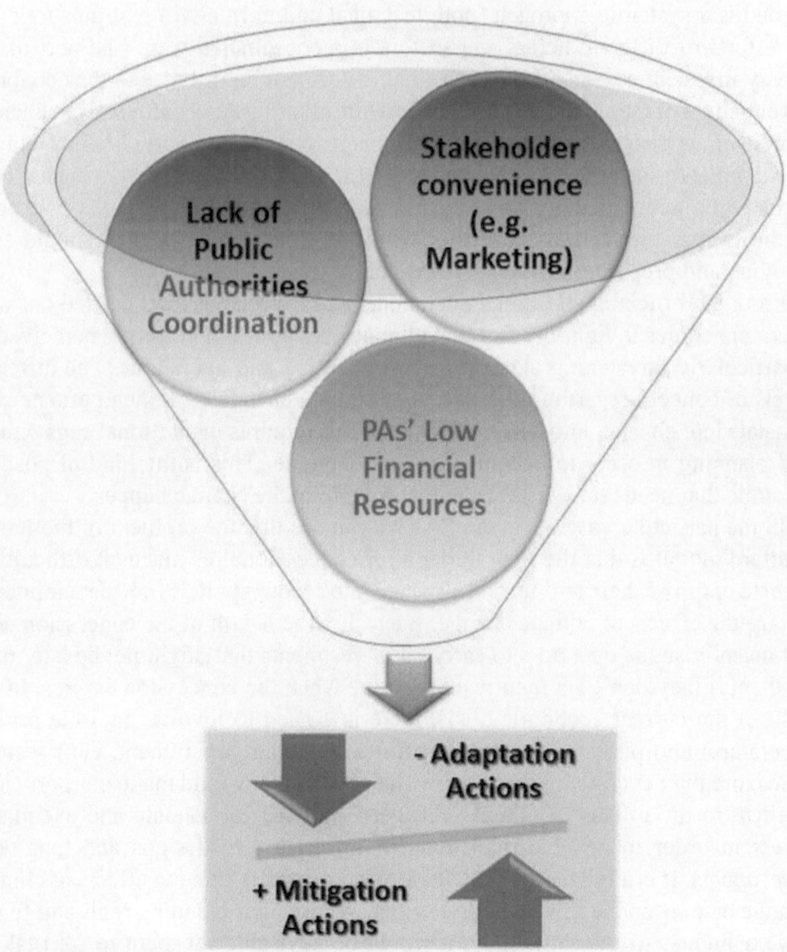

Figure 8.4 Barriers to the current development of adaptation measures.

time private port operators cannot benefit from these adaptation investments. In order to cope with this disincentive, port reform should introduce a concession renewal system able to guarantee investment returns even when the concession expiration is approaching, thus making this kind of investments more attractive for the private companies operating in the port.

Conclusion

Among all the uncertainties and mixed messages regarding the effects of climate change on Italian ports, one message is very clear: a business-as-usual approach is not a feasible option and waiting for a universal panacea to solve all our environmental problems is both unrealistic and a recipe for disaster. What is needed is a systemic approach (both technical and normative) that puts together the efforts of different actors and sectors in a coordinated way. The port of the twenty-first century should be a sustainable port that contributes to the economic development of the surrounding territory while assuring environmental excellence. Moreover, as a consequence of climate change, coastal and port cities around the world prone to suffer the consequences of SLR would have a deep impact not only in the port operations, but also in the transport infrastructural networks. The ports of the twenty-first century, together with the cities' governments, should start studying and preparing resilience plans before it is too late.

From the official institutional documents and from interviews carried out with operators comes to light the fact that climate change is not currently perceived as a particularly threatening element to port planning and operations. The different levels of concern regarding the effects of climate change on Italian ports need to translate into change, into a new reality, and this requires institutional engagement and planning in order to become a policy outcome. This is the kind of positive dynamic that needs to be established in order to make change happen.

In the particular case of Italian PAs we can see that the problem of the Italian landlord model is that the time horizon for concessionaires makes it difficult for them to optimise their activity and also take into account policies and developments to face the effects of climate change. Basically, the length of the concession does not incentivise the operators to carry out investments that might not be later used by them if they don't get their residual value when the concession expires. In the light of the current economic crisis, there is a need to involve both the private (operators) and public (PAs, municipalities, regional government, etc.) sectors, now more than ever, to invest in plans and actions that would make our port cities resilient to the effects of climate change. PAs need to evaluate and eventually invest in order to achieve robustness of infrastructure designs and long-term investments. It is advisable that all the PAs start considering the effects of climate change on port operations and port territory within their planning tools and future developments. At the same time, it would be advisable for them to start taking these kind of plans into consideration when assessing the different bids in their concession awarding processes.

One key element that requires further consideration is the necessity to coordinate climate-change resilience policies between the port and the city. A robust urban planning process that puts together a coordinated effort on behalf of all actors and institutions is the only way to face the consequences of climate change and this step would be easier to reach if there were stronger integrated guidelines, which are currently lacking.

The way forward for the ports of the twenty-first century is to devise adaptation strategies to short-term climate variability and extreme events that would serve as a starting point for reducing vulnerability to longer-term climate change. Our study underlines the lack of integrated policies between all the players involved in port-city development at local (e.g. PA, municipal government, port operators) and regional (district and regional governments, Environmental Authorities) levels, which is negatively affecting both the effectiveness of the (few) foreseen interventions and the possibility of introducing long-term investments due to the limited size of the single actors.

References

Acciaro, M., Ghiara, H. and Cusano, M. I. (2014). 'Energy management in seaports: A new role for port authorities', *Energy Policy*, 71: 4–12.

Antonioli, F., and Silenzi, S. (2007). *Variazioni relative del livello del mare e vulnerabilità delle pianure costiere italiane*. Rome: Quaderni della Società Geologica Italiana.

APICE. (2012). Project reports. www.apice-project.eu.

Assoporti. (2014). Association statistical database. www.assoporti.it.

Autorità Portuale di Genova. (2012). *Triennial Operative Plan*. Genoa, Italy.

Autorità Portuale di Trieste. (2013). *Port Regulatory Plan*. Trieste, Italy.

Autorità Portuale di Venezia. (2013). *Triennial Operative Plan*. Venice, Italy.

Becker, A., Inoue, S., Fischer, M. and Schwegler, B. (2012). 'Climate change impacts on international seaports: Knowledge, perceptions, and planning efforts among port administrators', *Climatic Change*, 110: 5–29.

Bosello, F., Nicholls, R. J., Richards, J., Roson, R. and Tol, R.S.J. (2012). 'Economic impacts of climate change in Europe: Sea-level rise', *Climatic Change*, 112: 63–81.

Brignardello, M. (2013). 'Le concessioni di aree e di banchine portuali ex art. 18 l. n. 84/1994: problemi interpretativi e prospettive di riforma', *Rivista di Diritto della Navigazione*, 34: 467–498.

Carbone, S. M. and Munari, F. (2006). *La disciplina dei porti tra diritto comunitario e diritto interno*. Milan: Giuffré Editore.

Cusano, M. I., Li, Q., Obisesan, A., Urrego-Blanco, J. R. and Wong, T. H. (2013). *Coastal City and Ocean Renewable Energy: Pathway to an Eco San Andres*. Southampton: Lloyd's Register Foundation Collegium Book Series.

Daamen, T. A. and Vries. I. (2013). 'Governing the European port–city interface: institutional impacts on spatial projects between city and port', *Journal of Transport Geography*, 27: 4–13.

Esteban, M., Webersick, C. and Shibayama, T. (2009). 'Estimation of the Economic Costs of Non Adapting Japanese Port Infrastructure to a Potential Increase in Tropical Cyclone Intensity', *IOP Conference Series. Earth and Environmental Science*, 6 (32).

European Commission. (2002). *Recommendation of the European Parliament and of the Council Concerning the Implementation of Integrated Coastal Zone Management in Europe*. COM 2002/413/CE.

European Commission. (2009). *Protocol on Integrated Coastal Zone Management in the Mediterranean*. Official Journal of the European Union, Barcelona, 2008.

European Environment Agency (EEA). (2014). *Global and European Sea-level Rise (CLIM 012)*. http://www.eea.europa.eu/data-and-maps/indicators/sea-level-rise-3/assessment (accessed 22 August 2014).

Ferrari, C. and Musso, E. (2011). 'Italian ports: towards a new governance?', *Maritime Policy and Management*, 38 (3): 335–346.

Ferrari, C., Merk, O. and Tei, A. (2015). *The Governance and Regulation of Ports: The Case of Italy*. Paris: OECD/ITF Discussion Paper, 1-2015.

Furuichi, M. and Otsuka, N. (2015). 'Proposing a common platform of shipping cost analysis of the Northern Sea Route and the Suez Canal Route', *Maritime Economics and Logistics*, 17 (1): 9–31.

Hanson, S., Nicholls, R., Ranger, N., Hallegatte, S., Corfee-Morlot, J., Herweijer, C. and Chateau, J. (2011). 'A global ranking of port cities with high exposure to climate extremes', *Climatic Change*, 104: 89–111.

Intergovernmental Panel on Climate Change, Working Group (IPCC) (2013). *Contribution to the IPCC Assessment Report on Climate Change 2013: The Physical Science Basis.* 30 September 2013. Genève, Switzerland.

Lim, B. and Spanger-Siegfried, E. (2004). *Adaptation Policy Frameworks for Climate Change: Developing Strategies, Policies and Measures.* United Nations Development Programme. Cambridge: Cambridge University Press.

Medri, S., Venturini, S. and Castellari S. (2013). 'Overview of key climate change impacts, vulnerabilities and adaptation action in Italy', *Euro-Mediterranean Center on Climate Change (CMCC) Research Papers*, Issue n. RP0178, July 2013.

Nelson, D. R., Adger, W. N. and Brown, K. (2007). 'Adaptation to environmental change: Contributions of a resilience framework', *Annual Review of Environment and Resources*, 32: 395–419.

Nicholls, R., Hanson, S., Herweijer, C., Patmore, N., Hallegatte, S., Corfee-Morlot, J., Chateau, J. and Muir-Wood, R. (2007). *Ranking Port Cities with High Exposure and Vulnerability to Climate Extremes: Exposure Estimates.* OECD Environmental WP. Paris: OECD Publishing.

Organisation for Economic Cooperation and Development (OECD) (2011). *Environmental Impacts of International Shipping: The Role of Ports.* Paris: OECD Publishing.

Parola, F., Tei, A. and Ferrari, C. (2012). 'Managing port concessions. Evidence from Italy', *Maritime Policy and Management*, 39 (1): 45–61.

Policy Research Corporation (PRC) (2009). 'Italy', in *The Economics of Climate Change Adaptation in EU Coastal Areas, Country Overview and Assessment.* European Commission, Directorate-General for Maritime Affairs and Fisheries.

United States Environmental Protection Agency (USEPA). (2008). *Planning for Climate Change Impact at U.S. ports.* White Paper, Washington, DC.

9 Adaptation to an increase in typhoon intensity and sea level rise by Japanese ports

Miguel Esteban, Hiroshi Takagi and Tomoya Shibayama

1. Introduction

Climate change and sea level rise are expected to pose considerable challenges to ports in the course of the twenty-first century and beyond, particularly due to the effects of temporary and permanent flooding exacerbated by sea level rise, high winds and storm surges (Becker et al., 2013). Ports form a crucial component of international trade systems, and disruption to their activities can have serious second-order consequences for the regional and even global economy (Becker et al., 2013). These second-order consequences are especially important for those regions most integrated within the globalized economy, and it could be argued that the cities that surround Tokyo Bay are crucially important to this system. The total population of Tokyo is around 13 million inhabitants (Tokyo Metropolitan Government, 2012) and has the largest GDP output of any city in the world, exceeding even New York City, with a gross output of 1,479 billion dollars (PricewaterhouseCoopers, 2009). The Kanto region, however, encompasses what is called "greater Tokyo," which includes cities such as Yokohama and Kawasaki and has a total population of more than 35 million people—the largest megalopolis on the planet (Japan Statistics Bureau, 2010). A number of ports serve this massive population centre, which are particularly important given that it constitutes one of the world's greatest financial, commercial, industrial and transport hubs.

In this chapter, the authors describe the most significant consequences of a potential increase in typhoon strength and sea level rise to Japanese port infrastructure, though particular emphasis will be placed on the ports in Tokyo Bay (see Figure 9.1), given the disproportionate impact that they have on the country's economy. Also, a rough economic analysis of the cost of raising port areas in Tokyo Bay will be outlined, which can serve to quantify the magnitude of this aspect of the problem. The Port of Tokyo alone is one of the largest in the Japan and the entire Pacific Ocean basin, with an annual capacity of around 100 million tonnes of cargo and 4.5 million TEUs,[1] serving 32,000 ships a year and employing 30,000 employees.

Essentially the main impacts are likely to be related to a possible future increase in typhoon intensity and sea level rise (as summarised recently by the 5th Assessment Report of the Intergovernmental Panel on Climate Change, or IPCC

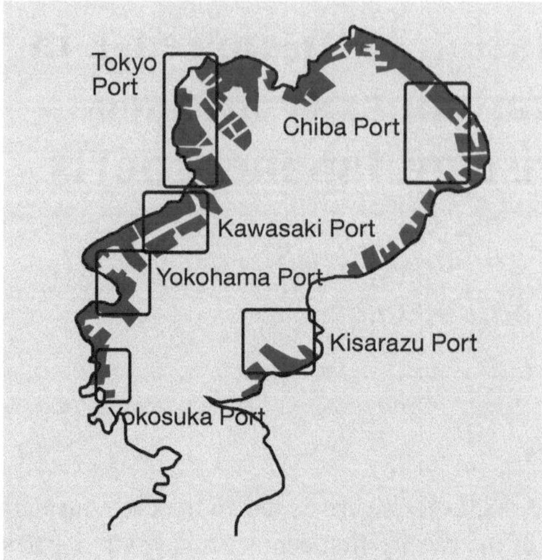

Figure 9.1 Main ports in the Tokyo area. Areas outside of continuous line indicate extent of reclaimed lands.

5AR). As a country, Japan is regularly subjected to a number of tropical cyclones, some of which have caused widespread damage, damaging breakwaters and other port installations (Takagi et al., 2013; Esteban et al., 2012a; Takagi et al., 2011; Esteban et al. 2009a), disrupting operations (Esteban et al., 2009b), and causing effects on the economy in general (Esteban and Longarte-Galnares, 2010; Esteban et al., 2009b).

2. Climate change, tropical cyclones and sea level rise

Japan is hit annually by a number of typhoons, which predominantly affect the southern islands of Okinawa and Kyushu, though the central island of Honshu is typically also hit by several of these weather events. Tropical cyclones form in subtropical latitudes with high surface sea temperatures, using the heat from the evaporation of sea water to maintain or increase their strength, and then propagate away to higher latitudes. Essentially, heat is released when moist air rises from the sea, resulting in condensation of the water vapour contained in this moist air. Aside from damaging winds, these weather systems also generate powerful waves and storm surges, which can inundate coastal areas and lead to the destruction of property and the loss of lives, as shown by the devastating damage caused by typhoon Haiyan in the Philippines in November of 2013 (Takagi et al., 2014).

As tropical cyclones feed on ocean heat, it appears logical that global warming caused by increasing concentrations of greenhouse gases in the atmosphere could in turn lead to an increase in the future intensity of tropical cyclones. The 5th Assessment Report of the Intergovernmental Panel on Climate Change (IPCC 5AR) highlighted how during the course of the twentieth century the oceans have continued to warm up, and it is expected that this trend will continue in the twenty-first century (a conclusion reached by many authors such as Knutson and Tuleya,

2004; Elsner et al., 2008; Landsea et al., 2006; Webster and Holland, 2005). However, currently there is little consensus on the future frequency of these events, and it is also believed that there could be a certain cyclical variability in the natural frequencies and intensities of such phenomena, possibly attributed to short-scale decadal cycles such as the El Niño-Southern Oscillation, or ENSO (Mousavi et al., 2011). Knutson et al. (2010) summarised the most important work on tropical cyclone simulations, which appears to suggest that their intensity could increase by between 2 and 11 percent by 2100, depending on the simulation used and the part of the world concerned. Knutson et al. (2010) find that the higher resolution models predict higher increases in intensity than those with lower resolution, and thus it is possible that earlier research on the subject relying on global-scale models could be underestimating potential future problems.

2.1 Climate change and typhoons in Tokyo Bay

For the case of Tokyo Bay, sea defences have been designed against the largest historical typhoons (namely the Isewan and Muroto typhoons), which could be equivalent to a once-in-100-years event. Thus, dyke design was not necessarily based on a thorough statistical analysis because of the shortage of available data (e.g. typhoon track, pressure, wind speed) at the time of the design (Kawai et al., 2006). Instead, a numerical simulation was performed for a number of possible typhoon tracks for these two typhoons, assuming their intensities and following some historical events during the last century (Miyazaki, 2003).

One of the most important storms to hit Japan during the twentieth century was the 1959 Isewan Typhoon (Vera), which caused a 3.5m storm surge in Ise Bay in Japan (Kawai et al., 2006). The storm breached the weak dikes present in the area and flooded low-lying lands behind them, destroying numerous buildings and infrastructure. Following this event the Japanese Government determined that defences around Japan should be designed to cope against another typhoon equal to the Isewan typhoon (which was thus designated as the "standard typhoon" against which defences should be designed), and undertook extensive efforts to construct coastal defences (Kawai et al., 2006). As a result of this typhoon, the essential philosophy underpinning the design of storm surge defences in Japan is that the design water level should be determined by one of the two following criteria (Kawai et al., 2006):

- The sum of the mean spring high tide level and the maximum storm surge recorded at a tide station or simulated assuming this "standard typhoon" (1959 Isewan typhoon); or,
- The highest tidal level recorded at a tide station in the area concerned.

The first criterion was adopted for major bay areas with a large population, such as Tokyo, Ise and Osaka Bays, while the second criterion is used for the central region of the Seto Inland Sea (Kawai et al., 2006). However, design tidal levels are still established in a deterministic way and there are a number of problems in attempting to ascertain the return period of storm surges (Kawai et al., 2006).

While the defences constructed around Tokyo Bay should protect the area from an event equal to the 1959 Isewan typhoon, such a typhoon has not actually affected the Tokyo area since modern records began. Thus, in order to provide a conservative analysis, it is worthwhile to focus on the worst typhoon to have actually taken place in Tokyo in the last 100 years. In October 1917 (6th year of the Taisho period) a typhoon made landfall in Tokyo Bay with a lowest pressure recorded during its passage of 952.7hPa, resulting in a +3.0m storm surge,[2] according to Miyazaki (1970). However, if similar typhoon events increase in intensity in the future as a consequence of climate change, it is likely that the associated storm surge would also be larger.

To understand how future increases in typhoon intensity will affect storm surges it is necessary to estimate how much stronger typhoons can become. Hoshino et al. (2011) used the research of Yasuda et al. (2010a) to determine that this typhoon (the worst historical typhoon in the 20th century in Tokyo Bay) in the year 2100 would have a minimum central pressure of 933.9hPa instead of the historically recorded value of 952.7hPa, which could result in an amplification of the expected storm surge height by +0.5m with respect to the historical event. Despite this analysis, the possibility of an event similar to the 1959 Isewan typhoon causing +3.5m of surge in Tokyo Bay should not be dismissed entirely. Design of defences should also consider the combination of an event similar to the 1959 Isewan typhoon with the effects of climate change, which could create an even bigger storm surge than that for which existing defence structures were designed for, highlighting the potential perils of anthropogenic climate change.

2.2 Sea level rise

Typhoons intensified by climate change are not the only factor that could increase inundation risks in low-lying coastal areas. There is a high level of confidence that sea level rise will be seen over the course of the twenty-first century (IPCC AR5). In fact, during the twentieth century, global average sea level rose by around 1.7mm on average per year and the pace of rise appeared to increase towards the end of the century, with satellite observations showing average increases of 3mm/year since 1993 (IPCC AR4). According to the IPCC AR5, it is believed that sea level is likely to rise in the range of 26 to 82 cm by 2100, substantially higher than the 18 to 59cm projection given in the IPCC 4R. Estimates have improved thanks to collection of more data on ice loss, as well as more research on how ice sheets react to climate change, which improved the modelling of thermal expansion. In fact, the IPCC 4AR assumed that ice was accumulating over the Antarctic ice sheet, though it is currently losing mass as a consequence of dynamic processes, as shown in Allison et al., (2009). As global CO_2 emissions continue to increase it appears likely that global temperatures will also continue to rise, and hence a significant amount of level of sea rise is inevitable due to thermal expansion and melting of glaciers and ice sheets/shelves. So-called "semi-empirical methods" (see IPCC 5AR) such as those by Vermeer and Rahmstorf (2009) indicate that for the future global temperatures scenarios given in the IPCC 4AR, the projected sea level rise for the period 1990–2100 could be in the 0.75 to 1.9m range, over twice the upper range indicated in the IPCC 5AR. The rate of increase will vary by region (IPCC 5AR); thus, from an economic, engineering and social point of view, it will be critical to predict the specific regional increases in sea levels, which could differ significantly from the global average (Magnan et al., 2011). However, for the case of Tokyo Bay, sea level rise is not expected to significantly deviate from global means.

3. Consequences of sea level rise on port installations

Sea level rise will eventually result in permanent flooding of the working areas of many ports at high tides. Exactly how quickly such situations will arrive will depend on the rate of this rise as outlined in the previous section. A useful indicator for the consequences that this sea level rise, could have were illustrated by the land subsidence caused by the 2011 Tohoku earthquake and tsunami (see Mikami et al., 2012). Following this earthquake, the level of the ports in many areas was lowered, resulting in flooding during high tides. To mitigate the impacts of flooding on their operations, the ports raised existing surfaces with gravel and sand to regain access for vehicles and equipment (see Figure 9.2).

Figure 9.2 Potential consequence of land subsidence on port areas. Flooding of a car park area following subsidence during an earthquake; on the left the area has been raised to enable use.

Table 9.1 Low-lying port areas in Tokyo, Kawasaki and Yokohama

	Tokyo	*Kawasaki*	*Yokohama*
Area (km^2)	11.9	17.6	8.5
Storm surge height +SLR (T.P., in m)	4.5	4.0	3.9
Cost (bn yen)	19.5	67.7	34.5

The effect illustrated is comparable to the more severe scenarios of Vermeer and Rahmstorf (2009) of 1.9m sea level rise by 2100, indicating the scale of adaptation actions which will be required. Hoshino (2013) calculated that the cost of the materials that would be required for raising the ground could amount to 19.5bn yen for the case of Tokyo Port, while raising other ports around the Bay could prove significantly more expensive (Figure 9.3 shows that 67.5 and 34.5bn yen could be needed for other ports and the low-lying areas of Kawasaki and Yokohama, respectively). Such costs are not directly proportional to the land area as they depend on existing ground level and by how much it should be raised, among other factors. Nevertheless, such calculations may be overly conservative, as they do not include the cost of demolishing and rebuilding affected buildings and structures (as these arguably have a design life of 15–20 years and would probably be rebuilt several times in the next century).

The table indicates the extent of each area, the expected height of a combined storm surge due to a 933.9hPa typhoon and a 1.9m sea level rise by the year 2100 (i.e. T.P.) at each location,[3] and the adaptation cost to raise lands outside the leveed areas to preserve risk levels constant to what they are nowadays by the year 2100 (see Hoshino, 2013).

4.Adaptation of breakwaters to an increase in typhoon intensity and sea level rise

The stability of breakwaters in Japan is likely to suffer as a consequence of the potential future increase in typhoon intensity and sea level rise. Sea level rise and other effects of climate change, such as an increase in tropical cyclone intensity (Knutson et al., 2010), could alter future wave patterns (Mori et al., 2010) and lead to increased damage to coastal defences. However, it is not necessarily clear to what degree these influences will impact on the stability of these structures and whether they should be strengthened accordingly. Such effects are likely to cause an increase in the required materials (and thus the cost) to build and maintain these and other types of sea defences. Strong winds due to typhoons generate high waves, with the size of the waves being proportional not only to the wind speed but also to the duration of the wind and the fetch. More intense typhoons are typically bigger, have higher maximum wind speeds, and act over a longer duration (see Esteban et al., 2009b).

Although recently sea level has started to be considered during breakwater design in some countries, under traditional design methods it was not included, something that could lead to them being under-designed towards the end of their lives, especially for a rapid increase in sea level scenario. It is important to note that these structures typically have a relatively long design life, usually of 30–50 years, though many continue to serve their purpose for longer periods due to the great expense involved in their construction and their resilience to major deterioration if well-built and maintained. Traditionally these structures are designed by looking at historical records of wave conditions over an area, which are assumed not to change over time. Coastal engineers typically use the significant wave height (H_s)[4] to estimate the size of the various elements of a given structure, with the two most common examples of structures being rubble mound or caisson type breakwaters. Generally speaking, rubble mound breakwaters are installed in shallower water than caisson breakwaters; these two typologies will be discussed separately in the next subsections.

4.1 Rubble mound breakwaters

Rubble mound breakwaters are typically made up of a mound with some porosity that is covered by a sloping armour layer consisting of large rock or concrete units, often tetrapods in the case of Japan (see Figure 9.3). These armour units are the crucial part of the structure, as their task is to resist the force of the wave, and once they are removed the breakwater is considered to have failed because the underlayer can then be quickly removed by the effect of wave action. Figure 9.3 shows a typical cross-section of a rubble mound breakwater. The weight of the armour units is generally proportional to the size of the waves (with higher waves requiring larger and heavier armour units).

Figure 9.3 Tetrapod armour units at Ishihama port in Japan.

These breakwaters are typically located in shallower waters than caisson breakwaters, due to the larger cross-section that characterises them, which makes them uneconomical in larger water depths. For the case in which these structures are located in typhoon-prone countries, the "Limiting Breaker Height" (H_b) parameter is more relevant than H_s for design. Essentially, as waves approach the coastline they gradually increase in size until the movement at the top becomes too extreme and the wave breaks. The location where this happens depends on the bathymetric profile at each point along the shore. The term "Limiting Breaker Height" is often used, as there is an upper limit to the waves physically possible at a certain water depth for a given wave period.

This breaking wave effect can be used to calculate the impact of sea level change on rubble mound breakwater design. As sea levels rise, the armouring requirements will increase substantially for breakwaters situated in shallow depths because the Limiting Breaker Height H_b parameter will increase and hence higher waves will reach the breakwater.

Esteban et al. (2014b) calculated the structural design changes due to sea level rise for rubble mound breakwaters with a variety of geometries, bathymetries, and structure configurations. These were computed for four different sea level rise scenarios over a period of 50 years,[5] and compared to a control scenario where conditions remained the same as at present (i.e. no sea level rise):

- Scenario 1: 0.15m increase, which would correspond to an annual increase of 3mm per year
- Scenario 2: 0.44m increase
- Scenario 3: 0.9m increase
- Scenario 4: 1.3m increase, similar to that calculated by Vermeer and Rahmstorf (2009) for the period 2050–2100.

For the case of rubble mound breakwaters another important parameter to keep in mind is the height of the structure, which will depend on a parameter referred to as the run-up. The crest of the structure must be high enough to prevent the waves from overtopping, but not high enough for it to be uneconomical or aesthetically unpleasing. As wave heights increase due to a higher *Hb*, the potential run-up on the breakwater will also increase and require them to be built with higher crests so that there is not significant overtopping. It appears that for the more extreme cases of sea level rise (Scenario 4 with a sea level rise of 1.3m) a breakwater designed in 2050 would require between 8 percent (for the deeper sections) and 66 percent more materials (for the shallower sections) than one designed in the twentieth century, not taking into account sea level rise (see Esteban et al., 2014b), also significantly impacting the cost of the structure.

Such measures are compatible with guidance in other countries. For example, the USA (USACE, 2011) requires consideration of three sea level rise scenarios, where the highest assumes a global sea level rise of 1.5m during the twenty-first century. Figure 9.4 shows the likely increase in the size of the required armour units for breakwaters situated in relatively shallow depth of 5 to 11m, showcasing

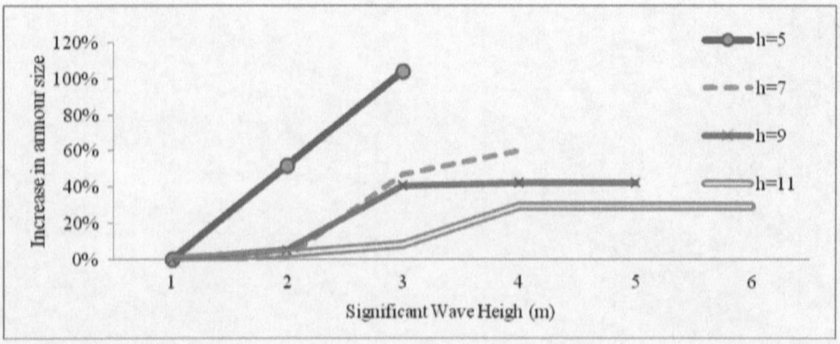

Figure 9.4 Increase in armour size for Scenario 4 for a variety of significant wave heights and water depths (water depths *h*=5–11m).

how for higher H_s values the increase in armour size plateaus, due to the effect of H_b. Nevertheless, accounting for sea level rise will represent an important increase in the cost of building and designing such breakwaters.

4.2 Caisson breakwaters

The evaluation of the risks climate change poses to caisson breakwaters is more complex than for rubble mound breakwaters, given that Japanese engineering codes of practice currently recommend the use of computationally complex third-order reliability design methods. Takagi et al. (2011) proposed a methodology to evaluate the sliding stability of breakwaters for future climate change scenarios based on the probabilistic method outlined by Takagi et al. (2007, 2008). This included the caisson breakwaters located at the Shibushi Ports.

The Shibushi Ports are located in the southern part of Kyushu Island and are regularly subjected to very large waves, particularly during the typhoon season (July–October). The offshore breakwater that protects the Shibushi Ports is 2570 m long and has eight different cross-sections of upright caissons, as shown in Figure 9.5. Among these caissons, those of section types I and VIII are equipped with wave-dissipating blocks in front of the caisson wall, whereas the other sections (II to VII) are simple upright caisson type breakwaters without blocks (Takagi et al., 2008).

Simulations of the wave climate using a third-generation spectral wave model (*SWAN)* found that a 10 percent increase in the potential future wind speed of tropical cyclones in the Asia-Pacific Region (Knutson et al., 2010) could correspond to a 21 percent increase in the peak significant wave height at this location. The study calculated that the combined damage from sea level rise and wave height increase would be greater than if they are calculated separately (see Figure 9.6), as sea level rise will increase the Limiting Breaker Height (*Hb*) and result in greater wave pressures than when sea level rise is ignored. For the case of the Shibushi breakwaters, the expected sliding distance[6] under the future climate scenario could increase to more than five times that under the present climate (see Figure 9.6). Hence, designers should take into account the possibility of decreased stability of caisson breakwaters due to climate change when they design new breakwaters or renovate existing ones.

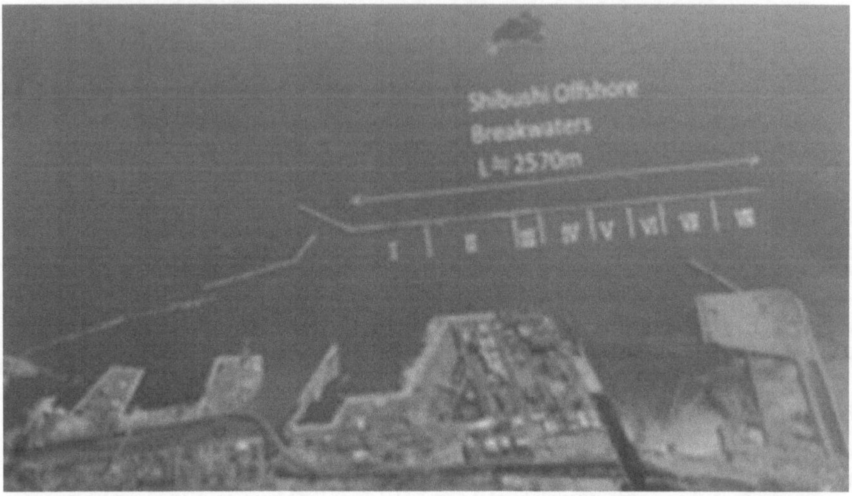

Figure 9.5 Aerial photo showing location of breakwaters at Shibushi port in the southern part of Japan.

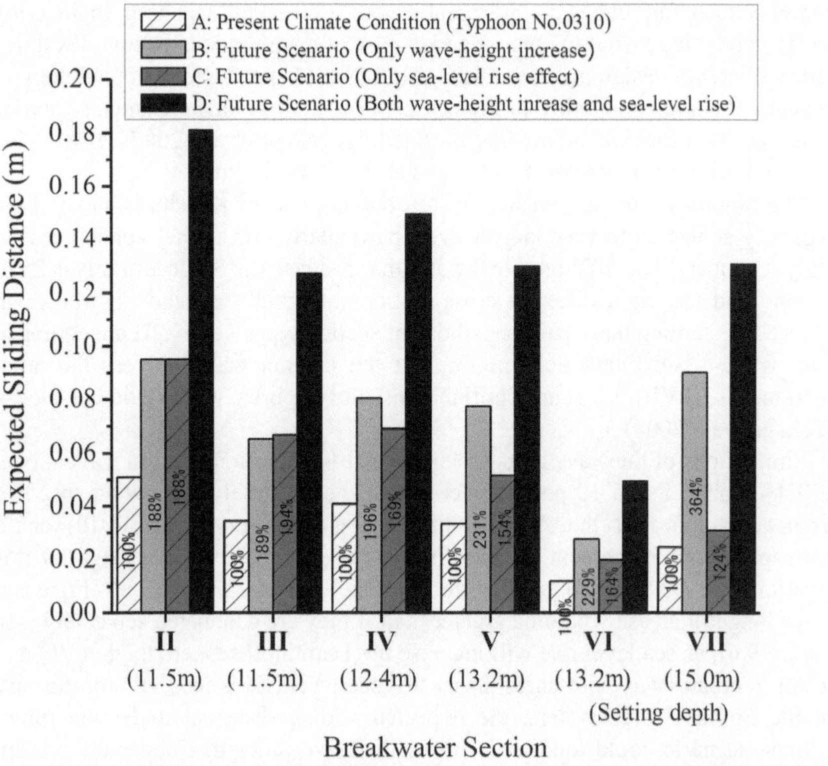

Figure 9.6 Comparison of the expected sliding distances between present climate and future climate conditions (after Takagi et al., 2011).

4.3 The move towards "adaptable" designs

Because sea level rise and climate change will be gradual processes, many authors have already started to call for "adaptable designs", allowing for breakwaters to be gradually modified, accommodating increasing wave heights during their design life (Headland et al., 2011). However, financial considerations about when to undertake such adaptation strategies is still not clear, along with challenges in setting aside appropriate funding mechanisms to perform the various stages of structure raising. While some level of understanding exists concerning the stability of breakwaters vis-à-vis changes in sea levels and wave climate, limited research has been carried out on what triggers will determine the timing for reinforcing these structures to respond to climate projections (Headland et al., 2011). Uncertainty in decision-making will therefore have to be factored into risk assessment at the local levels, something which was arguably less the case during the twentieth century. In view of the complexity of some of the issues at stake, targeted port case studies may significantly contribute to improved risk and vulnerability assessments (UNCTAD, 2011).

5. Indirect economic damage due to downtime in ports

The economic impact of tropical cyclones in Japan depends on several factors such as the location of economic activity, number of storms, intensity of storms, and the topography of the affected region and other geographical attributes, such as land-use patterns. Increased typhoon intensities will not only increase the direct damage to ports, but also the indirect economic effects caused by them having to temporarily suspend their operations during the passage of a given weather system. As there is a general relationship between the maximum sustained wind speed and the size of a typhoon (Esteban et al. 2010), stronger wind events will increase typhoon size and the damage caused by them. This could cause operational delays of billions of dollars per day and may have important second-order consequences, not only for the regional economy and the quality of life of those who depend on the port's functionality, but also for the operation of global supply chains (Hallegatte et al., 2011)

Esteban et al. (2009) calculated the expected downtime in the year 2085 based on the results of Knutson and Tuleya (2004). According to their results, Figure 9.7 show the average hours that representative Japanese ports can be expected to be affected by 30 knot winds for both the control scenario and future climate change scenarios. On average, each of the ports can be expected to be affected by 30-knot winds 18 to 43 percent more of the time in the future. The southern parts of Japan typically suffer the effect of typhoons more than the northern parts, and this is likely to continue to be the case in the future. Another consideration is the seasonality of downtime. In Japan, the typhoon season usually lasts from around July to September, and hence most of the downtime usually occurs during these months. On average, the climate change scenarios represent an 18 percent increase in downtime during the summer months, which could greatly disrupt port activities, especially for the southernmost regions (Esteban et al., 2009b).

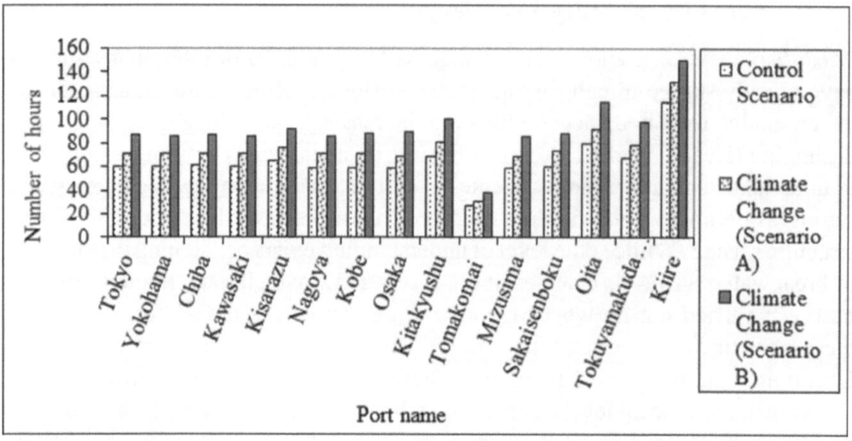

Figure 9.7 Expected annual port downtime at different ports in Japan at present and under two different climate change scenarios by the year 2085 (Esteban et al., 2009b).

Due to the fact that Japan's economy is heavily trade-orientated, the expansion of the country's ports has been indispensable for its growth. Kawakami and Doi (2004) established a direct correlation between the natural logarithm of the Real Port Capital Stock (*RPCS*)—defined as the total value of all port infrastructure and stock in Japan's ports—and the growth in Japanese GDP. It is clear that for the Japanese economy to grow there must be a continuous expansion of *RPCS* to deal with a greater volume of exports.

To obtain the cost of adaptation, it is necessary to calculate the additional *RPCS* needed due to future climate change and its effect on port downtime, for which ports must be expanded to increase their capacity so that they can quickly catch up with the downtime suffered during a typhoon. If RPCS is not increased to compensate for the increases in downtime this will have knock-on effects on supply chains and exports, damaging the country's economy. To establish the capital investment needed, Esteban et al. (2009) analyzed four scenarios, depending on rate of economic growth (1 or 2 percent) and the relationship between maximum wind speed and typhoon area. According to their results, a total of between 30.6 to 127.9 billion additional yen could potentially be needed by the year 2085 to expand all ports in the country to cope with the increase in downtime associated with the increase in typhoon intensity given in Knutson and Tuleya (2004), as shown in Figure 9.8 (for reference, current GDP of Japan is 484,382 billion yen). Failure to spend this money could reduce potential future GDP by between 1.5 and 3.4 percent by 2085, due to delays in exports.

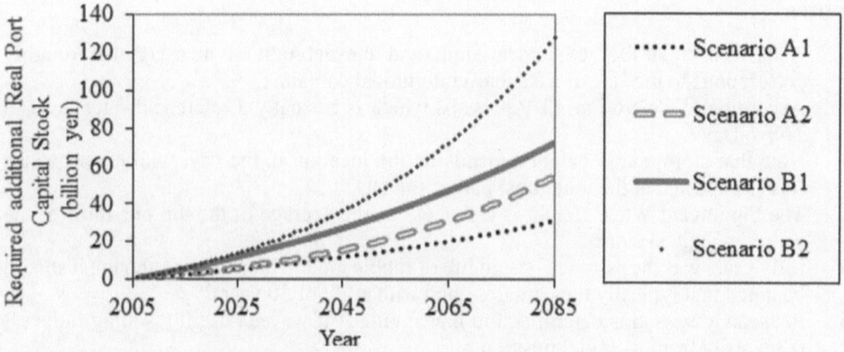

Figure 9.8 Increased estimated required additional real capital stock for all Japanese ports by the year 2085.

Conclusions

The combined effect of an increase in typhoon intensity and sea level rise could pose significant challenges to ports around Japan. Essentially, stronger tropical cyclones and their associated storm surges, together with the long-term effects of sea level rise could lead to the inundation of many port areas. This would require the elevation of many of them, with the extra materials needed in this elevation costing in excess of 120bn yen for the case of the ports around Tokyo Bay, not taking into account the extra costs of demolishing and rebuilding port installations and buildings. An alternative to this could be the construction of a storm surge barrier or permanent lock at the entrance of Tokyo Bay, which could somehow reduce the extent of this problem by allowing water levels to be controlled though also slowing the passage of ships (Esteban et al., 2014a).

Breakwaters will also have to be considerably strengthened in the future, as stronger waves generated by stronger typhoons will require larger cross-sections across all types of breakwaters. As these structures are typically in use long after their designated lives have expired, this would eventually probably require the strengthening of many existing structures. Given the considerable number of ports existing in Japan, the costs involved in doing this are considerable, and will be particularly challenging giving the aging nature of Japanese society.

Finally, stronger typhoons will also lead to increased downtime and bottlenecks in supply systems, which could substantially harm the economy of a country heavily dependent on the import of raw materials and exports of manufactured goods, such as Japan. This would require extra port handling capacity to be built, to be able to rebound from these longer periods of port shutdown.

While all these challenges are not insurmountable, especially for a rich developed country such as Japan, the costs associated with them clearly indicate that mitigating the effects of climate change is preferable to adaptation costs (Esteban and Longarte-Galnares, 2010).

Notes

1 A TEU is a 20-foot equivalent unit, and inexact unit of measure that roughly corresponds to the size of a standard intermodal container.
2 According to Tokyo Pail (T.P.) levels, which is basically a reference water level in Tokyo Bay.
3 Note that storm surge height depends on the location in the bay, with higher storm surges expected at the innermost part of the bay.
4 The Significant Wave Height is estimated as the average of the top one-third of the waves in a given storm.
5 Fifty years was the assumed design life of rubble mound breakwaters, though it should be noted that typically they are designed with a life of 30 years.
6 In Japan a caisson is determined to have "failed" if waves push it ("sliding failure") 0.3m away from its initial position.

References

Allison, I., Bindoff, N. L., Bindschadler, R., Cox, P., Noblet-Ducoudre, N. et al. (2009). *The Copenhagen Diagnosis, 2009: Updating the World on the Latest Climate Science.* http://www.copenhagendiagnosis.org/read/default.html (retrieved January 26, 2010).

Becker, A., Acciaro, M., Asariotis, R., Cabrera, E., Cretegny, L., Crist, P., Esteban, M., Mather, A., Messner, S., Naruse, S., Adolf, K. Y., Stefan Rahmstorf, S., Savonis, M., Song. D. W., Stenek, V. and Velegrakis, A. F. (2013). "A note on climate change adaptation for seaports: A challenge for global ports, a challenge for global society", *Journal of Climatic Change*, DOI 10.1007/s10584-013-0843-z.

Elsner, J. B., Kossin, J. P. and Jagger, T. H. (2008). "The increasing intensity of the strongest tropical cyclones", *Nature*, 455: 92–94.

Esteban, M. and Longarte-Galnares, G. (2010). "Evaluation of the productivity decrease risk due to a future increase in tropical cyclone intensity in Japan", *Journal of Risk Analysis*, http://onlinelibrary.wiley.com/doi/10.1111/risk.2010.30.issue-12/issuetoc.

Esteban, M., Takagi, H. and Shibayama, T. (2009a). "Methodology for the simulation of the construction of a breakwater taking into account climate and construction accident risks", *Coastal Engineering Journal (CEJ)*, 51 (1): 49–68.

Esteban, M., Webersik, C. and Shibayama, T. (2009b). "Methodology for the estimation of the increase in time loss due to future increase in tropical cyclone intensity in Japan", *Journal of Climatic Change*, 102 (3–4): 555–578.

Esteban, M., Takagi, H. and Shibayama, T. (2012a). "Modified heel pressure formula to simulate tilting of composite caisson breakwater", *Coastal Engineering Journal*, 54 (04). DOI: 10.1142/S0578563412500222

Esteban, M., Zhang, Q. and Longarte-Galnares, G. (2012b). "Cost-benefit analysis of green electricity system in Japan considering the economic impacts of tropical cyclone intensity", *Journal of Energy Policy*, 43: 49–57.

Esteban, M., Mikami, T., Shibayama, T., Takagi, H., Jonkman, S. N. and Ledden, M. V. (2014a). "Climate change adaptation in Tokyo Bay: The case for a storm surge barrier", Proceedings of the 34th International Conference on Coastal Engineering (ICCE), June 15–20, 2014, Seoul, Korea.

Esteban, M., Takagi, H. and Nguyen, D. T. (2014b). "Tropical cyclone damage to coastal defenses: Future influence of climate change and sea level rise on shallow coastal areas in southern Vietnam". In: Nguyen, D. T., Takagi, H. and Esteban, M. (Eds.), *Coastal*

Disasters and Climate Change in Vietnam: Engineering and Planning Perspectives. Amsterdam: Elsevier.

Hallegatte, S., Ranger, N., Mestre, O., Dumas, P., Corfee-Morlot, J., Herweijer, C., Wood, R. M. (2011). "Assessing climate change impacts, sea-level rise and storm surge risk in port cities: A case study on Copenhagen", *Climate Change*, 104 (1): 113–137.

Headland, J. R., Trivedi, D., and Boudreau, R. H. (2011). "Coastal structures and sea level rise: Adaptive management approach". In: Magoon, O. T., Noble, R. M., Treadwell, D. D. and Kim, Y. C. (Eds.), *Coastal Engineering Practice*. Reston, VA: ASCE, pp. 449–459.

Hoshino, S. (2013). *Estimation of Storm Surge and Proposal of the Coastal Protection Method in Tokyo Bay*. MSc thesis, Waseda University.

Hoshino, S., Esteban, M., Mikami, T., Takabatake, T. and Shibayama, T. (2011). "Effect of Sea Level Rise and Increase in Typhoon Intensity on Coastal Structures in Tokyo Bay". Proceedings of Coastal Structures Conference, Yokohama, September 6–8, 2011.

Japan Statistics Bureau. (2010). http://www.e-stat.go.jp/SG1/estat/XlsdlE.do?sinfid=000008640423 (accessed September 15, 2012).

Kawai, H., Hashimoto, N. and Matsuura K. (2006). "Improvement of stochastic typhoon model for the purpose of simulating typhoons and storm surges under global warming", Proceedings of the 30th International Conference on Coastal Engineering (ICCE, 2006), Vol. 2, pp. 1838–1850.

Kawakami, T. and Doi, M. (2004). "Port capital formation and economic development in Japan: A vector autoregression approach", *Papers in Regional Science*, 83: 723–732.

Knutson, T. R. and Tuleya, R. E. (2004). "Impact of CO_2-induced warming on simulated hurricane intensity and precipitation sensitivity to the choice of climate model and convective parameterization", *Journal of Climate*, 17 (18): 3477–3495.

Knutson, T., McBride, J., Chan, J., Emanuel, K., Holland, G., Landsea, C., Held, I., Kossin, J., Srivastava, A. and Sugi, M. (2010). "Tropical cyclones and climate change," *Nature Geoscience*, 3 (3): 157–163.

Landsea, C. W., Harper, B. A., Hoarau, K. and Knaff, J. A. (2006). "Can we detect trends in extreme tropical cyclones?", *Science*, 313 (5786): 452–454.

Magnan, A., Duvat, V., Pirazzoli, P. and Woppelmann, G. (2011). "In light of climate change, can coral archipelagos be defined as vulnerable "resource systems"?", 4th Congress of the Asia and Pacific Network, September 14–16, 2011, Paris, France (in French).

Mikami, T., Shibayama, T., Esteban, M. and Matsumaru, R. (2012). "Field survey of the 2011 Tohoku earthquake and tsunami in Miyagi and Fukushima prefectures", *Coastal Engineering Journal (CEJ)*, 54 (1): 1–26.

Miyazaki, M. (1970). *Tsunami Storm Surge and Coastal Disasters*. Edited by K. Wadachi. Tokyo: Kyouritsu Shuppan (in Japanese).

Miyazaki, M. (2003). *Study on Storm Surge*. Tokyo: Seizando Publishing (in Japanese).

Mori, N., Yasuda, T., Mase, H., Tom, T. and Oku, Y. (2010). "Projection of extreme wave climate change under global warming", *Hydrological Research Letters*, 4: 15–19.

Mousavi, M. E., Irish, J. L., Frey, A. E., Olivera, F. and Edge, B. L. (2011). "Global warming and hurricanes: The potential impact of hurricane intensification and sea level on coastal flooding", *Climatic Change*, 104: 575–597.

Pricewaterhouse Coopers. (2009). http://www.pwc.com/ (accessed September 15, 2012).

Takagi, H., Shibayama, T. and Esteban, M. (2007). "An expansion of the reliability design method for caisson-type breakwaters towards deep water using the fourth order approximation of standing waves," *Proceedings of Asian and Pacific Coasts 2007*, APAC, pp.1723–1735.

Takagi, H., Esteban, M. and Shibayama, T. (2008). "Proposed methodology for evaluating the potential failure risk for existing caisson-breakwaters in a storm event using a level III reliability-based approach," Proceedings of 31st International Conference on Coastal Engineering, ASCE, pp. 3655–3667.

Takagi, H., Kashihara, H., Esteban, M. and Shibayama, T. (2011). "Assessment of future stability of breakwaters under climate change", *Coastal Engineering Journal*, 53 (1): 21–39.

Takagi, H., Esteban, M., Shibayama, T., Mikami, T., Matsumaru, R., de Leon, M., Thao, N. D. and Oyama, T. (2014). "Track analysis, simulation and field survey of the 2013 Typhoon Haiyan storm surge", *Journal of Flood Risk Management*, DOI:10/1111/jfr3/12136.

Tokyo Metropolitan Government. (2011). Bidding information: http://bidfind.openbeat.org/ (accessed January 15, 2013).

United Nations Conference on Trade and Development Main Outcomes and Summary of Discussions (2011) In: *Climate Change Impacts and Adaptation: A Challenge for Global Ports,* Ad Hoc Expert Meeting, Geneva, Switzwerland, September 29–30, 2011. doi:UNCTAD/DTL/TLB/2011/3.

United States Army Corps of Engineering (USACE) (2011) *Sea-Level Change Considerations for Civil Work Programs.* Circular No. 1165-2-212.

Vermeer, M. and Rahmstorf, S. PNAS 2009;106:21527–21532.

Webster, P. J., Holland, G. J., Curry, J. A. and Chang, H.-R. (2005). "Changes in tropical cyclone number, duration, and intensity in a warming environment", *Science*, 309 (5742): 1844–1846.

Yasuda, T., Mase, H. and Mori, N. (2010a). "Projection of future typhoons landing on Japan based on a stochastic typhoon model utilizing AGCM projections", *Hydrological Research Letters,* 4: 65–69.

10 Modeling and evaluation of green port development

A case study on Tianjin Port

Di Zhang, Chengpeng Wan and Zaili Yang

1. Introduction

The present observation and studies show that over the past century, the global climate has been undergoing significant change as a main characteristic of the global warming (VijayaVenkataRaman et al., 2012). Problems caused by climate change such as high temperature, sea level rising and extreme weathers have aggravated the loss of life and property around the world, and caused severe impacts on both nature and society. As a crucial node in comprehensive transport networks, ports play a significant role in promoting a country's domestic and foreign trade development. For example, in China they counted for around 93 percent of Chinese international trade, 95 percent of crude oil trade and 99 percent of iron ore trade in 2012 (Wu and Ji, 2013). Meanwhile, billions of US dollars has been spent on port infrastructure construction, reconstruction and maintenance for the rapid development of ports in China. Thus, the negative effects of climate change will greatly influence port facilities and operations in China in the coming decades. Although estimates of the extent to which climate change and extreme weather will cause damage to ports are debatable, efforts are still needed to gain a better understanding as to how ports will be affected by climate change and what risk-control measures should be taken, in order to ensure the resilience of port facilities and personnel, and provide reliable service to customers.

In spite of the fact that a port is not a direct site for production processing, nor does it have a large amount of material consumption, it is, however, an important distribution center for various goods, allowing a large number of cars and ships to be engaged in transport operations, which can be a source of contamination (discharge of waste gas and rubbish) (Chen, 2009). Apart from these traffic conveyances, there is pollution from the goods themselves such as coal dust, dangerous materials and chemicals. A Chinese study (Ma et al., 2014) showed that in 2011, emissions of carbon dioxide (CO_2), nitrogen dioxide (NO_2) and dust from harbor districts throughout the country reached about 127 thousand tons, 146 thousand tons and 1.2 million tons respectively, contributing immensely to environmental deterioration in China. Pollution from port operation will not only damage the ecological balance of nature and the urban environment, but also cause adverse effects on global climate change, which further increases risk in port operation processes. The development of a low-carbon economy is considered to be a fundamental way to solve the problems caused by climate change. Nevertheless, ports and shipping are still lacking effective controls for emissions of greenhouse gas and the importance of sustainable development and climate change is still being ignored by most port management authorities (Wang, 2014). In view of this, the concept of fifth-generation ports, namely green ports (or low-carbon ports), was proposed at the United Nations climate change conference in 2009 (Wu and Ji, 2013). Therefore, the construction of green ports is a basic way to cope with global climate change and the key to addressing the problems encountered in the process of port development. On the basis of the organic combination of port development, utilization of resources and environmental protection, green ports refers to ports characterized by healthy ecological environments, reasonable utilization of resources, low energy consumption and low pollution (Chen, 2009).

Started initially in the USA, Japan and other developed countries, prominent achievements have been obtained by actively exploring and implementing the planning and construction of green ports (Gupta et al., 2005; Barney, 1988). As one of the advocates of green ports, the Port of Long Beach has made remarkable achievements. A "green port" policy was launched in the Port of Long Beach in January 2005 with a series of environmental protection plans established for seven aspects, namely water quality protection, improvement of air quality, soil conservation, wildlife and habitat protection, alleviating traffic pressure, sustainable development and community participation (Lv, 2005). Since the implementation of the above environmental protection plans, the water quality at Long Beach has been improved. In the meantime, Sydney Harbor carried out Green Port Guidelines for other aspects, paying more attention to the importance of the quality of the water and air, biological diversity, noise control, rubbish and dangerous cargo management, environmental education and training, etc. (Lu and Hu, 2009). Strengthening legislation and enforcement was the main focus for Sydney Harbor. In Italy, a shore power supply system was equipped in both the Venetian Harbor and Port of La Spezia in 2010, resulting in about 30 percent reduction of CO_2 emissions, 95 percent reduction of nitric oxide (NO) emissions, and significant noise reduction as well (Cai, 2010). Tokyo Harbor tried to improve the environment through planting trees and other green techniques. When planning the layout of the port, its influence on the environment should be considered including both ecological and living environments. It is also required that port construction projects and environmental protection planning should be implemented simultaneously (Liu, 2004). The aforementioned countries apply "green" to their daily operation and future design of port construction to strengthen port infrastructure and the emergency response capability.

Research on green ports started relatively late in China (Song, 2011). Shanghai Port took the lead in the development of green port technology in China. It actively explored environmental protection measures for the administration of Shanghai Port and listed the construction of an ecological port in Shanghai as an important research subject. Based on the research of ecological port development and countermeasures for Shanghai Port in 2006, an evaluation index system of green port development was developed (Lin, 2010). Tianjin Port officially launched a research project into green port development in October 2007, elaborating the concrete measures of development of green ports from aspects of environmental protection infrastructure, environmental pollution control, environmental risk prevention and management, development of environmental management systems as well as the construction of green logistics networks. The practice showed that by the end of 2010 in Tianjin Port, water quality, air quality, noise control, industrial water recycling rate, intensity of chemical oxygen demand (COD) emissions, intensity of sulfur dioxide (SO_2) emissions and sewage treatment rate all met the requirements of the assessment system of Tianjin eco-city construction (Wan et al., 2011). Lianyungang Port benefitted from the advantages of shore power technology in the control of pollutant emissions, energy saving and noise reduction, which brought considerable economic and social benefits (Yu, 2012). A variety of measures have been taken in Tangshan Port including partition management of the yard, arrangement of dust-proof nets, weight and speed limit of vehicles, improvement of production process, increasing the virescence area proportion, establishment of specialized management pattern and building information systems for port environmental management. These measures enhanced the recycle rate of resources and improved the port's environmental protection (Li et al., 2010). Qingdao Port introduced new equipment and new technologies imported from developed countries to improve working efficiency and to reduce energy consumption (Yu and Liu, 2012). Compared to developed countries, problems associated with green port development in China mainly include (Song, 2011):

a) There are a lot of old ports in China that came into service a few decades ago. Generally, theses ports lack funds and are under management with outdated concepts and techniques.
b) Low consciousness of environmental protection and energy saving has resulted in the lack of systematic and comprehensive planning and design when developing green ports.
c) The lack of sound evaluation criteria for the development of green ports has caused a certain blindness in green port development and has seriously affected the sustainable development of port resources, environment and economy.

The main aim of this chapter is to establish a model for the assessment of green port development through the identification of the key indexes in the model. The analytic hierarchy process (AHP) method is applied to determine the weight of each index in the model and the estimations are aggregated to obtain the overall development status of green ports using the Evidential Reasoning (ER) approach. This chapter investigates assessment grades for each criterion, converts quantitative criteria to qualitative ones by employing a fuzzy membership function and applies the ER approach to synthesize the assessment results. A case study on the Tianjin Port is used to demonstrate the applicability of the proposed assessment model.

2. Modeling of green port development

The port assessment model is a complex system involving many indexes from various aspects. These indexes included in the model should be comprehensive, quantitative and exclusive without overlaps with each other. Besides, these indexes should reflect real-time changes accordingly to adapt to the needs of social development, while maintaining a relative stability for the evaluation of green ports in different periods.

The assessment model developed in this chapter consists of three levels. The top level reflects the model's goal, which is to assess the development level of green ports.

The second level (criteria level) is constructed according to the DPSIR (Drivers, Pressures, States, Impacts, and Responses) framework which was introduced in the late 1990s and then applied in the evaluation of sustainable development (Carr et al., 2007). This framework enabling the integration of different types of indexes concerning environmental, social and economic issues, has been successfully implemented in the evaluation of the modernization of inland port and shipping management (Wen and Chen, 2013). Thus, the DPSIR framework is utilized in this research and the indexes in Criteria Level are set to be Drivers (B1), Pressures (B2), States (B3), Impacts (B4), and Responses (B5).

The indexes in the bottom level (index level) are chosen in terms of their associated elements in the upper level. "Drivers" here refers to the factors that show the interests of major stakeholders, which may lead to environmental pressures if over demanded. The throughput of the port and its growth rate are important quantitative indexes in the evaluation of the production and business activities of a port. They also impact on the economic performance of the port city (Zhang and Liu, 2006). "Pressures" indicates the stress on the environment of the port area due to daily port operations. There are lots of factors influencing the port environment, with dust pollution, water pollution, solid waste pollution and noise pollution are revealed as primary pollution sources by a recent study (Jiang, 2014). Accordingly, the indexes in "States" can be determined to reflect the condition of the port environment. Indexes in "Impacts" should be able to address not only the operational efficiency but also the safety degree of the port. These two important aspects in the development of green ports have been examined in a recent study (Dong et al., 2011). Despite society's responses, indexes in "Responses" here focus more on the actions taken by port authorities to release the pressures in the development of green ports, among which the construction of infrastructure, information platform and management systems, as well as emergency response capacities are of great significance to the safety and efficiency of port operations (Ouyang and Wang, 2014).

Hence, 18 indexes are selected in the index level (as seen in Figure 10.1) with respect to their significant roles reflecting the development level of green ports. These indexes are further verified through extensive discussions with domain experts. The detailed information of experts is shown as follows:

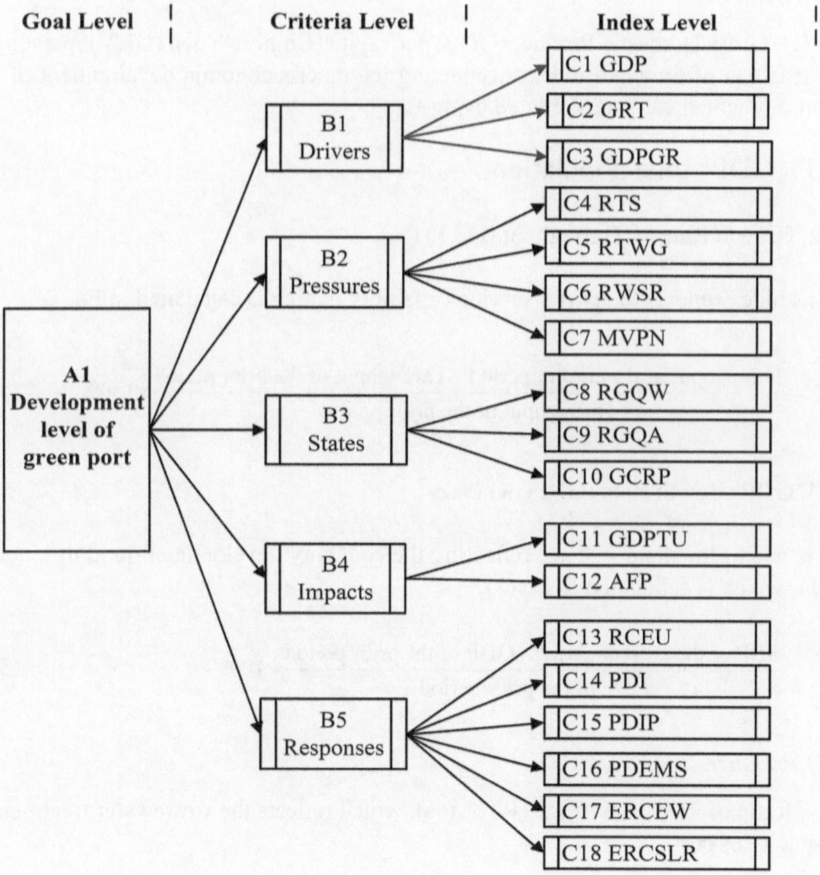

Figure 10.1 Assessment model of green port development.

- Expert No. 1: An experienced port manager with more than eight years' experience of safety management in port operations.
- Expert No. 2: A professor engaged in port performance research for more than ten years with particular reference to environmental effects, who is also a consultant for the development of Tianjin green port.
- Expert No. 3: A senior officer in charge of environmental protection from Tianjin Port Holdings Company Limited.

Interpretations of each index in the index level are provided as follows:

(1) Drivers:

C1, Gross Domestic Product (GDP) per capita (Chinese Yuan (CNY)/ person):

It is one of the main indexes reflecting the macroeconomic development of a port city which can be calculated using Eq (1),

$$C1 = GDP / \text{total population} \quad (1)$$

C2, Growth Rate of Throughput (GRT) (%):

This index reflects the port's service capability, which is calculated in Eq (2),

$$C2 = \frac{\text{Throughput of the current period - Throughput of the prior period}}{\text{Throughput of the prior period}} \times 100\% \quad (2)$$

C3, GDP Growth Rate (GDPGR) (%):

It is one of the main indexes reflecting the economy development trend of a port city, which is calculated in Eq (3),

$$C3 = \frac{\text{GDP of the current period - GDP of the prior period}}{\text{GDP of the prior period}} \times 100\% \quad (3)$$

(2) Pressures:

C4, Ratio of Treated Sewage (RTS) (%), which reflects the wastewater treatment capacity of port.

C5, Ratio of Treated Waste Gas (RTW) (%), which reflects the waste gas treatment capacity of port.

C6, Rate of Solid Wastes Recycling (RSWR) (%), which reflects the solid wastes treatment capacity of port.

C7, Mean Value of Port Noise (MVPN) (Decibel), which reflects the quality of production environment of port.

(3) States:

C8, Rate of Good-Quality water (RGQW) (%), which reflects the water quality of port.

C9, Rate of Good-Quality Air (RGQA) (%), which reflects the air quality of port.

C10, Green Coverage Ratio of Port (GCRP) (%), which reflects the green degree and environmental quality of port.

(4) Impacts:

C11, GDP per Throughput Unit (GDPTU) (CNY/ ton), which reflects the operational efficiency and professional ability of port.

C12, Accident Frequency of Port (AFP) (accident number/year), which reflects the degree of port safety.

(5) Responses:

C13, Rate of Clean Energy Utilization (RCEU) (%), which reflects the green degree and environmental quality of port.

C14, Perfection Degree of Infrastructure (PDI) (%), which refers to the mechanical equipment and facilities of the port. It is the guarantee of safe and efficient production at the port.

C15, Perfection Degree of Information Platform (PDIP) (%), which is another guarantee of efficient production and safety at the port.

C16, Perfection Degree of Environmental Management System (PDEMS) (%),

the establishment of the organization of environmental management, rules and regulations as well as reward and punishment measures is included in this index, which is the key to energy conservation and emissions reduction.

C17, Emergency Response Capacity of Extreme Weather (ERCEW), which refers to the port's emergency response ability to flooding, storm surge, heavy precipitation and other extreme weather due to climate change.

C18, Emergency Response Capacity of Sea Level Rising (ERCSLR), which reflects the ability of existing port facilities in coping with sea level rising.

3. Methodology

The following steps were developed in order to carry out an estimation of green port development.

Step 1: Carry out the pairwise comparisons in each level of the hierarchical structure in terms of the relative importance of the identified indexes to the development of a green port and calculate the weighting vectors of the indexes in the corresponding level.

Step 2: Develop a set of grading evaluation criteria and fuzzy membership functions to transform quantitative criteria into qualitative ones using an information transformation technique.

Step 3: The ER algorithm is used to carry out the assessment for the synthesis of basic criteria in the hierarchical structure.

Step 4: The results are prioritized and compared by using utility values for obtaining the development levels of a green port.

Step 5: The proposed model is validated through analyzing the research findings and some countermeasures are discussed.

3.1 Analytical hierarchy process (AHP)

AHP was developed by Satty to solve complex multi-criteria decision problems (Satty, 1980). AHP requires the decision makers to supply judgments about the relative importance of each criterion and then specify a preference for each decision alternative against each criterion. AHP is especially appropriate for complex decisions which involve the comparison of decision criteria that are difficult to quantify (Pillay and Wang, 2003). It is based on the assumption that when dealing with a complex decision the natural human reaction is to cluster the decision criteria according to their common characteristics. Since AHP was introduced three decades ago, many useful applications have been seen in the literature, including but not limited to, evaluation of green ports (Maritz et al., 2014), transportation system studies (Shang et al., 2004), risk and safety assessments (Sii and Wang, 2003), industrial engineering applications (Yang et al., 2003) and many more.

3.2 Evidential reasoning (ER)

ER was developed in the 1990s to deal with Multiple Criteria Decision Making (MCDM) problems under uncertainty. The ER algorithm is based on decision theory and the D-S (Dempster–Shafer) theory of evidence, which is well suited for handling incomplete assessment of uncertainty (Yang, 2001; Yang and Singh, 1994). The algorithm can be used to aggregate criteria of a multilevel structure. ER is widely used in many applications such as engineering design, system safety, risk assessment, organizational self-assessment and supplier assessment (Chin et al., 2009; Liu et al., 2008; Ren et al., 2008).

The set $S(E) = \{(H_n, \beta_n), n = 1, \ldots, N\}$ represents a criterion E which is assessed to grade H_n with degree of belief β_n, $n = 1, \ldots, N$. Let $m_{n,i}$ be a basic probability mass representing the degree to which the ith basic criterion e_i supports the hypothesis that the criterion y is assessed to the nth grade H_n.

To obtain the combined degrees of belief of all the basic criteria, $E_{I(i)}$ is firstly defined as the subset of the first i basic criteria as follows:

$$E_{I(i)} = \{e_1, e_2, ..., e_i\} \tag{4}$$

Let $m_{n,I(i)}$ be a probability mass defined as the degree to which all the i criteria in $E_{I(i)}$ support the hypothesis that E is assessed to the grade H_n and let $m_{H,I(i)}$ be the remaining probability mass unassigned to individual grades after all the basic criteria in $E_{I(i)}$ have been assessed. Eq. (5) and Eq. (6) are obviously correct when $i = 1$.

$$m_{n,I(1)} = m_{n,1},\ n = 1, 2, ..., N \tag{5}$$

$$m_{H,I(1)} = m_{H,1} \tag{6}$$

By using Eq. (5) and Eq. (6), Eq. (7) can be constructed for $i = 1, 2, ..., L$-1 to obtain the coefficients $m_{n,I(L)}$, $\bar{m}_{H,I(L)}$ and $\tilde{m}_{H,I(L)}$ (Yang and Xu, 2002):

$$K_{I(i+1)} = \left[1 - \sum_{t=1}^{N}\sum_{\substack{j=1 \\ j\neq t}}^{N} m_{t,I(i)} m_{j,i+1}\right]^{-1} \tag{7}$$

$K_{I(i+1)}$ is a normalizing factor.

$\{H_n\}$:

$$m_{n,I(i+1)} = K_{I(i+1)}[m_{n,I(i)} m_{n,i+1} + m_{H,I(i)} m_{n,i+1} + m_{n,I(i)} m_{H,i+1}]\quad n = 1, 2, ..., N \tag{8}$$

$$\tilde{m}_{H,I(i+1)} = K_{I(i+1)}[\tilde{m}_{H,I(i)} \tilde{m}_{H,i+1} + \bar{m}_{H,I(i)} \tilde{m}_{H,i+1} + \tilde{m}_{H,I(i)} \bar{m}_{H,i+1}] \tag{9}$$

$$\bar{m}_{H,I(i+1)} = K_{I(i+1)} \bar{m}_{H,I(i)} \bar{m}_{H,i+1} \tag{10}$$

$\{H\}$:

$$m_{H,I(i)} = \tilde{m}_{H,I(i)} + \bar{m}_{H,I(i)},\ i = 1, 2, ..., L - 1 \tag{11}$$

At last, the combined degrees of belief of all the basic criteria for the assessment to criterion E are calculated by:

$$\{H_n\}\text{: } \beta_n = \frac{m_{n,I(L)}}{1 - \bar{m}_{H,I(L)}},\ n = 1, 2, ..., N \tag{12}$$

$$\{H\}\text{: } \beta_H = \frac{\tilde{m}_{H,I(L)}}{1 - \bar{m}_{H,I(L)}} \tag{13}$$

The ER approach is used in Step 3 of the proposed methodology for synthesizing the evaluations of the basic criteria in the hierarchical structure.

3.3 Degree of evaluation index membership

As both the quantitative and qualitative indexes are included in the evaluation system, they should be transformed and presented in certain grades respectively, as shown in Section 3.3.1.

3.3.1 Membership degree of quantitative indexes

This study divides the development of green port into three levels, namely, Good, Average, and Poor. Membership degrees of quantitative indexes can be obtained through the graph of function shown in Figure 10.2, which is composed of fuzzy triangular and trapezoidal distributions. Further explanations (Wang, 2014) of this membership function are given in Table 10.1.
Here, *a* refers to the most possible value of Good while *c* is the most possible value of Poor, and b represents the most possible value of Average. In this study, the grading for quantitative indexes is obtained from a recent study (Wang, 2014) and in-depth discussions with the experts listed in Section 2, as shown in Table 10.2.

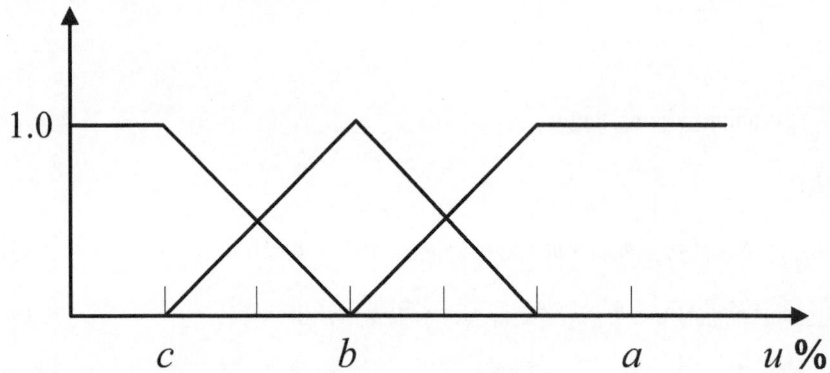

Figure 10.2 Graph of membership degree function.

Table 10.1 Membership functions for quantitative criteria

Grade	*Criteria*	*Membership function*	
Good	$u \geq b$	$(u-b)/(a-b)$	$b \leq u < a$
		1	$u \geq a$
		0	others
Average	$c < u < a$	$(u-c)/(b-c)$	$c \leq u < b$
		$(a-u)/(a-b)$	$b \leq u < a$
		0	others
Poor		$(b-u)/(b-c)$	$c < u \leq b$
	$u \leq b$	1	$u \leq c$
		0	others

Table 10.2 Grading for quantitative indexes

Quantitative index	*a*	*b*	*c*
C1 GDP (CNY/ person)	80000	70000	60000
C2 GRT (%)	30	20	10
C3 GDPGR (%)	30	20	10
C4 RTS (%)	85	75	65
C5 RTWG (%)	50	40	30
C6 RWSR (%)	95	85	75
C7 MVPN (db)	50	60	70
C8 RGQW (%)	90	70	50
C9 RGQA (%)	90	80	70
C10 GCRP (%)	20	15	10
C11 GDPTU (CNY/ ton)	3000	2500	2000
C12 AFP (accident number/ year)	0	1.5	3
C13 RCEU (%)	50	40	30

3.3.2 Membership degree of qualitative indexes

The grades of qualitative indexes are also described using Good, Average, and Poor. The evaluation results are obtained from questionnaires distributed to senior staff working in this area. Their judgments are merged according to their relative weights. Suppose M valid questionnaires are collected and there are altogether m feedbacks supporting the jth grade of ith criterion, then the membership of this criterion can be presented as $r_{ij} = m/M$. Criteria of each grade for qualitative indexes (China Water Transportation Construction Association, 2013) are shown in Table 10.3.

Table 10.3 Definition of each grade for qualitative indexes

Qualitative indexes	*Definition of each grade*		
	Good	*Average*	*Poor*
C14 Perfection Degree of Infrastructure	Sufficient equipment and port facilities	Moderate equipment and port facilities	Insufficient equipment and port facilities
C15 Perfection Degree of Information Platform	High information level	Medium information level	Low information level
C16 Perfection Degree of Management System	Mature rules and regulations for management	Newly-built rules and regulations for management	Lack of rules and regulations for management
C17 Emergency Response Capacity of Extreme Weather	Complete emergency response plans	Partially complete emergency response plans	No emergency response plans
C18 Emergency Response Capacity of Sea Level Rising	Existing port facilities can completely satisfy the requirement of sea level rising	Existing port facilities can partially satisfy the requirement of sea level rising	Existing port facilities cannot satisfy the requirement of sea level rising

3.4 Utility value

It is difficult to rank the development levels of green ports by using belief degrees associated with linguistic terms because they are not sufficient to show the difference between the assessments. Numerical values (crisp values) are therefore generated from the obtained distributed assessments. The concept of expected utility is used to obtain a crisp value for each alternative in order to rank them in terms of development levels.

Suppose the utility of an evaluation grade H_n is denoted by $u(H_n)$ and $u(H_{n+1}) > u(H_n)$ if H_{n+1} is more preferable than H_n (Yang, 2001). Therefore, the utility of the general criterion can be calculated using a linear distribution as Eq. (14) and Eq. (15):

$$u\left(H_n\right) = \frac{n-1}{N-1}, n = 1, 2, \ldots, N \tag{14}$$

where, N denotes the number of the linguist terms.

$$u\left(E\right) = \sum_{n=1}^{N} \beta_n u(H_n) \tag{15}$$

Thus, a crisp value can be calculated based on the distribution generated via the ER technique and thus a comparison between alternatives can therefore be carried out.

4. Case study

As an important waterway transport hub connecting north China with other regions, Tianjin Port has been developed into a considerable scale, three-dimensional traffic transportation system. It is the largest comprehensive port in north China and also one of the most advanced artificial deep-water ports in the world. Tianjin Port has a water area of 336 square kilometers and land area of 131 square kilometers. At present, the depth of the main channel has reached 21.0 meters with a coastline of 32.7 kilometers. In 2013, the cargo throughput of Tianjin Port reached 500 million tons and container throughput reached 13 million twenty-foot equivalent units (TEUs), making it one of the ten biggest container ports in the world. Thus, Tianjin Port is chosen as the case in this study due to its representative significance in the development of green ports in China. The time period for data gathering is set to be the recent three years considering the data availability.

4.1 Calculation of the weights of evaluation indexes

Three domain experts mentioned in Section 2 through in-depth interviews make pairwise comparisons and the weights of each index in the assessment model are obtained using the AHP method. Taking the criteria level as a demonstration, the following comparison matrix for this level can then be formed according to the judgments from the involved experts, as shown in Table 10.4. Since the knowledge and experience of all three experts involved are considered as equivalent, the normalized relative weight of every expert is equally assigned while combining their judgments.

The relative weights of each index in this level can be calculated based on the comparison matrix in Table 10.4. A similar process can then be implemented to lower levels and the weighting vectors of all pariwise comparison matrixes can be obtained, to represent the local importance degree of each index. The weights of each index are shown in Table 10.5.

Table 10.4 The comparison matrix of the criteria level

A	B_1	B_2	B_3	B_4	B_5
B_1	1	2	3	4	1
B_2	–	1	2	2	1/2
B_3	–	–	1	2	1/3
B_4	–	–	–	1	1/4
B_5	–	–	–	–	1

Table 10.5 Weights of each index of assessment model

Goal level	*Criteria level*	*Index level*	*Relative weights*	*Global weights*
Development level of Tianjin Port	Drivers 0.317	C1	0.521	0.165
		C2	0.338	0.107
		C3	0.141	0.045
	Pressures 0.174	C4	0.345	0.060
		C5	0.218	0.038
		C6	0.234	0.041
		C7	0.203	0.035
	States 0.120	C8	0.407	0.049
		C9	0.407	0.049
		C10	0.186	0.022
	Impacts 0.072	C11	0.333	0.024
		C12	0.667	0.048
	Responses 0.317	C13	0.203	0.064
		C14	0.051	0.016
		C15	0.170	0.054
		C16	0.170	0.054
		C17	0.203	0.064
		C18	0.203	0.064

4.2 Evaluation results of index level

Historical objective data used in the evaluation of quantitative indexes come from official statistics (Tianjin Transportation and Logistics Association, 2014), news, statistical yearbooks (Economic BBS by Renmin University of China, 2014) and research literatures (Wang, 2014). They are collected and shown in Table 10.6.

Grades of each quantitative index can be obtained based on the standards in Table 10.1 and those of qualitative indexes are determined according to the standards in Table 10.3 when subjective data is obtained from experts' judgments. Taking the development of Tianjin green port in 2011 as an example, the evaluation results are shown in Table 10.7.

Table 10.6 Values of quantitative indexes, 2011–13

Qualitative indexes	*Value*		
	2011	*2012*	*2013*
C1 GDP per capita (CNY/ person)	62574	72994	85213
C2 GRT (%)	7.1	8.4	9.7
C3 GDPGR (%)	–23.8	32.1	20.2
C4 RTS (%)	80.1	85.3	86.8
C5 RTWG (%)	36	43	54
C6 RWSR (%)	98.31	98.57	99.12
C7 MVPN (db)	67.7	67.7	67.5
C8 RQGW (%)	61.1	38.9	19.4
C9 RGQA (%)	84.1	84.4	87.7
C10 GCRP (%)	10	12.5	14.3
C11 GDPTU (CNY/ ton)	2226	2646	2756
C12 AFP (accident number/ year)	0	0	0
C13 RCEU (%)	19	32	45

Table 10.7 Evaluation of each index in 2011

Criteria level	*Index level*	*Grade*			*Global weights*
		Good	*Average*	*Poor*	
Drivers	C1 GDP		0.26	0.74	0.165
	C2 GRT			1	0.107
	C3 GDPGR			1	0.045
Pressures	C4 RTS	0.51	0.49		0.060
	C5 RTWG		0.6	0.4	0.038
	C6 RWSR	1			0.041
	C7 MVPN		0.23	0.77	0.035
States	C8 RQGW		0.56	0.44	0.049
	C9 RGQA	0.41	0.59		0.049
	C10 GCRP			1	0.022

Criteria level	*Index level*	*Grade*			*Global weights*
		Good	*Average*	*Poor*	
Impacts	C11 GDPTU		0.55	0.45	0.024
	C12 AFP	1			0.048
Responses	C13 RCEU			1	0.064
	C14 PDI	0.3	0.7		0.016
	C15 PDIP		0.9	0.1	0.054
	C16 PDEMS	0.2	0.8		0.054
	C17 ERCEW	0.2	0.8		0.064
	C18 ERCSLR		0.8	0.2	0.064

4.3 Evaluation of development of green ports

In this section, the IDS software (Xu and Yang, 2005) was used to compute the development level of Tianjin Port, employing the ER algorithm for synthesis of the criteria in the hierarchical structure. All the inputs with weightings of the relevant lowest level criteria are combined to determine the estimation of their corresponding higher level criteria. Based on the result in Table 10.7, the development level of the green port of Tianjin in 2011 can be calculated as shown in Figure 10.3.

It can be seen from Figure 10.3 that the development level of Tianjin green port in 2011 is evaluated as 40.16 percent poor, 39.62 percent average and 20.22 percent good. Thus, utility value of green port development in 2011 can be calculated as 0.4023 using Eq. (14) and Eq. (15). Similarly, the utility values of 2012 and 2013 can be calculated as shown in Figure 10.4.

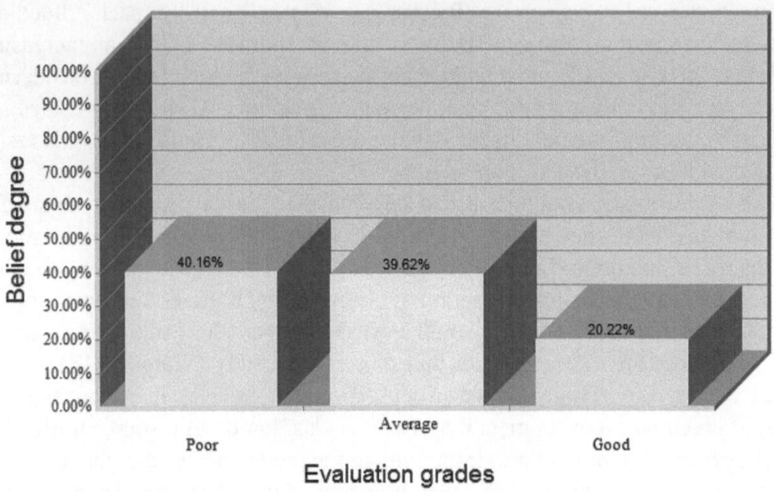

Figure 10.3 Development level of Tianjin green port in 2011.

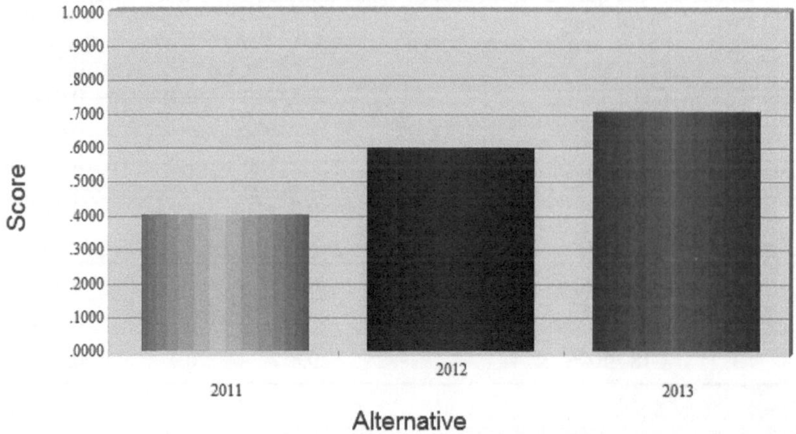

	2011	2012	2013
Average score	.4023	.6018	.7079

Figure 10.4 Utility value of development level of Tianjin green port, 2011–13.

4.4 Discussion and validation

It can be seen from Figure 10.5 that the indexes "Drivers" and "Responses" take the highest importance in criteria level than others sharing the same weight of 0.317. Within the "Drivers," "GDP per capita" (C1) is the most important index with a weight of 0.521, while "Emergency Response Capacity of Extreme Weather" (C17) and "Emergency Response Capacity of Sea Level Rising" (C18) are the two most important indexes in the "Responses." The variation trends of indexes in criteria level can be virtually presented as in Figure 10.6.

Though the utility values of "Pressures," "States," and "Impacts" fluctuate during the period, it is clear that "Drivers" and "Responses" keep in an increasing trend from 2011 to 2013, which is in accordance with the goal level. Considering the "Drivers" and "Responses" take the relatively higher weights and contribute the most to the improvement of the development level of Tianjin Port, this result is in line with the analysis of Figure 10.5.

From the evaluations above, it can be seen that the development level of a green port in Tianjin holds a growth trend in recent years, which is in harmony with the real situation in Tianjin Port in these years. This is evidenced by the fact that: a) there is an increasing utilization of clean energy; b) the enhancement of the emergency response ability of Tianjin Port. Moreover, the evaluation result of 2013 in this research accords with that of a recent study (Wang, 2014), which further verifies the validity of the proposed model. Indexes in the assessment model of green port development mainly reflect a situation of resource utilization and environmental protection, from which the deficiencies and the trend of development of green ports can be identified and analyzed as well. Therefore, the assessment model developed in this paper can be applied not only in the evaluation

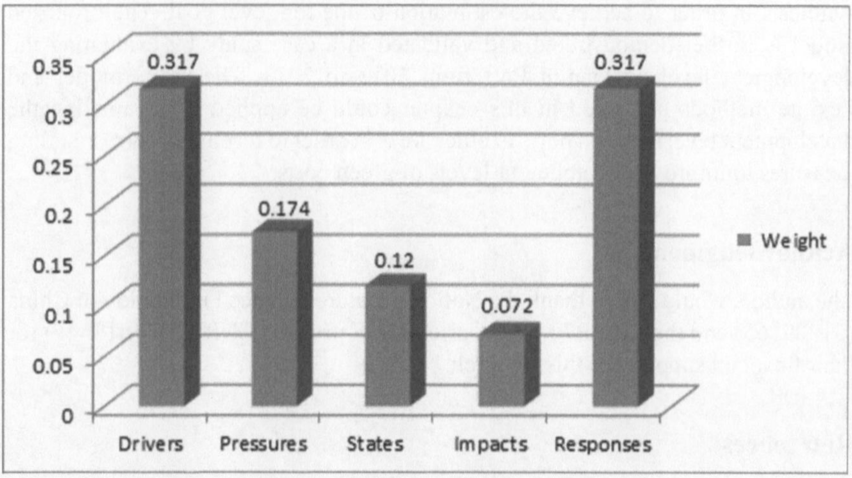

Figure 10.5 Weights of each index in criteria level.

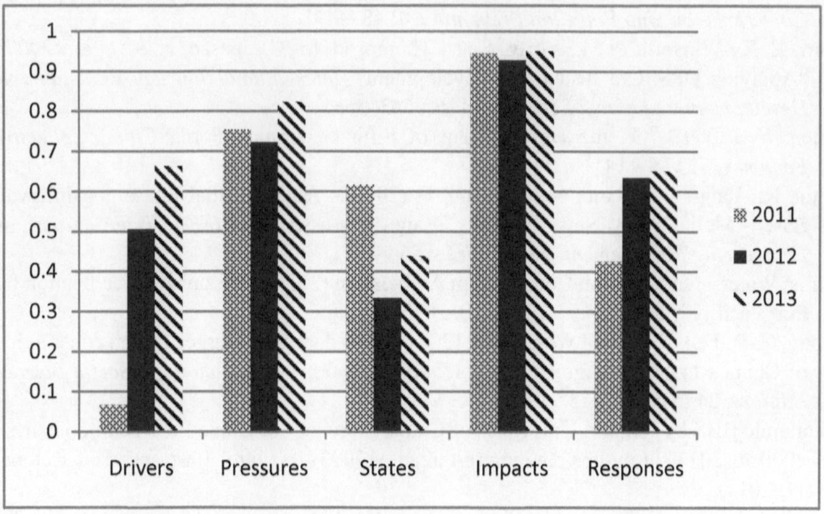

Figure 10.6 Utility values of indexes in criteria level from 2011 to 2013.

of the current situation of green port development, but also as a tool to provide port managers with certain insights on improving daily operations.

Conclusion

This chapter develops a hierarchical model for the evaluation of green port level from five main perspectives, namely, Drivers, Pressures, States, Impacts, and Responses. AHP and ER are used in the proposed model to calculate the relative importance of the relevant qualitative and quantitative criteria and deal with synthesis in order to achieve the estimation of the top level goal. The proposed model is further demonstrated and validated in a case study by evaluating the development level of Tianjin Port from 2011 to 2013. The novel model and flexible methods presented in this chapter could be applied for evaluating the development level of green ports in other areas in order to formulate corresponding measures to improve development levels of green ports.

Acknowledgment

The authors would like to thank the National Nature Science Foundation of China (51209165) and the EU FP7 Marie Curie IRSES project "ENRICH" (612546) for their financial support for this research.

References

Barney, W. (1988). "The Port Authority of New York–New Jersey", *Professional Geographer,* 40 (3): 297–306.

Cai, L. N. (2010). "Green port development in foreign countries", *Proceeding of Annual Conference on Ship Pollution Prevention,* 4: 487–491.

Carr, E. R., Wingard, P. M., Yorty, S. C., Thompson, M. C., Jensen, N. K. et al. (2007). "Applying DPSIR to sustainable development", *International Journal of Sustainable Development and World Ecology,* 14 (6): 543–555.

Chen, Y. (2009). "On the Development of Fifth Generation Port", *China Collective Economy,* 7: 113–114.

Chin, K., Yang, J. B., Guo, M. and Lam, P. (2009). "An Evidential-Reasoning-Interval-Based Method for New Product Design Assessment", *IEEE Transactions on Engineering Management,* 56 (1): 142–156.

China Water Transportation Construction Association (2013). "Technical Specification for Port Facility Maintenance", *JTS,* 310–2013, China.

Dong, G. S., Fan, H. M. and Wen, W. H. (2011). "The analysis of target system constitution of China's low carbon green port", *Wuhan University of Technology (Social Science Edition),* 24 (5): 672–675.

Economic BBS by Renmin University of China. Summary of China Port Yearbooks from 2000 to 2013. http://bbs.pinggu.org/thread-3170931-1-1.html (last accessed October 11, 2014).

Gupta, A. K., Gupta, S. K. and Rashmi, S. P. (2005). "Environmental management plan for ports and harbors projects", *Clean Technology Environmental Policy,* 7: 133–141.

Jiang, T. (2014). "Environmental pollution prevention analysis and governance at Tianjin Port", *Journal of Transport Information and Safety,* 32 (4): 102–108.

Li, S. P., Wang, L. J. and Shi, W. (2010). "Explore model of developing green ports", *World Shipping,* 2: 34–35.

Lin, Q. (2010). "The preliminary study of evaluation index system for the Shanghai green port", *Science & Technology of Ports,* 1:3–7.

Liu, J., Yang, J. B., Ruan, D., Martinez, L. and Wang, J. (2008). "Self-tuning of fuzzy belief rule bases for engineering system safety analysis", *Annals of Operations Research,* 163 (1): 143–168.

Liu, L. M. (2004). "Environmental construction in Tokyo Port", *Environmental Protection in Transportation,* 25 (3): 51–52.

Liu, L. M., Shao, C. F., Ju, M. T., Jiang, T. and Li, M. X. (2011). "Innovation exploration and practice of Tianjin green port construction", *Port Economy*, 1: 24–29.

Lu, Y. and Hu, H. (2009). "Sydney port's practice in green port development and China's inspiration from it", *Navigation of China*, 32 (1): 72–76.

Lv, H. (2005). "The US approach to green ports", *China Ship Survey*, 8: 42–44.

Ma, D., Ding, Y., Yin, H., Huang, Z. H. and Wang, H. L. (2014). "Outlook and status of ships and ports emission control in China", *Environment and Sustainable Development*, 39 (6): 40–44.

Maritz, A., Shieh, C. J. and Yeh, S. P. (2014). "Innovation and success factors in the construction of green ports", *Journal of Environmental Protection and Ecology*, 15 (3A): 1255–1263.

Ouyang, B. and Wang, L. (2014). "Study on green port development strategy in China", *China Harbor Engineering*, 4: 66–73.

Pillay, A. and Wang, J. (2003). *Technology and Safety of Marine Systems.* Elsevier Ocean Engineering Book Series. Oxford: Elsevier.

Ren, J., Jenkinson, I., Wang, J., Xu, D. L. and Yang, J. B. (2008). "A methodology to model causal relationships on offshore safety assessment focusing on human and organizational factors", *Journal of Safety Research*, 39 (1): 87–100.

Satty, T. L. (1980). *The Analytic Hierarchy Process*. New York: McGraw Hill International.

Shang, J. S., Tjader, Y. and Ding, Y. (2004). "A unified framework for multi-criteria evaluation of transportation projects", *IEEE Transactions on Engineering Management*, 51 (3): 300–313.

Sii, H. S. and Wang, J. (2003). "A design-decision support framework for evaluation of design options/proposals using a composite structure methodology based on the approximate reasoning approach and the evidential reasoning method", *Proceedings of the Institution of Mechanical Engineers, Part E (Journal of Process Mechanical Engineering)*, 17 (1): 59–76.

Song, X. B. (2011). "Talking about development status of green port in China and some suggestions", *Science & Technology of Ports*, 11: 17–20.

Tianjin Transportation and Logistics Association. (2014). *Overview of Tianjin Port Development.* http://www.tj.xinhuanet.com/web/jtxxw/tjzlmore.htm (last accessed August 30, 2014).

VijayaVenkataRaman, S., Iniyan, S. and Goic, R. (2012). "A review of climate change, mitigation and adaptation", *Renewable & Sustainable Energy Reviews*, 16 (1): 878–897.

Wan, J. H., Liu, L. M., Su, C. B. and Jiang, T. (2011). "Study on the environmental management system of Tianjin green port", *China Ports,* 11 (03): 19–21.

Wang, T. S. (2014). *Research on Green-Port Development of China.* Master thesis, Dalian Maritime University, Dalian, China.

Wen, X. F. and Chen, N. (2013). "Studies on evaluation of modernization of the inland port and shipping management based on DPSIR model and Gray correlation evaluation model", *Intelligent and Integrated Sustainable Multimodal Transportation Systems Proceedings from the 13th COTA International Conference of Transportation Professionals*, 96: 1792–1800.

Wu, X. D. and Ji, L. (2013). "Research on the impact of climate change on port of China and the countermeasures", *China Water Transport*, 13 (10): 116–118.

Xu, D. L. and Yang, J. B. (2005). "Intelligent decision system based on the evidential reasoning approach and its applications", *Journal of Telecommunications and Information Technology*, 3: 73–80.

Yang, J. B. (2001). "Rule and utility based evidential reasoning approach for multiattribute decision analysis under uncertainties", *European Journal of Operational Research,* 131 (1): 31–61.

Yang, J. B. and Singh, M. G. (1994). "An evidential reasoning approach for multiple attribute decision making with uncertainty", *IEEE Transactions on System, Man and Cybernetic*, 24(1): 1–18.

Yang, J. B. and Xu, D. L. (2002). "Nonlinear information aggregation via evidential reasoning in multiattribute decision analysis under uncertainty", *IEEE Transactions on System, Man and Cybernetic—Part A: Systems and Humans*, 32 (3): 376–393.

Yang, T., Lee, R. S. and Hsieh, C. (2003). "Solving a process engineer's manpower-planning problem using analytic hierarchy process", *Production Planning and Control*, 14 (3): 266–272.

Yu, M. and Liu, W. B. (2012). "Approach to the construction of inland green port", *China Ports,* 2: 59–60.

Zhang, S. Y. and Liu, S. G. (2006). "An econometrical analysis of relationship between the throughput of Qingdao Port and GDP of Qingdao", *China Water Transport*, 6 (11): 216–218.

11 Terminal Maritimo Muelles El Bosque, Cartagena, Colombia

Vladimir Stenek, Jean-Christophe Amado, Richenda Connell, Olivia Palin, Stewart Wright, Ben Pope, John Hunter, James McGregor, Will Morgan, Ben Staley, Richard Washington, Diana Liverman, Hope Sherwin, Paul Kapelus, Carlos Andrade, and José Daniel Pabón

Introduction

In 2008, International Finance Corporation (IFC), a private sector arm of the World Bank Group, initiated its Climate Risk and Adaptation Program[1] with a series of assessments of risks posed by climate change to private sector investments. The Program's overall goal is to develop sector-specific tools and methodologies that allow assessment of financial, environmental and social risks and opportunities resulting from the impact of climate change, and evaluation of appropriate adaptation responses. The studies' focus is private-sector investments but with a significant emphasis on the cooperation and synergies with the public sector, research institutions, and civil society.

This study explores general risks and opportunities for the ports sector, and examines specific implications for a private port company, Terminal Maritimo Muelles El Bosque (MEB), in Cartagena, Colombia. In 2007 IFC, the largest global developmental institution focused on the private sector, made a US $15 million dollar investment in MEB to partially finance the company's expansion program. The analyses and findings are based on the information available at the moment of its elaboration, in 2010.

Muelles El Bosque (MEB)

MEB port handles four types of cargo: containers, grain, bulk materials, and coke. Between 2005 and 2010 containers represented the largest share of MEB's revenue. In 2008, MEB moved 1 percent of Colombia's international trade (in tonnage). MEB currently employs approximately 250 people.

The 10ha terminal is located in a mixed industrial and residential zone of Cartagena. It is composed of two sites, an island named 'Isla del Diablo' (see Figure 11.1) and an adjacent mainland area linked via a causeway road.

Cartagena is too close to the Equator to be prone to tropical storms, and MEB is located in the natural harbor of the Bay of Cartagena, which is one of the most

Figure 11.1 MEB on Isla del Diablo (island site): 1) and 2) Quays; 3) Grain silos; 4) Coke storage area; 5) Container patio; 6) Part of the causeway that connects to the mainland site; and 7) the mangrove around the causeway.

secure on the Caribbean coast of Colombia. It is sheltered from high winds, waves, and storm surges. This natural resilience brings advantages compared to its competitor ports in Colombia.

Climate variability, climate change, and port operations

In Cartagena, average temperature varies little from month to month, generally lying in the range 27 to 29°C. Projections from ten General Circulation Models (GCMs) for three greenhouse gas emission scenarios (A2, A1B, and B2) point at future increases between 0.7 and 1.2°C and 1.2 and 2.2°C from the 1961–90 baseline to the 2020s and 2050s respectively. However, empirical downscaling of 14 GCMs for the A2 greenhouse gas emission scenario shows possible temperature increases of 6° by the 2050s in all seasons. For this study, a broad set of climate models and greenhouse gas emission scenarios were considered, to capture the range of uncertainty in future climate change.

Average precipitation on wet days has increased by 0.6 percent per year in Cartagena between 1941 and 2009. There is also evidence that precipitation is becoming more intense in some parts of Colombia.[2]

Making resilient decisions in the face of uncertainty requires managing gaps in knowledge and data. For example, there is poor agreement between climate models over future precipitation changes in Colombia, due to the complex topography and the lack of understanding of how tropical cyclones might change in the future. No credible daily or peak rainfall projections are available.

Thermal expansion of the oceans due to higher temperatures, and increased melting of glaciers, ice caps, and the Arctic and Antarctic ice sheets all contribute to sea-level rise. To capture the range of possible future sea-level rise over the twenty-first century, this study considered two scenarios: an observed sea-level rise scenario of 5.6mm per annum,[3] and an accelerated sea-level rise scenario (of up to 1.3m by 2100), starting from the current rate of sea level rise (5.6mm) and following a quadratic equation that approximates the exponential sea-level increase projected by many studies.

Winds are predominantly from the north and north-east in Cartagena and are either calm (21 percent of the time) or between 1.6 and 13.9m/s (69 percent of the time). These wind speeds are below the operating thresholds of MEB's cranes. However, no data is available on observed wind gusts.

Cartagena is located south of tropical cyclone tracks and is protected by its location in the north-east corner of the South American land mass. No significant change in wind speeds and storminess are projected by climate models, though this is an area of uncertainty.

Table 11.1 Summary of present-day conditions and future projections for the key climate variables considered in this study

Climate and climate-related variables	*Observed conditions recorded by meteorological stations and tidal gauges in Cartagena*	*Future scenarios from climate model projections*
Mean temperature	Between 26.7°C and 28.5°C (average monthly) No obvious trend in temperature over the last 70 years	Increases of 0.7 to 1.2°C by the 2020s, 1.2 to 2.2°C by the 2050s and 1.7 to 3.7°C by the 2080s (projected by an ensemble of GCMs) Potential increases up to 6°C by the 2050s (projected by an ensemble of downscaled GCMs)
Mean precipitation	Annual average rainfall was about 600mm per year in the 1940s and has risen steadily, to about 1,100 mm per year in the last decade Increase of 6mm per year in 1941–2009; corresponding to a 0.6% increase per year on wet days	Assumed yearly increase of 0.6% on wet days, based on continuation of observed trends[4] Climate models perform poorly at projecting future rainfall in Colombia
Sea-level rise	Rising at 5.6 mm per year (± 0.008 mm) (Source: Tidal gauge data)	'Observed sea level rise scenario': 5.6mm per year, i.e. 504mm by 2100 'Accelerated sea level rise scenario': 1,300mm by 2100
Wind	Calm or between 1.6 and 13.9m/s for most of the time	Increases by up to 0.2 m/s by the 2020s and 0.5 m/s by the 2050s and 2080s (projected by an ensemble of GCMs) Winds in the range 3 to 10 m/s could become more frequent (projected by the regional climate model PRECIS)
Storminess and storm surges	Not affected by tropical cyclones Storm surge height up to 171mm for a 1-in-300-year event	Little to no change

Port success criteria at risk from climate change

It is useful to take a generic view of the various success criteria for ports which can potentially be affected by climate change, as presented in Figure 11.2. These relate to the chain of external systems and internal assets and activities on which a port's commercial success relies:

- Trade levels and patterns and the consequent demand for ports' services.
- Navigation in and out of ports and ship berthing.
- Cargo handling inside ports.
- Movements of goods, vehicles, and people inside ports.
- Goods storage in ports.
- Inland transportation outside ports' fence lines.

In the case of MEB, the climate change risk analysis found that only a small number of these success criteria are likely to be significantly affected by climate change. These include:

- Reduced vehicle movements inside MEB due to increased seawater flood risk.
- Decreased global trade and US grain exports and associated shipping movements, with consequences for MEB's revenues.
- Increased risk of damage to goods stored inside ports due to seawater flooding, with potential reputational consequences.
- Degradation of the mangroves located around the causeway due to sea-level rise.

In the case of materials handling and surface flood risk, climate change does not appear to challenge the operational threshold of MEB's cranes or its drainage capacity, though there is some uncertainty about this. For instance, while precise data was available on the operational limits of MEB's cranes, information on observed extreme winds was uncertain and there were no credible future extreme wind projections. However, given that the climate models do not project significant changes in winds, and because Cartagena lies too close to the Equator to be significantly at risk from hurricanes, the study discarded this as a significant risk to MEB.

The absence of risk to navigation and berthing relates to MEB's favorable location on the Bay of Cartagena, where extreme climatic hazards are limited. The generally low risk of social or environmental issues driven by climate change (with the exception of impacts on the mangrove, see Figure 11.1) can primarily be explained by MEB's existing management systems.

In some cases, the available information was too limited to assess risk accurately. This was the case for assessing climate-change impacts on the transportation network outside MEB. Quality elevation data for the city of Cartagena could not be obtained, which led the study to consider potential sea-level thresholds above which flood risk of the roads used by MEB and its customers increases.

Figure 11.2 Conceptual model of a cargo handling port and the main success criteria that can be affected by climate change.

Demand, trade levels, and patterns

Commercial ports rely on the vitality and growth of the shipping industry, which is very sensitive to fluctuations in world and country GDP, as shown by the recent economic downturn.

For many products, climate change will influence market conditions which can translate either into risks of decreased trade through ports or opportunities for new exports or imports. For example, the supply of agricultural and forest products is very sensitive to climatic conditions.

Demand for a port's services can also be affected by population movements and changes in the locations of industrial centers, as well as by the way port customers perceive its reliability in the face of extreme weather events. Those that suffer major disruptions during storms or other climatic events may experience decreased throughput. For example, following Hurricane Katrina some customers shifted to alternative ports on the US East coast.

The potential impacts of climate change on international trade are difficult to quantify with confidence over coming decades, due to the interactions between future climate-change impacts, economic performance and trade, the number of other factors that significantly influence trade, and the unstable nature of trade itself. However, trade is a potentially significant climate-related risk area for ports, and merits further research.

The study adopted a practical approach to assessing the potential implications of climate change for MEB's imports and exports, aiming to provide insights into these potentially significant impacts to MEB.

For the period 2005–2010, MEB's revenue and world GDP appear closely related: on average, a 1 percent change in world GDP resulted in a 0.7 percent change in MEB's revenue. The study acknowledges that this relationship could be different in the future.

The Stern Review on the Economics of Climate Change estimated that the average costs of climate change could go up to 1.88 percent of global per capita consumption across different climate-change scenarios by the 2080s.[5] The implications of such global GDP reductions for MEB could amount to an annual loss of US $1.1 million by 2100. The effect on MEB's annual revenue of different climate change scenarios is shown in Figure 11.3.

Projected increases in imported grain prices and impacts on land suitability for the Colombian crops that are exported through MEB are not expected to affect MEB's revenue significantly.

Customer demand for MEB's services could benefit from climate change. When tropical cyclones disrupt other ports in the Caribbean, MEB has observed a 2 percent increase in its total income on average, due to its reputation as a safer hub than other Caribbean ports. Research has shown that tropical cyclone intensity may increase in the Caribbean (though not at Cartagena).

Compared to other major Colombian ports, MEB is not vulnerable to disruption from intense precipitation and fluctuations in water depth. In contrast, Buenaventura is known to face operational restrictions due to heavy rainfall, and shipping from Baranquilla is limited by depth. Therefore, future climate change could create a competitive advantage for MEB if it exacerbates existing problems in these competing ports.

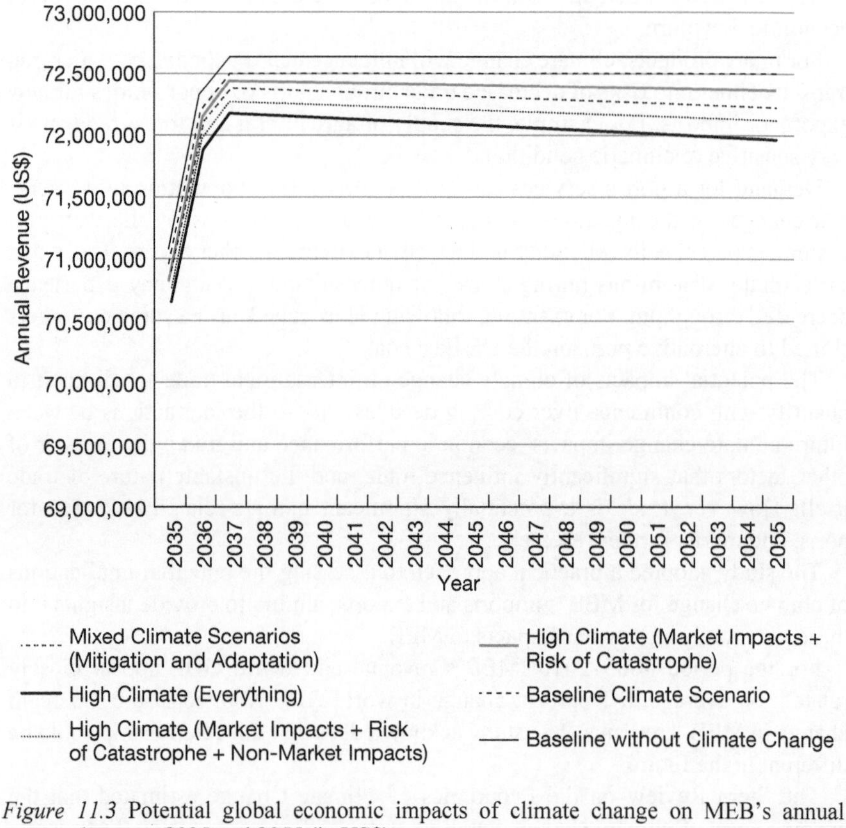

Figure 11.3 Potential global economic impacts of climate change on MEB's annual revenue between 2035 and 2055 (in US$).

Navigation and berthing

In general, as a result of sea-level rise, navigable water depths are likely to increase in many coastal ports and shipping channels. The increased available draft also generally equates to decreased dredging maintenance requirements. However, changes in rates of coastal erosion and deposition will affect the depths of some navigation channels, and lead to increased dredging costs. Furthermore, capital expenditure may be necessary in ports where sea levels rise above the operability range of infrastructure and equipment. For instance, reduced clearance under some bridges will restrict the low water level windows available to large vessels.

Climate change will lead to reduced river flows or lake levels in some areas with severe implications for navigation and port access. Increased shipping or improved inland transportation in areas where higher temperatures will increase ice-free periods is likely to benefit some ports. For instance, the opening of the Northwest Passage could provide a commercial alternative to using the Panama Canal, which would decrease shipping movements in Central America.

The Bay of Cartagena has limited openings to the sea via two sea channels: Bocachica and Bocagrande (Bocachia is the only entrance for commercial shipping) (see Figure 11.4). Seawater flows in and out of the bay through these two channels and there is a freshwater input from the Canal del Dique (see Figure 11.4). The bay offers protection against waves and storms surges; it is also characterized by low tidal ranges and little congestion or navigation difficulties.

Because of the characteristics of the Bay of Cartagena, climate change is not expected to represent a considerable risk. A number of issues were considered during the assignment:

- The top of MEB's quays and the operability range of cranes and fenders are able to cope with the rise in sea levels projected this century in both the observed and accelerated sea-level rise scenarios.
- MEB only dredges two short channels every five years approximately. The increased draft caused by higher sea levels is likely to reduce maintenance dredging requirements. A gross estimate of the total savings to the end of the century is from $325,000 to $400,000. MEB has no plans to accommodate vessels with a draft above 12m.
- There is no evidence that sedimentation in the bay will change. Further, because other port terminals located in the bay (SPRC and Contecar) are in charge of dredging the channels closest to the mouth of the Canal del Dique and have plans to accommodate Post-Panamax ships, the depth in the bay will be maintained at a level significantly greater than that required by MEB.
- Since wave height is limited in the Bay of Cartagena, any increases in wave height due to greater water depths (driven by sea-level rise) is unlikely to have a significant effect on MEB.

The municipality of Cartagena has been reviewing options to reduce the volume of sediment discharged by the Canal del Dique into the bay.

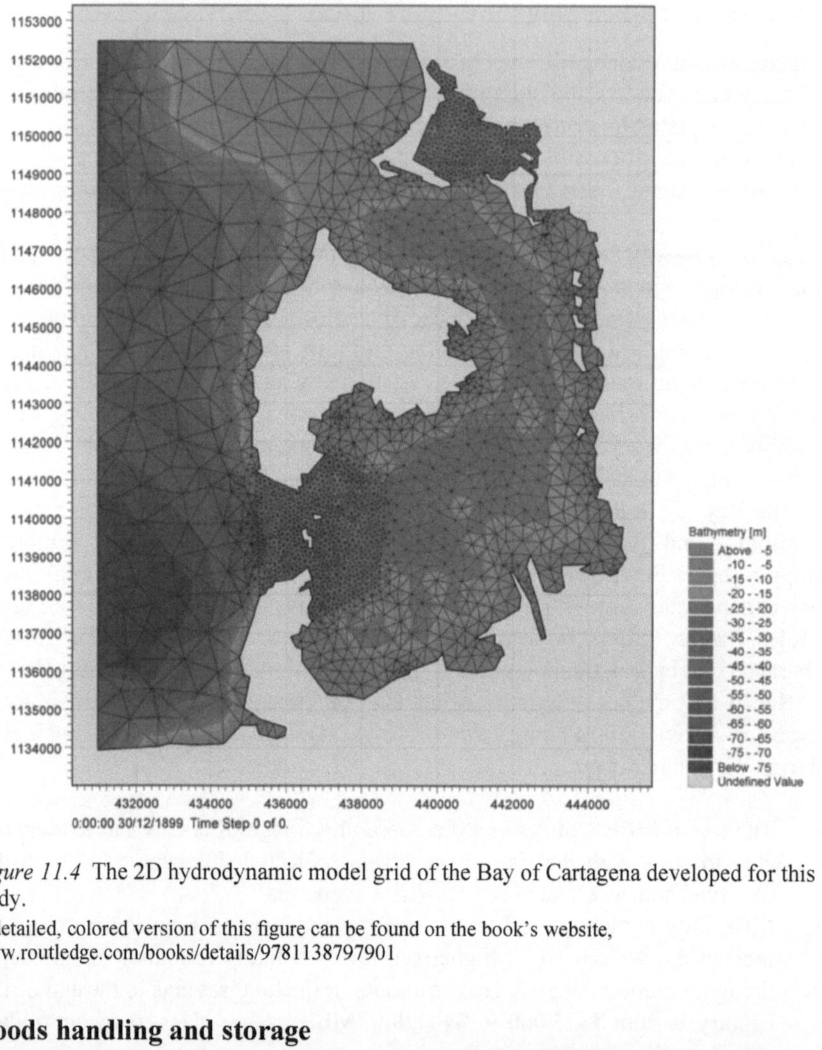

Figure 11.4 The 2D hydrodynamic model grid of the Bay of Cartagena developed for this study.

A detailed, colored version of this figure can be found on the book's website, www.routledge.com/books/details/9781138797901

Goods handling and storage

High winds, extreme rainfall and lightning can restrict port operations. For instance, cranes cannot be moved or used above certain wind speeds and heavy rain can affect crane's electrical systems. Some operations cease in times of strong winds, including tugboat and ferry movements. During heavy rainfall, the loading or unloading of weather-sensitive goods or material also ceases. Usually rainfall limits are set by ports on a case-by-case basis.

- Higher temperatures are expected to improve port operating conditions in cold regions.
- Risk of spoilage of goods stored is likely to increase because of increased temperature and changing rainfall and their effect on pests, diseases, rust and mould, as well as increased flood risk.

- Changing climate conditions can also have implications for port energy and water requirements.
- Risks of dust explosions associated with grain handling and storage may increase in areas which become hotter and drier.

The study assessed future surface flood risk on MEB's island site considering both changes in surface water runoff and sea-level rise which can lead to seawater surcharge of drainage pipes.

When rainfall is considered alone, the amount of rainfall required to overwhelm MEB's drainage pipes on the quay is 639mm per day. The maximum daily rainfall recorded between 2000 and 2008 in Cartagena was 158mm and daily rainfall only exceeded 50mm for about 1 percent of this time. Future precipitation changes in Cartagena are uncertain, so to assess the capacity of the drainage system to cope with future rainfall changes, the observed rate of increasing rainfall was extrapolated into the future and applied to the maximum daily rainfall observed between 2000 and 2008. Thus, considering a possible future scenario in which the observed 0.6 percent yearly increase in average rainfall during wet days continues, and peak rainfall increases by as much as average rainfall, no risk to MEB's required drainage capacity is expected, unless the future maximum daily rainfall calculated for 2100 (of 245mm in a day) falls within less than nine hours.

When the additional effect of seawater ingress is considered, surface flood risk could increase. For instance, mean sea level during the highest spring tide is expected to reach the critical limit above which the drainage system cannot cope with both seawater ingress and increased rainfall towards the end of the century in the accelerated sea-level rise scenario (estimated to be 1.7m above the port plan datum). If the effect of storm surges is added, seawater ingress is expected to be critical to drainage earlier in the century in the accelerated sea-level rise scenario.

No adaptation action appears to be necessary to prevent an increase of surface flood risk due to climate change. However, because of the uncertainty of future precipitation changes and the possibility of occasional flooding due to heavy rainfall events, a few steps are recommended including: regular maintenance, inspections, and fitting valves to all drainage sea outlets. Towards the end of the century, it could be necessary to raise the drainage outlets to avoid seawater surcharge, which would be best done when quays are replaced.

Water-sensitive storage areas at MEB (warehouses and the storage area on the mainland site) are likely to be flooded under the compound effect of rising sea levels and storm surges between 2050 and 2060 and 2030 and 2040 in the observed and accelerated sea-level rise scenarios respectively (see Figure 11.5 and Figure 11.6). Without knowing the value of the goods stored on the mainland site, it is difficult to assess the potential loss. However, it could represent a significant risk to MEB's reputation as a reliable port. To avoid such impact, MEB could add emergency flood protection or reorganize their storage areas in order to ensure that water-sensitive cargo is stored in less vulnerable areas. Raising the height of the mainland patio by 1.2m is financially beneficial from the 2080s and 2050s onward in the observed and accelerated sea-level rise scenarios respectively. The

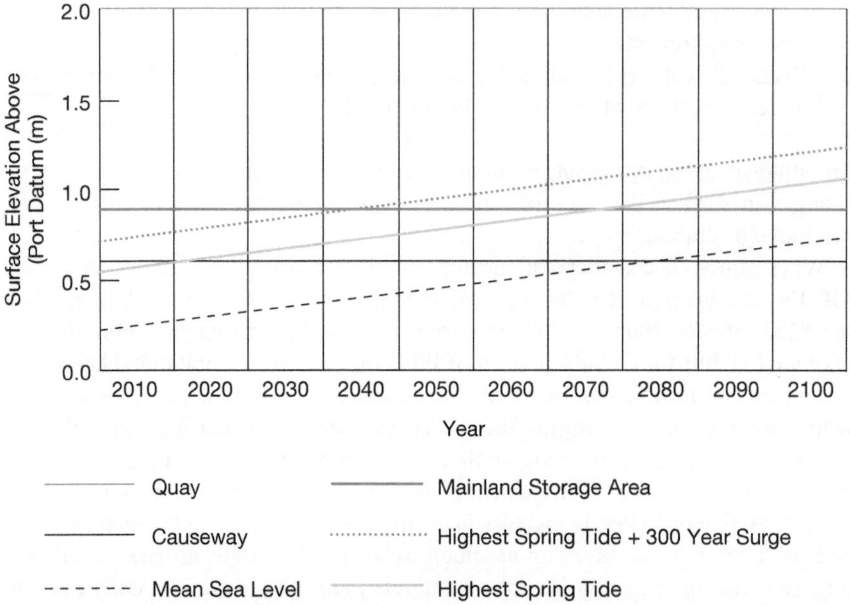

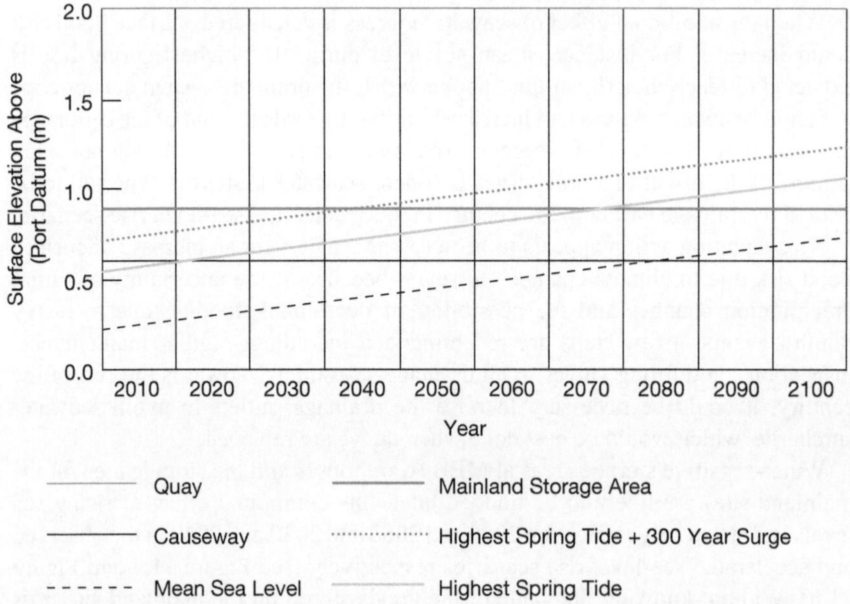

Figures 11.5 and 11.6 Projected sea level and flood risk in different parts of the port (including the mainland storage patio) under the observed sea-level rise scenario (Figure 11.5) and under the accelerated sea level rise scenario (Figure 11.6): 1) mean sea level; 2) mean sea level at the highest spring tide; 3) mean sea level at the highest spring tide and with a 1-in-300-year high storm surge. The horizontal lines represent the level of the causeway (red line), mainland storage area (blue line) and quay (green line).

total associated paving and drainage costs (approximately $2.2m) are lower when the patio is raised every ten years by 20cm between 2050 and 2100 compared to when the patio is raised by 1.2m only once in 2050 (for a 16 percent discount rate, see Figure 11.7).

It is possible that water consumption for coke spraying could decrease if the observed trend of increasing rainfall continues in the future. Savings appear to be modest and could amount to $14,000 per year. In contrast, higher temperatures will increase electricity consumption of refrigerated reefers; this could correspond to a 30 percent (or 84kW) increase in MEB's annual electricity consumption for a 6°C increase in temperature.

The risk of grain dust explosion is expected to be relatively low at MEB at present, provided grain dust and fire risks are adequately managed. Climate change is not expected to lead to any significant change in risk.

Hourly average wind speeds of between 13.9 and 17.1m/s, which prevent cranes from being moved, only occur about 2 percent of the time in Cartagena at present. Hourly wind speeds above cranes' operational mode threshold (20 m/s) are rare. Gusts can also affect port operations, but no data is available for Cartagena.

No significant risk to MEB's material handling and storage activities due to climate change is expected. Climate models project modest wind speed increases in Cartagena, at most 0.5m/s in July to August by the 2080s. Also, there is

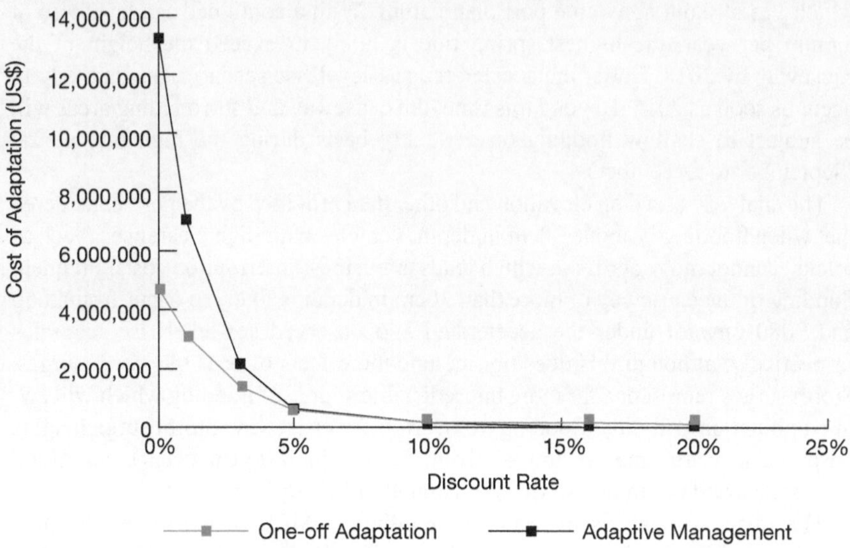

Figure 11.7 Costs of raising the height of the mainland storage patio by 1.2m for a range of discount rates. Two adaptation scenarios are considered: 1) an adaptive management scenario (blue line) whereby the patio is raised by 20 cm six times between 2050 and 2100; 2) a one-off adaptation scenario (red line) whereby it is raised all at once in 2050. (Although difficult to notice in this graph, the cost of adaptive management is lower than the one-off option for all discount rates higher than 10%. At DR=16%, it is approximately 60% of the one-off option cost.)

uncertainty about future increases in the intensity of tropical cyclones in the Caribbean; however Cartagena will most likely not be affected due to its proximity to the Equator.

Interruptions to loading or unloading of ships due to heavy rainfall seldom occur at MEB. Future precipitation increases are anticipated, but in the case of MEB it is not expected that the frequency of disruptions to materials handling will increase significantly, based on assuming the observed 0.6 percent yearly increase in average rainfall continues in the future.

Vehicle movements inside ports

- Increases in mean sea levels, storm surge heights, and changes in wave regimes, will increase flood risk for many coastal ports.
- Ports on rivers and lakes could also face increased flood risk in areas where river flows or lake levels will change.
- Ports' drainage systems can be overwhelmed by intense precipitation, leading to surface flooding.
- The levels of operation interruptions, road closures, and delays caused by flooding will depend on flood duration and depth.

MEB's lowest area is the causeway road linking its mainland and island sites, which lies at 0.6m above the port plan datum. With a continued sea level rise of 5.6mm per year, the highest spring tide is likely to exceed the height of the causeway by 2018. Under the accelerated sea-level rise scenario, this is likely to occur as soon as 2015. Beyond this time, the causeway and surrounding areas will be subject to shallow flooding on a weekly basis during the high tide period (September to December).

The analysis, based on elevation and other data provided by the port, considered that when flooding is above 30cm in depth, vehicles with high clearances (such as trucks) cannot move across it, which leads to business interruption costs. Frequent flooding of the causeway of more than 30cm in depth is likely to occur from 2050 and 2080 onward under the accelerated and observed sea-level rise scenarios respectively, although this does not include the effects of possible storm surges. Storm surges represent a far more unpredictable source of flooding which will act, in combination with tidal flooding, to overtop the causeway. Storm surge heights observed in Cartagena are low (171mm for a 1-in-300-year event). The flood analysis carried out in this study is illustrated in Figure 11.8.

The financial analysis of seawater flooding to MEB demonstrates that this threatens MEB's business continuity if no action is taken. By 2032 the cost of flooding of the causeway is estimated to be between 3 percent and 7 percent of MEB's annual projected earnings in the observed and accelerated sea-level rise scenarios respectively. In the accelerated sea-level rise scenario, all earnings are wiped out by about 2065 and in the observed sea-level rise scenario significant losses are incurred from 2060 onwards (representing more than 10 percent of MEB's annual projected earnings).

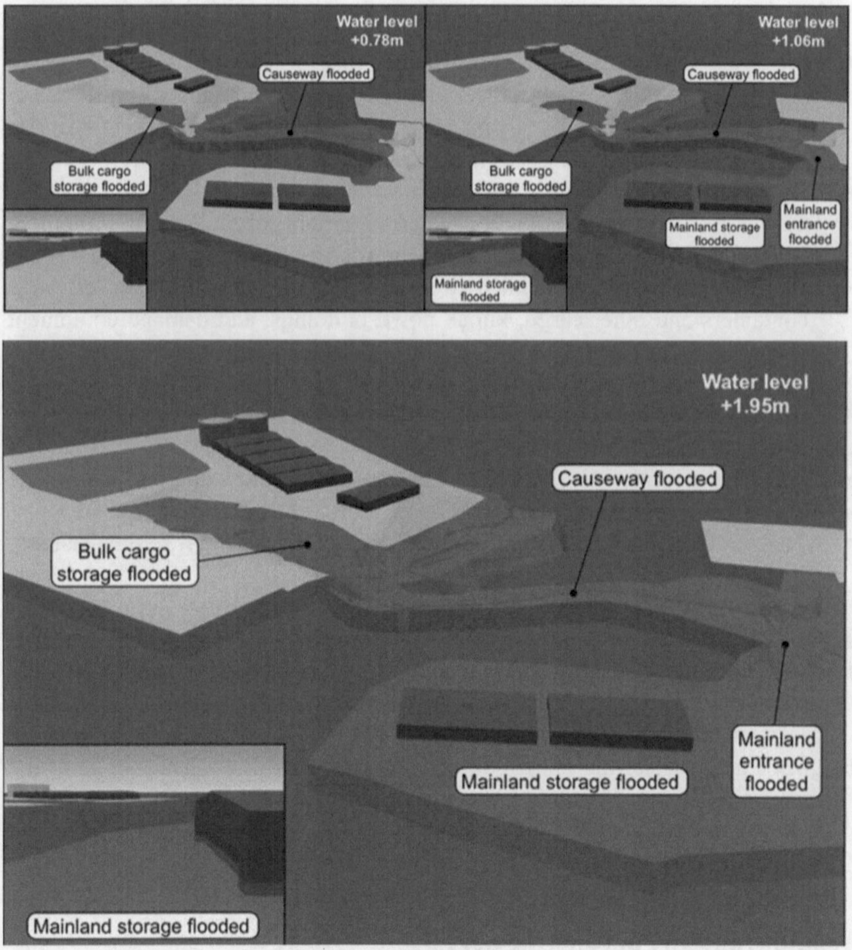

Figure 11.8 Projected seawater flooding during the highest spring tides (and the highest water level on the bay attributed to wind set up and rainfall). In 2050 under the observed sea-level rise scenario (left), in 2050 under the accelerated sea-level rise scenario (right), and by 2100 in the accelerated sea-level rise scenario and with the additional effect of a 1-in-300-year storm surge (bottom). Blue areas are underwater. In all cases, the causeway road is projected to be flooded.

To ensure that the causeway road is not flooded this century, raising its height by 60cm and 1.2m is necessary in the observed and accelerated sea-level rise scenarios respectively. It is expected that the cost is relatively small. MEB has two options: either raise the causeway by the required amount at once or raise it incrementally this century as required. Overall, it appears cheaper for MEB to raise the causeway in different increments under both sea-level rise scenarios for discount rates above 0.2 percent. This 'adaptive management' approach also has the advantage of allowing MEB to monitor the rate of rising sea levels and adjusts adaptation decisions accordingly.

Infrastructure, building, and equipment damage

- Increased flood risk will have implications for port infrastructure, building, or equipment damage. For instance, increased seawater flooding could lead to higher rates of metal corrosion on ports, especially as temperatures rise globally and salinity increases in some locations.
- While damage by shallow and temporary flooding is likely to be limited in most ports due to the construction materials used in ports, electrical equipment is very vulnerable with risks of arcing and short-circuits.
- Fast-moving water associated with storm-surge flooding can dislodge containers and other cargo, knock down buildings, and damage equipment and port infrastructures such as piers, pavements, and foundations. High winds can have similar impacts.
- Increased sea levels, changes in wave activity, and river flows can aggravate coastal and soil erosion.
- With rising sea levels, the standards of protection of natural and man-made sea defenses will be reduced.
- Temperature-sensitive port structures (such as cranes) will be increasingly stressed by higher peak temperatures.

In the case of MEB, it is not expected that climate change (including seawater flooding) will cause significant damage to its infrastructure, buildings, and equipment, or threaten its employees. This is due to both the types of assets and construction materials present on the port and the relatively low storm-surge heights. For instance, the electricity sub-station on MEB's island site is at the same level as the quays and is not projected to be flooded by 2100 (even by a 1-in-300-year storm surge). However, towards the end of the century, increased frequency of flooding could cause higher corrosion rates.

Inland transport beyond the port

Ports rely on the efficiency and resilience of inland transportation networks and assets to move import and export goods and connect to major economic centers.

Road, rail, and waterway transportation systems can be affected by climate change in a number of ways.

- Higher temperatures and longer periods of extreme heat can, among other things, affect road surfaces, cause rail tracks to buckle and damage overhead cables, and increase fire risk.
- In areas with frozen ground, rising temperatures will increase land instability. Where winters will be less snowy as well as warmer, transport system winter maintenance requirements could decrease.
- Changes in rainfall regimes and extreme rainfall, and the associated effects on soil moisture, will increase flood risk and impact on ground stability. This will have consequences for the structural integrity of assets such as

roads, bridges, and railways. Damage will increase in areas of increased hurricane intensity.
- In some areas, cargo transported on waterways will be affected by reduced flows in rivers or water levels in lakes due to the compound effect of higher rates of evaporation and plant transpiration, decreased surface water runoff, and increased droughts or dry spells associated with changes in land use.

All of these potential impacts on transportation can cause delays, traffic interruptions, or transportation restrictions on the transport routes taking goods and materials to ports. They can affect port revenues significantly.

Most goods to and from MEB are transported by truck, though waterway transport is also available. Transport cost is a decisive factor that influences decisions on which Colombian ports customers use.

Trucks coming in and out of MEB use a limited road network:

- Three two-lane residential streets oriented east–west connecting the port terminal gates and the highways.
- Two four-lane highways running into El Bosque going south (Avenida Crisando Luque) or north (Avenida Bosque).

Some flooding in Cartagena has been known to be caused by high sea levels. The roads neighboring MEB, assumed to be at 1.6m above the port plan datum, are projected to be at flood risk in the accelerated sea-level rise scenario by 2080 during the highest tides of the year and in the event of a 1-in-300-year storm surge. Flood maps of the city produced using a Light Detection and Ranging (LiDAR) elevation dataset (which was not available for use in this study) indicate that roads outside MEB are at flood risk when sea levels reach 1m.

Surface flooding is known to frequently affect parts of the city (Castillo-Grande, Bocagrande and the old part of Cartagena). However, it is not known how El Bosque, as well as the roads used by MEB and its customers, are affected. In 2008, the University of Cartagena found that El Bosque had high levels of material damming its drainage channels causing localized siltation and structural flaws. Due to the fact that the municipal drainage system in El Bosque is already experiencing strain and that increased heavy precipitation is expected, it would appear that surface flood risk of the local road network is increasing. It is recommended that MEB monitors progress of the various ongoing studies into the issue and supports better drainage maintenance by the City. It is worth noting that the large drainage channel built to the north of the port by MEB helps to manage surface flood risk.

Outside Cartagena, much of the road network is already in a state of disrepair. Over 50 percent of it is classified as "bad or very bad." The analysis of government data on hazards recorded in 2008 identified storms and floods as the major hazards affecting roads around Cartagena. Climate change could increase the rate of flood-related incidents.

Insurance availability and costs

- Ports may face changes in insurance terms and costs as the incidence of severe weather events increases and knowledge of climate change impacts in the insurance industry improves.
- For those ports that are most vulnerable to future extreme climatic hazards, it is likely that insurance premiums and deductibles will increase at least proportionally to port claim losses.
- The Association of British Insurers[6] (2009), using a typical insurance pricing formula, demonstrated the relationship between average annual loss (AAL) and premiums (P): $\Delta P = 1.1 * \Delta AAL/0.25$, where 1.1 represents fixed expenses and 0.25 variable expenses and profit load.
- MEB's insurance covers damage related to typical weather-related perils, such as floods, heavy rains, wind storms, tropical cyclones, and extreme temperatures.
- However, MEB does not have an 'ingress/egress' policy, so that the revenue consequences of business interruptions due to increased flooding (e.g. between 3 and 7 percent of MEB's projected earnings by 2032 because of flooding of the causeway road as outlined earlier) will not be covered, unless it relates to damage or loss of insured assets.
- Changes in MEB's insurance conditions because of climate change cannot be excluded since Colombia is already considered by the insurance market as "highly to very highly exposed" to a number of climate-related hazards which are projected to further increase (such as floods). Additionally, the study found that flood risk will considerably increase at MEB, unless adaptation measures are taken.
- In 2010, MEB's annual insurance premium amounted to US$122,877.

The study showed that raising the height of the causeway is economically beneficial from the 2040s and 2030s onwards for the observed and accelerated sea-level rise scenarios respectively. Discussion with insurers conducted for this study indicate that some insurers would consider offering more favorable insurance to customers who have undertaken similar actions that reduce future risk. It is recommended that MEB considers raising the causeway, rather than contracting an ingress/egress insurance policy.

Social performance

- Changing climatic conditions can aggravate or create additional safety risks to port workers related to cargo handling and use of machinery, vehicle movements, and flammable materials.
- Exposure to pollution may also increase under certain circumstances, for instance if changes in sea conditions affect the risk of chemical or oil spills. There could also be new pests and diseases present in ports.

- Increased tensions in the relations between ports and surrounding communities is possible if the combined effects of climate change and port activities have negative community impacts.
- Overall it is thought that climate change will pose little, if any, health risks to MEB's workforce. There is a small chance that climate change could lead to an increase in eye, skin, and cardiovascular complaints among the workforce due to higher exposure to ultraviolet radiation in sunnier conditions and increased peak temperatures in Cartagena. However, present-day costs to MEB for absenteeism due to these conditions are very small.
- Despite the limited risk, it would be useful for MEB to raise awareness among staff that certain health and safety conditions can be influenced by climatic changes.
- There are very few people living in the vicinity of the port and MEB itself has limited impacts on these surrounding communities. MEB's leadership is committed to good social management and maintains good relations with the surrounding communities. It has been found that climate change is unlikely to create new problems for these communities or aggravate existing ones. For instance, whatever precipitation changes occur in the future, MEB plans on paving areas of the port which will reduce the likelihood of mud spray on local roads.
- The unfounded community perception that MEB handles coal which produces dust detrimental to human health (it handles coke) will not be affected by climate change. In fact, if the observed trend of increasing rainfall continues, coke dust generation could be reduced.

Environmental performance

- By influencing port dredging requirements, climate change can disturb sea beds and areas important to marine life. Increased temperatures, sea levels and storminess, as well as modified salinity, sea currents and water runoff, will have implications for seabed conditions and sedimentation.
- Some of the major sources of port air pollution, as well as the ways port air pollution is dispersed into the atmosphere, will be affected by changes in factors such as wind, humidity, and temperature. Higher frequency and intensity of poor air quality episodes, which can lead to restrictions to port operations, could increase as a result.
- Future increases in precipitation intensity will increase the risk of on-site or off-site pollution due to pollutant runoff, in particular for ports with limited drainage capacity or with low levels of maintenance of sediment traps and oil/water separators.
- Waste reception and management facilities at ports could have limited capacity to cope with future climatic changes.
- Ports located near protected habitats or species could see their reputation challenged if changes in climate negatively impact these natural assets.

- Overall, port environmental impact assessment and environmental management plans which do not consider future changes in climate could be inadequate.
- MEB is certified according to ISO 14001, which demonstrates a commitment to environmental protection.

The environmental impact of MEB's operation is not expected to be significantly affected by climate change, except for a 30 percent increase in energy use for refrigeration and associated greenhouse gas emissions (assuming 6°C increase in average seasonal temperature as an upper-bound estimate); a possible increase of the risk of overflow of sediment traps and oil/water traps due to seawater flooding at the port, and a limited, but not impossible, increased risk of pollutant runoff during episodes of surface flooding.

In and around Cartagena, sea-level rise and rising temperatures are expected to further degrade corals, mangroves, and seagrass.

Works to raise the height of the causeway are likely to damage the surrounding mangrove. MEB could be required by the environmental agency of Cartagena (CARDIQUE) to pay for mangrove (re)generation elsewhere.

Sea-level rise is also expected to have an impact on the mangrove. Due to its location on the narrow tidal fringe against the port, it is not possible for the mangrove to retreat as sea level rises. Therefore it must move upward out of the water. To do this the sediment substrate must increase in height and it may be possible for MEB to assist this process, for example through the use of dredge material.

The analyses undertaken for MEB in this study have shown that climate change can have material business implications for ports. Though some of the risks analyzed in this study are likely to be specific to MEB and may not apply to other ports, a number of them will be of broad relevance to the industry around the world. To help port operators and their stakeholders to identify climate-related risks and adaptation options, a general ports climate risk and adaptation checklist was elaborated in the framework of this study and is available at www.ifc.org/climaterisks.

Adaptation investments

The study found that MEB is in a good position relative to some other ports, especially in Colombia. For instance, it is located in a sheltered bay which protects it from wave action and storm surges, it does not suffer from the high levels of precipitation experienced at Pacific ports in Colombia and is not as exposed to storms as other ports in the Caribbean. For these reasons, there are various climate-change risks which are generally significant to ports but have not been explored in depth in this assessment. However, the study did identify materially significant impacts in timeframes that are relevant for investment decisions, and proposed adaptation investments that were incorporated in the company's financial model.

Following the elaboration of the study in 2011, MEB announced a US $10 million investment in adaptation measures recommended in the study[7]; the final investment amounted to US $12 million.[8]

Notes

1 www.ifc.org/climaterisks
2 Aguilar, E., et al. (2005). "Changes in precipitation and temperature extremes in Central America and northern South America, 1961–2003", *Journal of Geophysical Research*, 110: 1–15.
3 Sutherland, M., Dare, P. and Miller, K. (2008). "Monitoring sea level change in the Caribbean", *Geomatica*, 62 (4): 428–436.
4 Due to the lack of good agreement between climate models on future precipitation changes, the study considered a future scenario whereby the yearly trend in average precipitation observed between 1941 and 2009 continues in the future.
5 Stern, N. 2006a. *The Economics of Climate Change: The Stern Review*. Cambridge University Press. [Online].
6 ABI (Association of British Insurers). (2009). *The Financial Risks of Climate Change: Examining the financial implications of climate change using climate models and insurance catastrophe risk models*. ABI Research Paper. No. 19. Report by Dailey, P., Huddleston, M., Brown, S. and Fasking, S.
7 Florez, G. (2011, April 27). Más inversión en 2 puertos. Portfolio.co. Retrieved from www.eltiempo.com/archivo/documento-2013/DR-14270.
8 A. Jimenez, personal communication, February 2, 2013.

12 Climate change adaptation in the Panama Canal

Michele Acciaro

1. Introduction

The Republic of Panama is well-known globally for the Panama Canal, which together with its financial sector has allowed the country to achieve a remarkable economic performance (ECLAC, 2012). Part of Panama's 3.5 million inhabitants, however, still live in poverty, especially in rural areas with over half of rural residents in 2004 classified as poor and 22 per cent as extremely poor (World Bank, 2011). In the last decade not much has changed, and the economic gap between the urbanized part of the country and the rural areas has increased notwithstanding the enormous resources in the country (Foster et al., 2011).

Poverty and inequality have also put pressure on the environment, contributing to depauperating its natural resources. This is exacerbated by an inequitable land tenure system and the difficulties for local farmers to access credit and compete on agriculture products (World Bank, 2007). Although forests cover 40 per cent of Panama's territory and more than 12 per cent of the country landmass is protected, poverty has an impact on wildlife and protected areas, and deforestation has been rampant.

The local tropical climate and the concentration of people in few densely populated regions also exposed the population to epidemics and natural disasters such as floods, heavy storms and droughts, especially in the poorest parts of the country. Panama ranks 14th among the countries most exposed to multiple hazards with 15 per cent of its total area and 12.5 per cent of its population exposed to at least two hazards (World Bank, 2011).

As in other developing countries, climate change constitutes a serious threat to Panama's population and the country's economic and environmental systems. In particular, agriculture, water resources, forestry, coastal maritime ecosystems and the health sector have been identified as the most vulnerable sectors (World Bank, 2011). Climate change is likely to affect the country through more intense storms, floods and droughts with the highest tolls on the poorest and most marginalized parts of society.

In addition to the impacts of climate change on the country, an issue of particular concern are the effects that climate change will exert on one of the main economic resources of the country: the Panama Canal. The Panama Canal is one of the most important arteries in world international transport. The Canal is transited daily by nearly 40 ocean-going vessels. Given its role, there are a large number of studies that have focused on its impact at a local and at a global level (e.g. CEPAL, 1965; Brandes, 1967; Nathan Associates, 2003). In addition, the ongoing expansion of the Canal has led the Panama Canal Authority (Autoridad del Canal de Panamá, ACP) to commission feasibility and economic studies (ACP, 2007; ACP and Nathan Asssociates, 2011). As Panama is a developing country, the importance of the Canal is even more critical (United Nations Economic and Social Council, 1979).

The Canal is a complex system of locks and artificial fresh water bodies that allows vessels of a maximum length of 294.13m, a width of 33.5 m and a draft of 12.56m – the so-called *Panamax* ships – to be elevated 26m above sea level on the Gatún Lake and make the 82km crossing of the isthmus of Panama. After the current expansion programme is completed, that, notwithstanding some delays, will allow the enlarged third set of locks to be operational and *New-Panamax* vessels of 366m length, 49m width and a draft of 15.2m will be able to transit through the Canal.

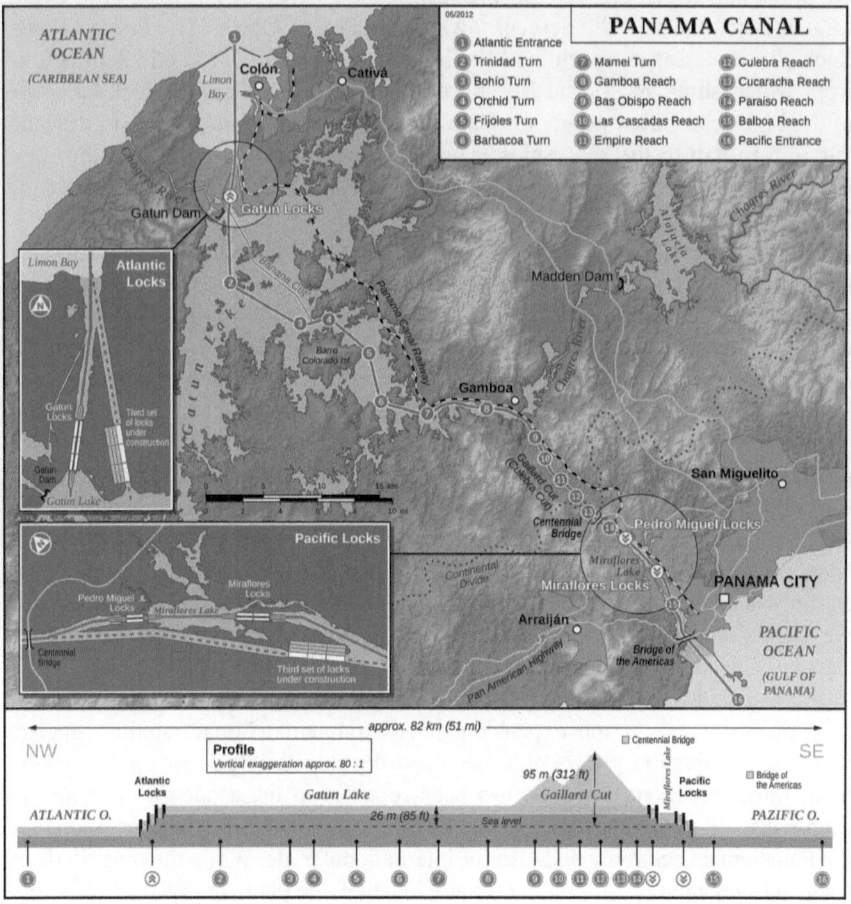

Figure 12.1 Overview of the Panama Canal.

Source: Thomas Römer/OpenStreetMap data

Figure 12.1 above shows an overview of the Canal, from the Atlantic Ocean to the Pacific Ocean cutting across the Panama Isthmus for 77.1 km from northwest to southeast. The Canal is a complex piece of engineering and is constituted of various fresh water lakes, connected by artificial channels and three sets of locks. In addition the Alajuela artificial lake is linked to the Canal system as it acts as a reservoir for the Canal.

From the Atlantic Ocean vessels approach the Canal at Limón Bay, sailing along to the deep-water port of Cristobal, and its intermodal facilities at the Colón Free Trade Zone. After approximately 8.5 km, they reach the 3.2 km approach channel to the Gatún Locks, that, with their three-stage, 1.9 km-long flight, allow vessels to be lifted 26.5 m above sea level to the Gatún Lake. This artificial lake constitutes the largest part of the Canal (24.2 km), and is emptied by evaporation and the lock operations. The lake is fed mainly by the Chagres River, along which ships will sail for another 8.5 km into the Culebra Cut, a 12.6 km passage through the mountain ridge. Vessels will then reach the single-stage 1.4 km-long Pedro Miguel Lock and start the descent towards the Pacific Ocean. The Pedro Miguel Lock allows vessels to reach the Miraflores Lake, which is situated 9.5 m below, from which ships access the Miraflores Locks. These two-stage locks are one of the best observation points for Canal operations and allow vessels to descend the 16.5 m gap to the sea. Ships leave the Canal after the port of Balboa and finally enter the Pacific Ocean in the Gulf of Panama, 13.2 km away from the Miraflores Locks.

The Canal is not only a complex technical system, but includes organizational, scientific, environmental, and political legislative components (Carse, 2012). This implies that such a large piece of infrastructure is bound to affect radically the environmental, social, political, and obviously economic contexts in the country. As a major manmade structure the Panama Canal has changed dramatically the local environment with the creation of a complex water management system, which includes two major artificial lakes and which required the displacing of the natural flow of rivers, and has since affected marine, coastal, and inland wildlife.

As nature-related complex socio-technical systems, it is not surprising that the Canal is vulnerable to climate-change-related risks. The impact of climate change became evident between 1997 and 1998, when the Niño phenomenon caused Canal earnings to drop, as trade for agricultural commodities shrivelled, and even required the Canal Authority to impose draft restrictions on the vessels sailing through the Canal as a result of low water levels at Gatún Lake (de Marucci, 2002).

The Panama Canal Authority included climate-change-related risk assessment procedures as part of the expansion of environmental impact studies, and an integrated adaptation programme has been developed as part of the expansion plan. Little attention, however, has been reported in the academic literature to such risks and to the consequences of any climate-change-related disruptions for the Panamanian economy and also for international trade. While the magnitude of such consequences is difficult to measure in detail, the Panama Canal offers a very illustrative example of the importance of adaptation measures and of the cost magnitude that the lack of adaptation could engender.

This chapter provides a contribution on the impacts of climate change that can affect Canal operation or result in the Canal's closure. This constitutes the background for any analysis aiming at quantifying the economic benefits of any further adaptation measures to be implemented in the Canal region. In addition to this introduction the chapter comprises of four more sections. Section 2 deals with the climate-change-related impacts relevant for the Panama Canal. This section will bring together the limited scientific evidence on the possible impacts of the Canal, highlighting the importance of further quantitative studies. Section 3 discusses the costs at a local and international level of eventual disruptions of Canal operations. Section 4 deals with current adaptation measures, and Section 5 concludes and provides some suggestions for further research.

2. Climate change impacts on Panama Canal operations

2.1 Climate baseline

Panama has a tropical climate characterized by two main seasons: a dry season (December to April) and a rainy season (May to November). The rainy season brings between 250–700 mm of rain across the country while the average rainfall is 1,900 mm annually. The climate is generally hot and humid, with average temperatures around 27°C, maximum temperatures between 31°C and 35°C, and minimum temperatures just above 20°C. There is some variability in rainfall and temperature depending on the region of the country and the altitude. The temperatures and average rainfall for the regions of Colón and Panama, where the Canal is located, are provided in Figure 12.2 below.

In the study of the environmental impacts of the ACP, data on rainfall, temperatures, humidity, wind speed, among other indicators are provided for three stations along the Canal (Balboa, Gamboa, and Limon Bay) that show a

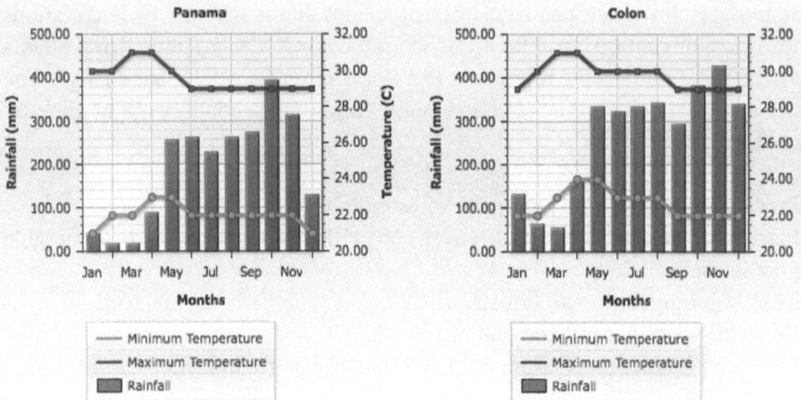

Figure 12.2 Rainfall, minimum and maximum temperatures in the Panama Canal region.
Source: World Bank Climate Change Knowledge Portal, Panama Dashboard

similar profile to the one reproduced in Figure 12.2. The assessment study (ACP, 2007: 62) highlights the importance of rainfall in the Canal watershed (Cuenca). In particular in the dry season, lack of rainfall is the result of the action of trade winds and of the movement of the Intertropical Convergence Zone (ITCZ) on the Isthmus.

The climate is mostly influenced by the ITCZ, which is responsible for the two-season regime in Panama and brings the summer rainfall. El Niño/La Niña is responsible for drought in the south and floods in the north of the country (de Marucci, 2002). Furthermore, the country benefits from the mitigating effects of the sea that contributes to regulating temperatures. The area of the Canal is affected by occasional tropical storms, with potential for landslides and heavy rain, but the country is out of hurricane paths. Droughts, therefore, are one of the most serious weather events that can affect the Canal area, mostly as a result of the lack of water in the Gatún Lake. The amount of water at Gatún Lake depends on multiple factors, among which rainfall in the watershed area is the most relevant, but temperatures and wind, through evaporation, and of course the Canal water management, are also of importance.

2.2 Climate change

Future climate in Panama is expected to remain humid with a general increase in rainfall of almost 80 percent by 2080 (World Bank, 2011), but the major increase would be observable in the months of January, April, and May. Sea-level rise will reach 35 cm by the end of the century, with important repercussions on estuaries and aquaculture. Dry season temperatures are projected to increase between 0.4°C and 1.1°C in 2020, 1°C to 3°C in 2050, and 1°C to 5°C in 2080. Uncertainty, however, is still substantial, and the magnitude of climate change in Panama, as in other parts of the world, could be underestimated. As indicted in Figure 12.3 below, the changes in rainfall in the various models show large variations.

The impacts, however, can be catastrophic, and that is why the United Nations Framework Convention on Climate Change (UNFCCC) has included Panama among the most vulnerable regions in the world in terms of climate change. The World Bank (2011) identified for Panama the following secondary major climate-change impacts:

- Increased incidence and intensity of crop failure.
- Increased intensity of heat stress on crop production and vulnerable population.
- Loss of biodiversity and forests.
- Reduced water quality and quantity.
- Increased incidence of climate-related human health impacts.

On the basis of the broader impacts of climate change on Panama, the following can be considered specific impacts on the Panama Canal operations and the Canal freight system.

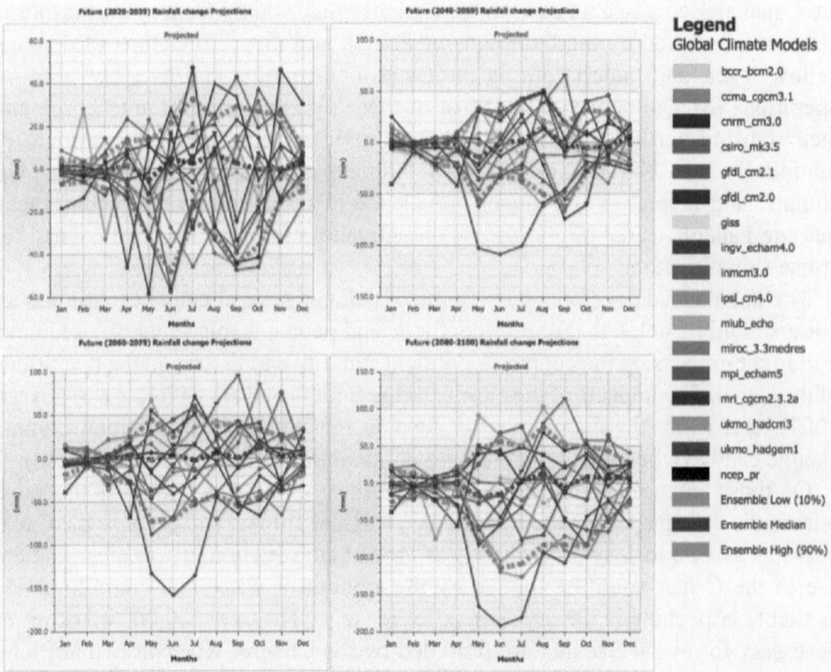

Figure 12.3 Comparison of rainfall change projections, 2020–2100.
Source: World Bank Climate Change Knowledge Portal, Panama Dashboard

- Reduction in cargo and ships transiting through the Canal, similar to the 1997–1998 Niño years, with a consequent loss of revenue for the Canal.
- Unrest and relocation of population.
- Heavy rainfall and gust winds which could affect logistics distribution areas, as well as reduce visibility or hinder Canal operations.
- Fog, which is a widespread phenomenon in the Canal areas, can cause disruptions in operations (Cheng and Georgakakos, 2011).
- Droughts, which can limit the navigability along the Canal, require restrictions on vessel drafts.
- Floods can affect the neighbouring areas to the Canal with possible population displacement and unrest (Shamir et al., 2013)
- Landslides can affect the Canal areas (Duncan, 2010; Berman, 1995; Stallard and Kinner, 2005).

A climate-change-induced cargo transit reduction through the Canal could have a substantial impact on the Canal, as in 1997–98. Climate change can dramatically affect agriculture production (e.g. Turral et al., 2011) and by modifying trade patterns also reduce the use of the Canal, as grain carriers constitute one-quarter of Canal ship passages. The attractiveness of the Canal for other types of ships can also be affected by climate change impacts. For example, Panama and specifically the Canal are long associated with tropical diseases, mainly from mosquito-borne viruses, as attested by the thousands of French and then US victims of dengue, yellow fever, and malaria during the Canal construction and its early years of operations (McCullough, 2001). Most of these diseases are mosquito-borne and their transmission rates are sensitive to climatic factors such as the amount of rainfall (Reiter, 2001). Links between climate change, a warmer and wetter climate, and increased and longer prevalence of disease-carrying vectors mean that the Panama Canal might lose its attractiveness for shipping, particularly for cruise ship operators.

The unbalanced development patterns of Panama have already been discussed, but it should be noted that unrest, conflicts, and people displacements could have important impacts on the country's political stability and in turn on the operations of the Canal. The impacts of the closure of the Suez Canal are well-known (Feyrer, 2009), and while the geopolitical context in Panama is very different, climate change can have bearings on increasing political risk (Oh and Reuveny, 2010).

Of those impacts related to the navigability of the Canal one of the most relevant is drought. Given the Panamanian tropical climate and because the Canal operates using a lock system, already at time of construction, it was clear that the use of the Canal would be limited by the amount of water that could be made available, especially in the dry season, for lockage. The over 200,000 m^3 of water necessary for every crossing are provided by the Chagres River in two artificial basins, but the sources of water, especially at Gatún Lake, are limited.

Multiple studies (e.g. World Bank, 2013) advance the possibility of a reduction in dry-season rainfall. In a recent study by DNV-GL (2014), the probability of water scarcity at Gatún Lake was predicted in absence of any adaptation measures (Figure 12.4). The study shows that water level reductions would be substantial without any adaptation measures.

Landslides have also been studied extensively (Duncan, 2010; Berman, 1995; Stallard and Kinner, 2005). While the impact of landslides is not specifically related to the infrastructure, it represents an obstacle to navigation when large trees are carried into Canal waters (Wohl and Ogden, 2013) with potential for ship allusion damage. The landslides are often associated with heavy rainfall and reduced visibility such as experienced in 2010 (Shamir et al., 2013). The manoeuvrability of vessels in enclosed waters is a particularly critical issue as vessel sizes increase following the Canal expansion and ship collisions or groundings would be catastrophic for Canal operations, requiring perhaps several weeks of work to remove the wreckage.

The increased severity of weather events such as rainfall, fog, and wind, which are typically associated with climate change, is known to dramatically impact ship navigation in enclosed waters (Becker et al., 2013). Fog, for example, has delayed the transit of over 428 vessels in 2014, 121 more than in 2013. Strong winds can dramatically impair the ability of a vessel to stay on course, as a consequence of the strong forces exerted on the ship's slab side of large container vessels, cruise ships, and car carriers (Athanasatos et al., 2014).

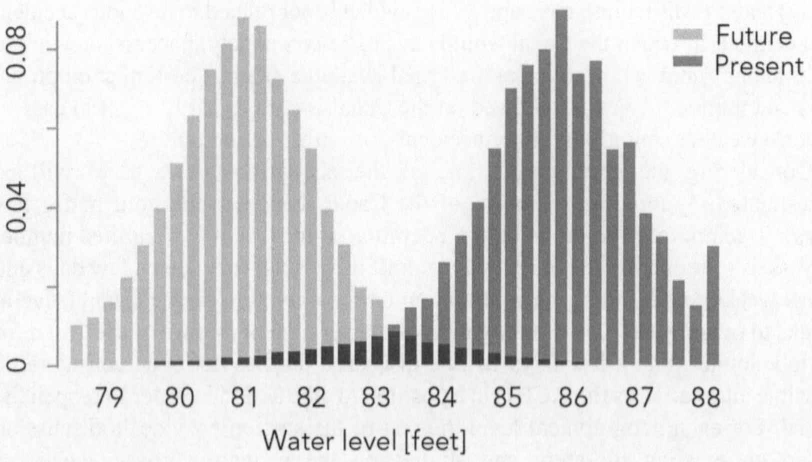

Figure 12.4 Probability distributions of water level at Gatún Lake observed in recent decades and projected for late this century assuming no adaptation.*

Source: DNV-GL, 2014: 60

* The estimate has been obtained as a hypothetical bottom line where no adaptation measures were taken. However, in the expansion plan adaptation measures are actually employed. The estimate is the result of both a tendency in projected decrease of precipitation and increased water use; the latter is imputable to the operations of the new locks and projected increase of water demand for municipal and industrial uses.

3. Economic consequences of disruptions in the Panama Canal operations

The economic consequences of disruptions in the operation of the Panama Canal depend on the gravity and type of the disruption as well as the length of time the disruption would affect operations. For this study consequences are assessed at three levels: the ACP level, the Panama level, and the global level. As far as the ACP impacts are concerned, the consequences investigated are specific to the ACP in terms mainly of loss of revenue. As far as the Panama level is concerned consequences relate to the Panamanian economy and the Canal-related and -induced economic activities. This would focus on job creation and value-added generation within the country of Panama. At a global level this study assesses how disruption in Canal's operations will impact the role that the Panama Canal plays in global supply chains and in particular with reference to the competition with the US land bridge connecting Asia to the East Coast of the USA and as a connection in South-South trades between Europe and North America with South America.

As far as the gravity of the disruptions is concerned, two alternatives will be investigated: complete closure of the Canal or limitations on the number and type of vessels using the Canal, based on vessel draft or size. The complete closure of the Canal is a rare event that has taken place in the past only in exceptional circumstances.

Typically such closure is limited to short periods of time (between a few hours and a day) and it is difficult to envisage a case, which is not related to a serious accident or a conflict, in which the Canal would have to be completely inaccessible without ACP taking rapid action to make the Canal available again. The limitation on the type and number of vessels allowed on the Canal is a more likely event in case of adverse weather conditions, minor accidents, or infrastructure failure.

Considering the gravity and time of the disruptions three cases will be investigated: 1) temporary closure of the Canal (between one hour and a few days); 2) temporary limitation on the operation of the Canal (e.g. limited number of vessels or reduction in the acceptable draft of vessels) (between a few days and a few weeks); and 3) long-term limitation on the operation on the Canal (several weeks to over a year). The investigation of the permanent closure of the Canal for periods longer than a few days will be discussed but has not been considered a possible alternative as the ACP is in a position to take action in order to reopen the Canal. For each geographical level the rest of the section provides a discussion based on existing literature and studies on the economic consequences of disruptions in the operations of the Canal.

3.1 Local level

One of the main sources of revenue for the ACP and Panama are Canal tolls.1 Tolls are set by ACP and are based on vessel type, size, and the type of cargo carried. For container ships, tolls are based on the ship's capacity in TEU (US $74 per TEU). In addition, a charge of US $8 per container is made for every loaded container transiting on board of the vessel. US $82 is charged for containers carried on board of ships other than cellular ships. Cruise ships pay a rate based on the number of passengers that can be accommodated in permanent beds. The per-bed charge is currently US $92 for unoccupied beds and US $134 for occupied beds and US $108 when transiting in ballast. Passenger vessels of less than 30,000 tonnes or less than 33 tonnes per passenger are charged according to the same per-tonne schedule as are freighters. Most other types of vessels pay a toll per tonnage. The toll is calculated differently for passenger ships and for container ships carrying no cargo (currently US $65.60 per TEU).

As of October 1, 2013, the toll ranges between US $4.29 and US $5.33 per tonne for the first 10,000 tonnes, between US $4.20 and US $5.22 per tonne for the next 10,000 tonnes, and US $4.12 and US $5.14 per tonne thereafter, as described in Table 12.1 below. As with container and passenger ships, a reduced toll is charged for freight ships transiting *in ballast*. Tolls for small vessels2 are calculated on the basis of the ship length. They range between US $1,300 for vessels below 50 ft and US $2,400 for vessel above 100 ft, with two intermediate tariffs of US $1,400 and US $1,500.

Table 12.1 Canal dues structure (2014)

Type of vessel	*Loaded*			*Ballast*		
	First 10,000 tons	*Subsequent 10,000 tons*	*Any 10,000 tons thereafter*	*First 10,000 tons*	*Subsequent 10,000 tons*	*Any 10,000 tons thereafter*
General cargo	5.10	4.99	4.91	4.07	4.00	3.93
Refrigerated cargo	4.29	4.20	4.12	3.43	3,36	3.30
Grains	5.06	4.89	4.81	4.04	3.90	3.85
Tankers	4.92	4.84	4.75	3.94	3.87	3.80
Chemical tankers	5.06	4.98	4.89	4.05	3.98	3.91
Gas carriers	4.99	4.91	4.82	4.07	4.00	3.93
Ro-Ro	4.40	4.31	4.24	3.52	3.45	3.40
Passenger ships	4.42	4.33	4.26	3.54	3.46	3.41
Other	5.33	5.22	4.14	4.27	4.18	4.12

Source: ACP official tariffs as in Regulation 1010.0000.

While the tolls represent the main source of revenue for the Canal, their increase cannot be unlimited given the existence of alternative routes available to the all-water route through the Panama Canal. Tolls per container vessels increased in 2009 from US $40 per TEU to US $72 and then to US $74 in 2011. Such increase already captured over 40 percent of the potential cost saving deriving from the economies of scale that can be exploited after the expansion (Rodrigue, 2010).

Total revenue for the ACP was US $1,962 million in 2010, US $2,319 million in 2011, US $2,411 million in 2012, US $2,411 million in 2013 and US $ 2,629 million in 2014, three-quarters of which were toll revenues (ACP, 2012, 2013, 2014). The remaining 25 percent are mostly deriving from transit-related services (US $395 million in 2012) and other revenues, i.e. interests, water, and electricity sales (US $163 million in 2012). Ninety-three percent of ACP revenues in 2012 were related to tolls or transit-related services performed by the company, showing the rather homogeneous line of activities that ACP performs.

Considering that the total vessel tonnage (measured in PC/UMS) was 332.5 million in 2012, which generated US $2,398.9 million, a simple calculation leads to an average of 910 thousand tonnes transported through the Canal a day, which amounts to roughly US $6.6 million revenue per day (roughly US $7.25 per tonne). Total expenses amount to US $682.6 million, just above US $2.05 per tonne. Net income for the Canal was US $964 million in 2010, and increased to US $1,229 million and US $1,258 million in 2011 and 2012 respectively, just above US $3.78 per tonne, or US $3.45 million per day using 2012 figures (the difference being taxes, amortization, and depreciation) (ACP, 2012, 2013, 2014).

In terms of number of vessels, in 2014 a total of 11,956 transits took place through the Canal. Panamax vessels (the largest) accounted for 56.3 percent of the total. In terms of shipping sectors, 36 percent of total transits are container vessels followed by dry bulk carriers (25 percent), tankers (15 percent) and vehicle carriers (11 percent). Containers accounted for 119.9 million tonnes (on average

35,000 tonnes per container vessel3). Considering that 3,331 container vessels transited the Canal in 2012, on average each container vessel delivered to the Canal's net income over US $132,000. Panamax vessels would allow on average for almost US $200,000 net income per transit. Since these are the vessels that would be the first to suffer in case of restrictions on the transit in the Canal, the importance of the expansion programme is quite evident. Post-Panamax container vessels would allow three times the carrying capacity of current Panama vessels (from 4,000 TEU to 12,000 TEU), which would make the transit of these new ships even more desirable through the new Canal.

From these simple calculations it is clear that the ACP earns in the range of US $150,000 per hour of operation and a disruption of any sort causes considerable losses. Even if we were to correct for the other sources of revenues and account for the possibility of delaying or charging higher tolls (e.g. through the auction system) for the vessels that would not be able to use the Canal as a result of the disruption, the losses for ACP would still be substantial.

These considerations are mostly based on unexpected disruptions that could potentially be resolved rapidly. Longer-term economic effects are more difficult to estimate as a result of the possible corrective action that ACP would take. In the specific case of climate-related disruptions, the profitability of the Canal is such to allow for operation for a relatively long period of time. Prolonged disruptions though would not only affect directly the revenue of ACP by reducing the total capacity of the Canal, but also by creating an incentive for cargo to move to other, more reliable corridors. Prolonged disruptions or an increase in the variability of the frequency of the disruptions would affect the Canal's competitiveness. This might have critical consequences on the Canal, considering that the expansion project, which will allow post-Panamax vessels to transit through the Canal in 2014, envisaged substantial growth rates in terms of vessel numbers and total tonnage (Pagano et al., 2012).

Using the estimated changes in Canal transit and volumes after completion calculated by Pagano et al. (2012), Canal transit would increase by 13.1 percent in terms of number of ships and in the range of 35 percent both in terms of tonnes and toll revenue by 2020, and increase further to 21.5 percent in terms of number of vessels, and 55 percent in terms of total tonnes transported and toll revenues by 2025. These increases rely on the additional cargo allowed by the transit of post-Panamax vessels. If by any chance the transit of these vessels would not be feasible, growth estimates would be affected more dramatically than in the past. So the increase in the size of the Canal increases also the variability of the revenues from the Canal, as those become more dependent on larger vessels.

3.2 Panama level

For Panama the benefits of the Canal are enormous as a result of the direct monetary transfers deriving from the Canal tolls and the importance of the Canal system for exports, which have accounted for over two-thirds of the Panama GDP in the last 15 years (ACP, 2014). More than 75 percent of the total exports are related to the activities of the cluster in the Interoceanic Transit Region, most of which are dependent on the Canal activity. Approximately 9,800 people work at the Canal and the contribution of the ACP to the Panamanian national treasury was just above 1 billion US$ in the last few years (ACP, 2012, 2013, 2014). The role of the Canal is important also for the induced and indirect effects of the economic and logistics activities connected to the Canal. ACP estimated that the Canal indirectly contributed to the Panamanian national budget over US $169.4 million by means of income tax, social security, and educational insurance. To these benefits, US $1,517.6 million from wages and the procurement of goods and services from local suppliers will be added (ACP, 2014).

From the perspective of the importance of the Canal for the state of Panama, a study performed in 2006 by the ACP estimated that the Canal expansion would allow cluster exports to triple by the year 2025 (ACP and Nathan Associates, 2011). The Canal expansion will allow Panama to attain a GDP of US $31,700 million by 2025 using 2005 as the base year. This represents almost 2.5 times the GDP of the country in 2005, and equals an average annual growth rate of over 5 percent for the next 20 years.

A more recent study (Pagano et al., 2012) confirms similar findings by assessing the overall contribution of the expansion of the Canal to the economy of Panama using an input-output approach and expanding the analysis by means of General Equilibrium Model (GEM). They observe that 82 percent of Panama's economy and 33 percent of its exports are services (excluding exports generated in the duty-free zone). Overall, the Canal GDP multipliers are estimated between 1.27 and 1.37 (Pagano et al., 2012).

The impact of a short-term closure of the Canal would entail a reduction in the treasury contribution. Since this is a matter of political discussion, an accurate forecast cannot be made. The minimum contribution that ACP has to transfer to the Panamanian government is calculated as US $568 million per year (ACP, 2014). A pro-rata quota calculation considering the treasury contribution of 2014 as a reference would result in a loss of US $2.82 million a day to the national treasury. If we consider indirect contributions, the ACP suggests it would rise to US $7.4 million a day (ACP, 2014), however, salaries and other indirect contributions would not be affected by short-term disruptions.

Reduction in service and operations of the Canal for a longer period of time would affect the contributions of the Canal to the local economy. In this case, the losses of the treasury would probably be more sizeable as long as they would be reduced pro-rata on the basis of the Canal service utilization. Assuming a reduction in the use of the Canal of 30 percent for ten days, it would be in the range of US $8.6 million. On more general terms, a one-point percentage reduction in the utilization of the Canal would cost in the range of US $30,000 per day. In terms of cargo, considering that the total amount of cargo transiting the Canal in 2014 amounted to 326.8 million tonnes, every tonne lost due to traffic restrictions would translate into a loss of 1.3 cents per tonne per day.

The Colón Free Trade Zone, as well as the port sector, are important drivers together with the Canal, of the Panamanian economy. While short-term disruptions in the performance of the Canal would probably not affect the development of the cluster,4 in the long term they might affect the competitiveness of the logistics cluster as cargo moves to other chains. The severity of such longer-term changes are difficult to asses as they depend among other things on the capacity of the cluster to attract cargo independently from the eventual operational restrictions on the Canal, and the growth of South-South trades.

3.3 Global level

The Panama Canal is considered one of the global choke points where international traffic converges, and in virtue of the limitations on capacity its infrastructural characteristics impact the global maritime circulation both in terms of volumes transported and transit times (Rodrigue, 2004). Disruptions in the operation of the Canal are likely to have a global impact. Maritime chokepoints have been looked at mostly in order to guarantee energy supply (Komiss and Huntzinger, 2011). In particular the expansion programme will allow post-Panamax vessels to transit through the Canal and reach the US East Coast ports. The expansion, however, will have little impact on oil transit, as most tankers will remain too large for the Canal. Thirteen per cent of Canal cargo is transit cargo between the West Coast South America/US East Coast route and the expansion is expected to increase the amount of trade within the region (Rodrigue et al., 2013). The Canal has a critical role for countries such as Peru, Chile, and Colombia, with over one-third of their imports and 20 percent of their exports transiting through the Canal. On the one hand, the expansion of the Canal will favour trade growth through Panama, on the other hand its success depends largely on the competitiveness of the Panama Canal option versus other transport options (Figure 12.5, left panel). The Canal is also used for South-South trades connecting North America and northern Europe with South America.

The most relevant trade for the Canal, however, is the link between the Asia Pacific Region and the American East Coast (approximately 40 percent of cargo transiting through the Canal). Of those trades, a bit more than half goes through the intermodal transport through Canada, the US West Coast and Mexico, while approximately 45 percent goes through an all-water route mostly via the Panama Canal (Rodrigue, 2010). It should be noted that the market share of the all-water route for cargo to the US via the Panama Canal has been increasing at the expense of the intermodal land bridge alternative. This is the result of capacity constraints in the US West Coast ports and railroad network (Fan et al., 2009) as well as of a logistics differentiation strategy performed by logistics service providers (Rodrigue, 2010). The increase in the evolution of the world's containerized carrying capacity, with the exponential increase of the post-Panamax fleet is one of the main determinants of the expansion project for the Canal. Container traffic is likely to remain the main beneficiary of the Panama Canal expansion as tankers still remain too large for making use of the enlarged Canal (Komiss and Huntzinger, 2011).

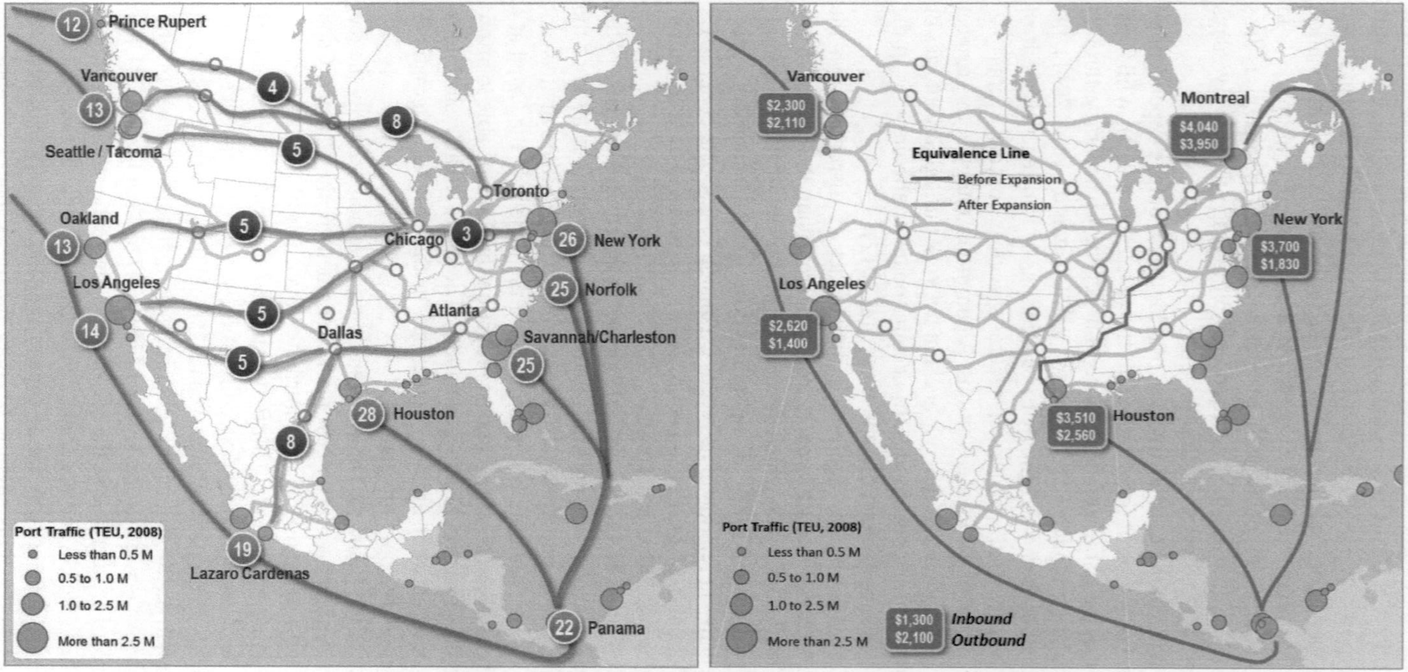

Figure 12.5 Standard transit times from Shanghai and North American routing options (in days) and shipping rate in US$ for a 40-foot container with Shanghai, selected port pairs, mid-2010.

Source: Rodrigue (2010)

The most important competitor of the Panama Canal for the international traffic between Asia and the US East Coast is the so-called US land bridge. US traffic through the west ports of Los Angeles/Long Beach, Oakland, Portland, and Seattle/Tacoma is loaded on freight trains or trucks to the US central distribution hubs of Chicago, Memphis, and St Louis. While the Panama Canal option is in general cheaper in terms of costs with respect to the land bridge, it is less competitive in terms of transit times as it requires on average three to five days more (Corbett et al., 2012; Fan et al., 2009).

It should be noted that the competitiveness of a logistics route is dependent on three main factors: costs, transit time, and reliability (Rodrigue, 2010). While cost and transit time can be estimated with a certain degree of accuracy through empirical and simulation models (e.g. Fan et al., 2009; Tavasszy et al. 2011), the reliability of a corridor is more difficult to estimate as it is often the result of variables that can be predicted with limited accuracy, such as political events (e.g. wars, strikes), accidents, or weather conditions. A common strategy in this respect is building sufficient resilience in the supply chain network through inventories or differentiated routes or transport modes, to accommodate erratic changes in the network flows due to supply chain disruptions (Lam, 2012).

There are no specific studies addressing the impact of disruptions on the Panama Canal on international trade apart from research done in relation of the Torrijos-Carter Treaties of 1977 which focuses more on political than climate risks (Elton, 1987; Randolph, 1988). While studies do exist on the competitiveness of the Canal, especially in relation to its recent expansion (e.g. Costa et al., 2012; Fan et al., 2009; Mitchell, 2011; Pagano et al., 2012), the findings of such studies can only partially be used for our purposes.

A recent study by Ungo and Sabonge (2012) looks at the competitiveness of the Canal routes vis-à-vis other international routes. They give an indication of the dependence of the world international trade on the Canal. Although it is not possible to determine an accurate cost estimation of the benefit of the Canal at a global level, it is possible to assess what cost reduction the Canal is able to afford international shipping (Figure 12.5, right panel). These are clearly short-term effects and might not account for the fluctuations in the costs of international shipping due to demand-supply imbalances.

The importance of the Canal for container shipping is demonstrated by the fact that the container segment is the largest market segment in the Panama Canal measured by revenues. Container ships contributed over half of total tolls and 34 percent of ocean-going transits in 2014. Between 1997 and 2007, the number of vessels crossing the Canal increased steadily, mostly as a consequence of the expansion of liner shipping networks between Asia and the US East Coast. Such growth came to a halt as a result of the 2008 recession, and the number of liner services through the waterway decreased from 43 in 2007 to 33 in late 2010 (Ungo and Sabonge, 2012). The decrease in the use of the Canal is the result of the drop in cargo demand from the US and the dramatic increase in available capacity, also due to the exponential fleet growth. Large capacity surplus encourages carriers to make use of longer routes in the attempt to reduce the total capacity available.

Table 12.2 Summary of the effects of disruptions of operations in the Canal

	Local level (ACP)	*Panama level*	*Global level*
Temporary closure of the Canal (between one hour and a few days)	Temporary but substantial loss of revenue	Limited effects mostly as a result of reduced transfer to the government treasury	Limited effects, as global supply chains absorb temporary shocks Temporary episodes of congestions
Temporary limitation on the operation of the Canal (between a few days and a few weeks)	Temporary but substantial loss of revenues	Limited effects mostly as a result of reduced transfer to the government treasury	More sizeable effects as longer severe restrictions on the use of the Canal would put under pressure the existing transportation systems
Long-term limitation of the operation on the Canal (several weeks to over a year)	Substantial loss of revenues. Demand for the Canal could be diverted to other corridors	Potential loss of competitiveness from the Canal	A modification of global supply chain patterns would occur

3.4 Summary of the economic consequences of disruptions of operations in the Canal

The economic consequences deriving from disruptions in the operation of the Panama Canal and the various geographical levels are summarized in Table 12.2 above.

An important dimension that is not discussed in Table 12.2 refers to the reliability and the frequency of disruptions. While an occasional disruption in the operations of the Canal might not have long-term effects on the competitiveness of the Canal, a more frequent or unpredictable occurrence of such disruptions could potentially compromise the attractiveness of the Canal as an alternative to the US land bridge.

4. Adaptation measures

Panama has been active in developing responses to climate change through the implementation of mitigation measures and adaptation. Mitigation measures have been extensively discussed in particular with reference to emission reduction (e.g. Mulligan and Lombardo, 2011; Corbett et al., 2012). Adaptation measures instead are only incidentally discussed. Those measures are the human responses to foreseeable climate-change impacts at a local, national, and global level. As discussed in Section 2, climate-change impacts affect Panama on multiple fronts. Secondary major climate-change impacts have the possibility of affecting the operations of the Canal as a result of people displacements, unrest, and political instability. As the assessment of these impacts on the Canal is very challenging, adaptation measures will be dealt with, only briefly, mostly focusing on agricultural impacts and coastal management.

Among the adaptation measures suggested to deal with agricultural impacts on the region there is the improvement of irrigation efficiency, the introduction of new crops less vulnerable to climate change, the development of controls against pests and crop diseases, and the subdivision of the agricultural zone taking into account the increase in weather variability due to climate change. Coastal management is another area where adaptation measures can have positive effects. Among the measures suggested, there is the implementation of an Integrated Coastal Zone Management program, which would include the improvement of fish and shrimp farming, water flux and reflux in mangrove areas, as well as better urban plans (construction lines, residual waters treatment, sanitation and pluvial systems).

These measures would contribute to the reduction of political and societal risks although direct assessment of costs and impacts of adaptation measures would require further study. As far as health-related risks, such as mosquito-borne diseases, are concerned, specific actions have not been taken to account for the impact of climate change.

Most relevant for the Canal are adaptation options in terms of water management. The Canal in fact is only, albeit important, user of freshwater, and an integrated approach to water resource management would be advisable. In the World Bank adaptation report (2011) the following measures are suggested for water management:

- Increasing water supply, e.g. by using groundwater, building reservoirs, improving or stabilizing watershed management, desalination.
- Decreasing water demands, e.g. by increasing efficiency, reducing water losses, water recycling, changing irrigation practices.
- Improving or developing water management.
- Developing and introducing flood and drought monitoring and control systems.
- Strengthening of water and weather station networks to better predict future changes in the water regime, including floods and droughts.
- Developing new irrigation technologies.
- Promoting conservation and rational use of water resources.

The ACP has already taken into consideration in the expansion plan some adaptation measures to ensure improvements in water management. The Panama Canal expansion program includes raising the maximum operating level at the Gatún Lake by 45 cm and the widening and enlargement of the navigation channels. In order to ensure the water availability in the Canal system during the dry season months, the ACP has also planned for the development of water-saving basins for the new locks (Simonit and Perrings, 2013a).

While the increase in water depth at the Gaún Lake and the widening of navigation channels are operationally driven in order to allow the transit of larger vessels after the expansion, the development of the new lock system specifically accounted for the water-saving measures. While works on the lock segments are still ongoing, in the planning phase because of the lock size and the need to increase lock performance, efforts have been made to optimize filling and emptying times (Perez and Newbery, 2014). The lock control system has been built to ensure high levels of reliability and redundancy (99.6 percent).

Among the most important issues identified in several studies is the specific role that watershed management has to play to ensure the operability of the Canal. The topic of watershed management in the Panamanian context has been the subject of academic research highlighting the role of the environment in Canal operations (Carse, 2012). Although the impact on water preservation of the reforestation is ambiguous (see for example Simonit and Perrings, 2013a; 2013b; Ogden and Stallard, 2013), it has allowed for reducing the environmental impact of the Canal expansion project.

When it comes to navigability, climate-change-related risks are difficult to disentangle from operational risks. ACP has been very active in developing traffic control systems, expandimg pilotage and tugging capacity, and improving weather monitoring (ACP, 2014). Already in 2014 and in preparation for the Canal expansion, ACP modernized the infrastructure of the Maritime Traffic Control Center. The creation of berthing layup pockets dispersed along the Canal length could represent a viable operational risk reduction strategy, although in some parts of the Canal (e.g. Culebra Cut) costs would be prohibitive.

In recent years, the ACP implemented a Comprehensive Risk Assessment System and a Canal Business Continuity Plan. In addition, pursuant to the provisions of the Torrijos-Carter Treaties in 1987, the US Congress authorized the acquisition of a comprehensive multi-peril insurance that protected the Canal against catastrophes that might have a significant impact on Canal revenues. It is recommended that these tools would account specifically for climate-change-induced risk changes.

Conclusions

This paper provided an overview of the impact of climate change in the Panama Canal systems, mostly focusing on the costs associated with disruptions in Canal operations. The vulnerability of the Canal to climate-change disruptions is substantial and further studies should aim at quantifying the risks and the associated climate-change costs. The most relevant climate-change risk in the Canal system seems to be associated with droughts. Prolonged drought periods in the dry season could reduce the level of water at Gatún Lake, which could potentially require the ACP to impose restrictions on vessel draft or could even cause disruptions in Canal operations. The expansion programme aimed already at addressing water scarcity and an extensive reforestation programme is ongoing. While there seems to be consensus on the environmental benefits deriving from a better watershed management, little attention so far has been paid to other possible climate-change-related hazards.

The substantial costs associated with the partial closure of the Canal, or even with restrictions on traffic, justify a careful consideration of climate-change impacts. Climate-change impacts are dealt with operational and market risk adaptation measures, but in some cases, such as water management or weather-related navigational hazards, specific account of the impact of climate change on the risk level should be made. At the moment, although climate-change-related risks seem rather pervasive, there does not seem to be a comprehensive approach aimed at identifying and mitigating their impacts on the Canal system.

Further areas of investigation include:

- The impact of crop failures and agriculture trade flow changes on the Canal revenues.
- The impact of specific weather conditions on Canal navigability (e.g. fog, heavy rain) as well as the assessment of the risk of landslides.
- The impact of climate change on the port and logistics areas around the Canal.
- The impact of political unrest, famine, and population movements within Panama.

The potential for climate-change-related disruptions calls for further investigation for the implementation of adaptation measures on the Canal system.

Acknowledgments

Although the content of the chapter is original and the result of research work performed by the author in 2013 and 2014, some of the ideas in the chapter emerged from discussions when the author was employed at DNV-GL in the period 2010–12 and was involved in the ADAPT project. The author would like to acknowledge the contribution of Luca Garré, Peter Friis Hansen, and Elzbieta Bitner-Gregersen in these discussions. The author would like to also thank Ana E. Bucher and Jean-Paul Rodrigue for granting permission to reproduce some of the illustrations in this chapter. The author would like to acknowledge the suggestions and insightful comments of two anonymous reviewers. A preliminary version of this chapter has been presented with the same title at the 33rd World Congress of the World Association of Waterborne Transport Infrastructure (PIANC) in San Francisco, June 1–5, 2014. The usual disclaimers apply.

Notes

1 The information on Canal tolls is obtained from the ACP website. Where there are discrepancies between the English and the Spanish versions the most recent information is considered correct. When the time of publication of the information is the same or unknown, precedence is given to the Spanish version.

2 Small vessels are vessels up to 583 PC/UMS net tonnes, the system used to measure the size of vessels crossing the Panama Canal, when carrying passengers or cargo, or up to 735 PC/UMS net tonnes when in ballast, or up to 1,048 fully loaded displacement tonnes.

3 These are rough calculations, but 35,000 tonnes per container vessel is a reasonable estimate of the average size of the container vessels transiting the Canal. Panamax vessels typically have a DWT between 65,000 and 80,000 tonnes. It should be noted that Panamax vessels are not able to carry more than 52,000 tonnes through the crossing due to draft limitations in the Canal.

4 It could, on the contrary, even increase the cluster activity as disruptions in the Canal operations could potentially require more trans-shipment, intermodal transport, etc.

References

ACP. (2007). *Estudio de Impacto Ambiental Categoría III Proyecto de Ampliación del Canal Tercer Juego de Esclusas*, ACP, Panama City, Panama. Miami, FL: Autoridad del Canal de Panamá (ACP).

ACP. (2012). *Annual Report 2012,* ACP, Panama City, Panama. Miami, FL: Autoridad del Canal de Panamá (ACP).

ACP. (2013). *Annual Report 2013,* ACP, Panama City, Panama. Miami, FL: Autoridad del Canal de Panamá (ACP).

ACP. (2014). *Annual Report 2014*, ACP, Panama City, Panama. Miami, FL: Autoridad del Canal de Panamá (ACP).

ACP and Nathan Associates. (2011). *Re-estimation of Economic Impacts of Canal Related Activities, Programa De Dimensión del Impacto del Canal de Panamá Sobre la Economía del País, Panama City, Panama*. Miami, FL: Autoridad del Canal de Panamá (ACP).

ACP, Intracorp and Asesores Estratégicos. (2006). *Impacto Económico del Canal de Panamá en El Ámbito Nacional*, ACP, Panama City, Panama. Miami, FL: Autoridad del Canal de Panamá (ACP).

Athanasatos, S., Michaelides, S. and Papadakis, M. (2014). "Identification of weather trends for use as a component of risk management for port operations", *Natural Hazard*, 72 (1): 41–61.

Becker, A. H., Acciaro, M., Asariotis, R., Cabrera, E., Cretegny, L., Crist, P. ... and Velegrakis, A. F. (2013). "A note on climate change adaptation for seaports: a challenge for global ports, a challenge for global society", *Climatic Change*, 120 (4): 683–695.

Berman, G. (1995). "Landslides on the Panama Canal". In: *Energy and Mineral Potential of the Central American-Caribbean Region.* Berlin, Heidelberg: Springer, pp. 391–395.

Brandes, E. M. (1967). *Analysis of Panama Canal Traffic and Revenue Potential.* Prepared for Panama Canal Company. Menlo Park, CA: Stanford Research Institute.

Carse, A. (2012). "Nature as infrastructure: Making and managing the Panama Canal watershed", *Social Studies of Science*, 42 (4): 539–563.

CEPAL. (1965). *Estudio Sobre las Perspectivas del Actual Canal de Panamá*, Spanish version. Santiago, Chile: Comisión Económica para América Latina y el Caribe.

Cheng, F. Y. and Georgakakos, K. P. (2011). "Statistical analysis of observed and simulated hourly surface wind in the vicinity of the Panama Canal", *International Journal of Climatology*, 31 (5), 770–782.

Corbett, J. J., Deans, E., Silberman, J., Morehouse, E., Craft, E. and Norsworthy, M. (2012). "Panama Canal expansion: Emission changes from possible US West Coast modal shift", *Carbon Management*, 3 (6): 569–588.

Costa, R., Rosson, C., Robinson, J. and Fuller, S. (2012). *The Impacts of the Panama Canal Expansion on World Cotton Trade*. 53rd Annual Transportation Research Forum, Tampa, Florida, March 15–17, 2012.

De Marucci, S. (2002). *Impact of El Niño Southern Oscillation Phenomenon on the Panama Canal and its Markets*. International Association of Maritime Economists Annual Conference, Panama, November 13–15, 2002.

DNV-GL. (2014). *Adaptation to a Changing Climate*. Høvik: DVN-GL.

Duncan, J. M. (2010). "Managing unstoppable landslides in the Panama Canal", *Geo-Strata—Geo Institute of ASCE*, 14 (6): 12–13.

ECLAC. (2012). *Preliminary Overview of the Economies of Latin America, 2012*. Santiago, Chile: Comisión Económica para América Latina y el Caribe.

Elton, C. (1987). "Studies of alternatives to the Panama Canal", *Maritime Policy and Management*, 14 (4): 289–299.

Fan, L., Wilson, W. W. and Tolliver, D. (2009). "Logistical rivalries and port competition for container flows to US markets: Impacts of changes in Canada's logistics system and expansion of the Panama Canal", *Maritime Economics and Logistics*, 11 (4): 327–357.

Feyrer, J. (2009). *Distance, Trade, and Income – The 1967 to 1975 closing of the Suez Canal as a Natural Experiment* (No. w15557). Cambridge, MA: National Bureau of Economic Research.

Foster, W., Valdés, A., Davis, B. and Anríquez, G. (2011). "The constraints to escaping rural poverty: An analysis of the complementarities of assets in developing countries", *Applied Economic Perspectives and Policy*, 33 (4): 528–565.

Komiss, W. and Huntzinger, L. (2011). *The Economic Implications of Disruptions to Maritime Oil Chokepoints*. Arlington, VA: Center for Naval Analysis.

Lam, J.S.L. (2012). "Risk management in maritime logistics and supply chains". In: Song, D. W. and Panayides, P. M. (Eds.), *Maritime Logistics: Contemporary Issues*. Bingley, UK: Emerald, pp. 117–132.

McCullough, D. (2001). *The Path between the Seas: The Creation of the Panama Canal, 1870–1914*. New York: Simon and Schuster.

Mitchell, C. (2011). *Impact of the Expansion of the Panama Canal: An Engineering Analysis*. Master's Thesis, University of Delaware.

Mulligan, R. F. and Lombardo, G. A. (2011). "Panama Canal expansion: alleviating global climate change", *WMU Journal of Maritime Affairs*, 10 (1): 97–116.

Nathan Associates, Inc. (2003). *Transportation Study on the Dry Bulk Market Segment and the Panama Canal*, Volume I, Draft Final Report. Arlington, VA: Nathan Associates Inc.

Ogden, F. L. and Stallard, R. F. (2013). "Land use effects on ecosystem service provisioning in tropical watersheds, still an important unsolved problem", *Proceedings of the National Academy of Sciences*, 110 (52): E5037–E5037.

Oh, C. H. and Reuveny, R. (2010). "Climatic natural disasters, political risk, and international trade", *Global Environmental Change*, 20 (2): 243–254.

Pagano, A. M., Light, M. K., Sánchez, O. V., Ungo, R. and Tapiero, E. (2012). "Impact of the Panama Canal expansion on the Panamanian economy", *Maritime Policy and Management*, 39 (7): 705–722.

Panama Canal Authority (ACP). (2012). *Annual Report 2012*. Miami, FL: Autoridad del Canal de Panamá (ACP).

Perez, R. and Newbery, M. (2014). *Innovation in Large-Scale Civil Works Design/Build – The Third set of Locks Experience*. 33rd PIANC World Congress, June 1–5, San Francisco, CA, USA.

Randolph, F. F. (1988). *The Strategic Value of the Panama Canal: Value Versus Cost.* Monterrey, CA: Naval Postgraduate School.

Reiter, P. (2001). "Climate change and mosquito-borne disease", *Environmental Health Perspectives*, 109 (Suppl. 1): 141.

Rodrigue, J. P. (2004). "Straits, passages and chokepoints: a maritime geostrategy of petroleum distribution", *Cahiers de géographie du Quebec*, 48 (135): 357–374.

Rodrigue, J. P. (2010). *Factors Impacting North American Freight Distribution in View of the Panama Canal Expansion.* Calgary, Canada: Van Horne Institute.

Rodrigue, J. P., Comtois, C. and Slack, B. (2013). *The Geography of Transport Systems.* New York: Routledge.

Shamir, E., Georgakakos, K. P. and Murphy Jr, M. J. (2013). "Frequency analysis of the 7–8 December 2010 extreme precipitation in the Panama Canal watershed", *Journal of Hydrology*, 480: 136–148.

Simonit, S. and Perrings, C. (2013a). "Bundling ecosystem services in the Panama Canal watershed", *Proceedings of the National Academy of Sciences*, 110 (23): 9326–9331.

Simonit, S. and Perrings, C. (2013b). "Reply to Ogden and Stallard: Phenomenological runoff models in the Panama Canal watershed", *Proceedings of the National Academy of Sciences*, 110 (52): E5038–E5038.

Stallard, R. F. and Kinner, D. A. (2005). "Estimation of landslide importance in hillslope erosion within the Panama Canal watershed". In: *The Río Chagres, Panama.* Netherlands: Springer, pp. 281–295.

Tavasszy, L., Minderhoud, M., Perrin, J. and Notteboom, T. (2011). "A strategic network choice model for global container flows: Specification, estimation and application", *Journal of Transport Geography,* 19 (6): 1163–1172.

Turral, H., Burke, J. J. and Faurès, J. M. (2011). *Climate Change, Water and Food Security.* Rome: Food and Agriculture Organization of the United Nations.

Ungo, R. and Sabonge, R. (2012). "A competitive analysis of Panama Canal routes", *Maritime Policy and Management*, 39 (6): 555–570.

United Nations Economic and Social Council. (1979). *Panamá: Consideraciones Para la Formulación de un Programa de Desarrollo de la Subregión Canalera*, Documento de Trabajo. New York: United Nations, Economic and Social Council.

Wohl, E. and Ogden, F. L. (2013). "Organic carbon export in the form of wood during an extreme tropical storm, Upper Rio Chagres, Panama", *Earth Surface Processes and Landforms*, 38 (12): 1407–1416.

World Bank. (2007). *Country Partnership Strategy for the Republic of Panama.* Washington, DC: World Bank.

World Bank. (2011). *Vulnerability, Risk Reduction, and Adaptation to Climate Change: Panama.* Washington, DC: World Bank.

World Bank. (2013). *Panama: Climate Baseline.* Washington, DC: World Bank.

13 The impact of climate change on Australian ports and supply chains

The emergence of adaptation strategies

Stephen Cahoon, Shu-Ling Chen, Benjamin Brooks and Greg Smith

1. Introduction

Australia is the sixth-largest country in the world with a total land area of 7.69 million square kilometres, and is completely surrounded by water with a total coastline of 59,736 km, comprising a mainland of 35,877 km and offshore islands of 23,859 km (Blewett, 2012). It lies between latitudes 10° and 43° South and longitudes 113° and 153° East and experiences a variety of climatic zones due to its size, including equatorial, tropical, subtropical, desert, grassland and temperate (Figure 13.1) (Bureau of Meteorology, 2014). With such diversity, understanding the impact of climate variability and change on Australia's water resources, agricultural production, coastal communities and natural ecosystems presents a multi-levelled challenge because adaptation strategies in one region may not apply to another.

In Australia, the climate risks of greatest concern appear to be rising sea levels, extreme rainfall and wind, increasing intensity of cyclones and tropical storms, and higher temperatures (Commonwealth Scientific and Industrial Research Organisation, CSIRO, 2007). These risks have already resulted in coastal recession, flooding of essential public infrastructure in low-lying areas, loss of transport access due to coastal inundation, heatwaves, and an increasing frequency of bush fires (Steffen et al., 2014; Kinrade and Justus, 2008; Queensland University of Technology, QUT, 2010). Steffen et al. (2014) indicate that over half the Australian coastline (about 31,000 km) is vulnerable to recession from rising sea level, for example, 80 per cent of the Victorian and 62 per cent of the Queensland (see Figure 13.1 for the location of each Australian State) coasts are at risk. It is anticipated that the intensity of cyclones and storms will increase, with a 60 per cent to 140 per cent increase of category three to category five cyclones being predicted by 2030 and 2070 respectively (CSIRO, 2007). The coastal areas of Queensland, Northern Territory and Western Australia for example, are particularly vulnerable to cyclones. Exceptional heatwaves have also occurred in recent years resulting in record-breaking temperatures. The heatwave in southern

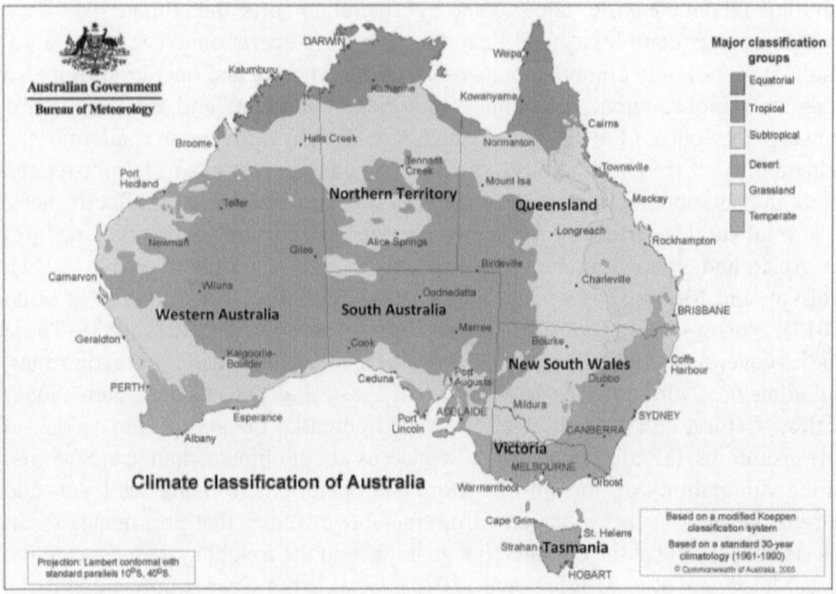

Figure 13.1 Climate of Australia.
Source: Adapted from Bureau of Meteorology (2014)

Australia in January to February 2009 for instance, led to extreme temperatures (45.7°C and 46.4°C in Adelaide and Melbourne respectively) and the resulting catastrophic Black Saturday bushfires on 7 February (QUT, 2010). In addition, an exceptionally extensive heatwave affected large parts of Australia in late December 2012 and the first weeks of January 2013 where 48°C and 49°C were recorded in New South Wales and South Australia (Bureau of Meteorology, 2013). The CSIRO also predicts that the number of days with maximum temperatures over 35°C may increase by 25 per cent by 2030 and double by 2070 in both Melbourne and Adelaide (CSIRO, 2007).

The above climate risks have implications for the operations of major Australian industries such as forestry, agriculture, mining, and natural resource-based tourism sectors (Maunsell, 2008). In particular, the important income earners including tourism and the prime resource industry are vulnerable. Additionally, these risks have impacts on critical infrastructure such as electricity transmission networks, water supply, transport, coastal buildings and settlements, human health, ecosystems and biodiversity, and indirectly on international trade (Maunsell, 2008; Steffen et al., 2014).

An important user of critical infrastructure and connectors between sea and land transportation for international and regional trade are the 70 ports located around the geographic outline of Australia (Figure 13.2). Of them, 65 are regional ports performing approximately 85 per cent of the total national cargo handling task (Anderson, 2010) and five are capital city ports located in Sydney, Brisbane, Adelaide, Melbourne and Fremantle (Perth) that mostly engage in container cargo handling. Given the wide scope served by Australian ports, the climate risks have brought greater complexity and uncertainty to port operations. The uncertainty results from not only climate threats to port infrastructure and operations but also from the regional impact of climate change on industries and freight systems utilising the ports. For instance, changing freight movement patterns, the vulnerability of freight facilities to heat and flooding, and risks to freight-related safety and port operations may impact on supply-chain efficiency and effectiveness.

Several studies pertaining to climate-change impacts and adaptation strategies for Australian ports have been undertaken recently by Chhetri et al. (2012); McEvoy and Mullet (2013); McEvoy et al. (2013); Maunsell (2008); Ng et al. (2013); Nursey-Bray et al. (2013); QUT (2010) and Scott et al. (2013). These studies cover nine Australian ports in five States located in various climatic zones, including the Northern Territory, Western Australia, Queensland, New South Wales, Victoria and Tasmania, and revealed potential threats of climate risk to port operations. Although most ports' concerns about climate change are centred on the vulnerability of port infrastructure and operations to rising sea levels and extreme weather events, it appears from the above studies that port management has started looking at the climate-change impact on the freight system through the port (McEvoy et al., 2013; Ng et al., 2013). What is not clear is how much focus is placed on adaptation strategies by port managers.

This chapter explores the climate-change risks to which Australian ports are exposed and, based on secondary data and by synthesising the findings of the above-mentioned major studies, discusses existing and potential adaptation strategies used by ports. Further, this chapter explains the results of a case study

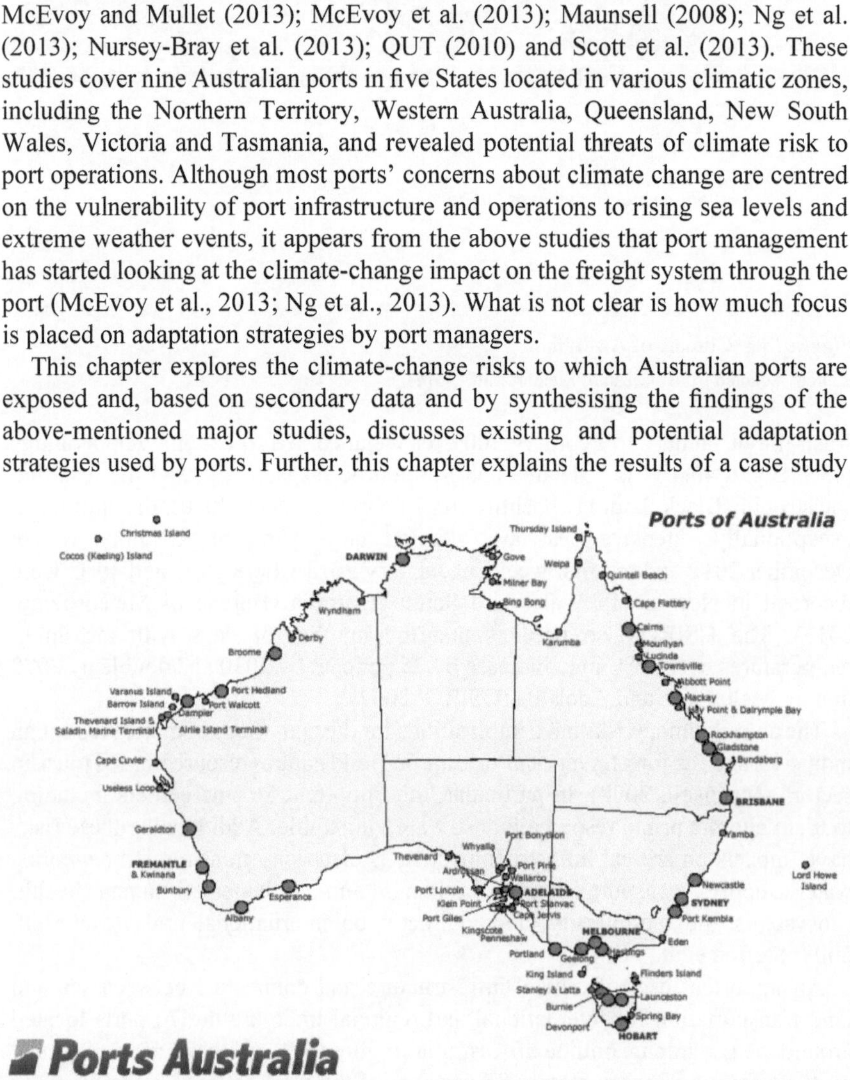

Figure 13.2 Ports, Australia.

Source: Ports Australia (2014)

conducted via interview that investigates how climate change is and may impact on a coal supply chain from the coal mines through to the port terminal. The chapter concludes with proposed future directions that highlight the need for a coordinated approach by port supply chains to climate-change adaptation.

2. Climate change risks to Australian ports and supply chains

Among the climate-change risks mentioned in Section 1, the increasing intensity of tropical cyclones is likely to be the key climate risk affecting port infrastructure and operations in the States of Western Australia, Northern Territory and Queensland. Southern States are not immune to this issue with summer low-pressure systems impacting the NSW coast and extreme cold fronts affecting South Australia and Victoria. Storm surges resulting from tropical cyclones are another important key climate-change risk, of which the result is an increase in coastal water levels for periods of several hours, and affect the environment and perimeter of the coastline. Consequently, port operations and maintenance practices are impacted. Increasing extreme weather events, in particular heavy rainfall and heatwaves, are already of wide concern to ports across the States of Western Australia, Northern Territory, Queensland, New South Wales and Victoria. For example, the Queensland floods in 2011, caused by excessive rainfall due to an extremely strong La Niña effect and powerful cyclone Yasi, shut down the Port of Brisbane for ten days, costing AUD50 million and decreasing total throughput by 6.4 per cent in the Queensland port system in 2010/11 (Chhetri et al., 2012). Rising sea levels are also of concern for ports across Australia. For example, since 1993, on the north and northwest coasts, sea levels are rising at the rate of 7 to 11 mm per year, which is two to three times the global average of 3 mm per year between 1993 and 2011. On the central east and southern coasts of the continent, the rise in sea-level rates is similar to the global average (CSIRO and Bureau of Meteorology, 2012).

The climate-change risks impose potential threats to infrastructure, superstructure, functions and services provided to cargo and ships, and to people (e.g. workforce and passengers) in the port system. The port system includes operations and activities occurring at the approach from the sea, quayside (interface), landside and further to the hinterland. Each of these subsystems is further explored below in relation to the potential threats at both subsystem and port-system levels.

2.1 Approach from the sea

Within the sea approach to the port, the most observed impacts of rising sea levels are coastal erosion at or adjacent to the port and deposition and sedimentation along the port channels (Ng et al., 2013; Nursey-Bray et al., 2013). Similarly, intense rainfalls causing floods are also affecting port channels and shipping berths, for example, the Queensland floods in 2011 deposited an extra 1.2 million m^3 of sediment and material into the Port of Brisbane's channel and shipping berths reducing the channel depths by up to one metre (Chhetri et al., 2012). The sea approach has also been affected by increasing storm intensity and high winds that create risks of vessels being taken off course in the channel (Nursey-Bray et al., 2013) and thus the need for additional tug boats. Evidence is indicating that difficulties at the berths from storm surge are resulting in operational downtime or port closure (Scott et al., 2013).

2.2 Quayside

At the quayside, rising sea levels have brought concerns about higher salt-water splash zones for port superstructure (Ng et al., 2013). Increased extreme climatic events including strong cyclones, extreme wind, flash floods and heatwaves can also damage port superstructure such as berths, yards, handling equipment and cargo, and operations, creating delays in ship discharging, increasing days of operation and risks for cargo stevedoring. The effect of these events is port downtime causing a loss in productivity and an increase in operational expenditure (Maunsell, 2008). For example, a port in Northern Territory suspended operations for two days due to cyclones in the past two years, compared with only four days in the past decade. Similarly a major iron-ore exporting port in Western Australia suspended operations due to five cyclones in six weeks, resulting in increased ship queuing that cost AUD3 billion to the Australian economy (Ng et al., 2013). In a personal communication in 2014 with a port manager from Western Australia, he explained that an increasing intensity of tropical cyclones and occasional supercell storms has the following implications for operations and costs in his port:

> Cyclone downtime and clean-up/repairs shared between port actors are perhaps AUD250,000 per cyclone. Because of port evacuation the vessel operators incurred losses for vessel daily hire. For example three platform supply vessels might be visiting the port at any one time and clearing the port results in operator loss of AUD40,000 daily hire per vessel. Other losses include cruise ships bypassing the affected storm area and port call, and fuel tanker and break bulk shipping delays.

In terms of the impact of intense rainfalls, in 2011 the Port of Botany in New South Wales was closed due to flash flooding after three days of intense rainfalls even though several truck-loads of sand were brought into the terminal area to prevent straddle slippage due to flooding distributing oil throughout the terminal (Scott et al., 2013). The impact of high winds and heatwaves on port operations at the quayside was also evidenced recently when in October 2012, gusty winds blew 45 full containers over at Port of Botany, causing operations to cease for three days (McEvoy et al., 2013).

High temperatures also affect machines, the surface of terminal yards and the workforce. In 2009, 5 per cent of the 36-hectare terminal at the Port of Melbourne was unable to be used due to the tarmac melting, and consequently port capacity was reduced (QUT, 2010). However, there is also a human cost, and thus increased operational costs due to heatwaves. For example, as a result of industrial agreements with trade unions and port hot weather policies, Australian port workers now have additional breaks or even a suspension of operations once the temperature gets to a certain degree (Ng et al., 2013). This is evident at DP World in the Port of Melbourne which has a heat agreement with the Australian Maritime Union (AMU) stating when the temperature exceeds 35°C, outdoor workers can take a 15-minute break per hour and when it is above 38°C workers can stop working until the temperature cools down (QUT, 2010). This reduces productivity and delays the process of loading and unloading ships, resulting in vessel delays and potentially affecting national and international supply chains. The loss of productivity is evident when considering that a heatwave in January 2009 at the DP World terminal resulted in a loss of 72 crane hours (QUT, 2010).

2.3 Landside

The purpose of port landside logistics is to provide multi-modal transport infrastructure and superstructure for shipping cargo to and from hinterlands. Climate impacts such as flooding due to intense rainfalls, storms and cyclones, and heatwave damage to road and railway lines servicing ports, disrupt supply chains and increase maintenance and replacement costs. For instance, the heatwave in 2009 caused buckling of the rail tracks at the Port of Melbourne (Chhetri et al., 2012). In another example (see Ng et al., 2013), a port manager in Western Australia explained that cargo movements are also affected by heavier than normal precipitation. The port manager was concerned that the live cattle industry would be heavily affected because many regional routes would become impassable, therefore affecting planned throughputs via his port. Bulk ports, which are important for Australian exports, are even more vulnerable to this threat. In Queensland, for instance, the flood in 2011 stopped coal being transported to the Port of Gladstone due to damage to the rail line and flooding at coal mines (McEvoy et al., 2013). A case study identifying the vulnerability of coal supply chains to climate-change risks will be illustrated later in this chapter to highlight the inter-connectedness of the climate-change risks and level of complexity that port and supply chains deal with on a daily basis.

2.4 The hinterland

The fourth port subsystem impacted by climate change is regional sectors/industries in the port's hinterland, each of which may affect port throughput positively or negatively. According to Maunsell (2008) agriculture is the key industry in Australia to be affected by climatic conditions, followed by tourism and mining. Ng et al. (2013) explain that the predicted drier climate and high temperatures will adversely affect crop and beef production thus mainly decreasing trade throughput for the southern, northern and western Australian ports. The same authors suggest that on the positive side, opportunities for increased crop production in Tasmania are likely because the State will be less affected by climate change, thereby representing an opportunity for pasturelands, situated in its port's hinterland, to become the 'new food bowl' for Australia.

Cruise tourism is an example of a key industry sector that contributes significant revenues to Australian ports and their respective States. The threats imposed by climate change such as extreme hot temperature and reduced road access to major attractions due to storm or flooding will make regional cruise supply chains vulnerable. The study by Ng et al. (2013) identifies that an increase in the number of extremely hot and high humidity days, particularly in the Northern Territory and Western Australia, has a potential disadvantage to cruise tourism meaning that the ports may become less attractive destinations for cruise passengers in the future. A respondent in their study (p. 191) said:

> When the cruise ships come here, and the passenger profile is normally fairly senior, the one thing that faces them is that when they get out of the air-conditioned ship they get into the air-conditioned bus and they get off-loaded in the middle of the town and all of a sudden they are in temperatures in the high thirties and very high humidity and nowhere cool or shady for them to go to sit down and recover.

Ng et al. (2013) further suggest that the offshore oil and gas industries in Australia would face interruptions due to more extreme cyclones affecting operations.

3. Current climate change adaptation actions and measures taken by Australian ports

As indicated in the previous section, research findings reveal that several Australian ports are aware of climate-change impacts. In addition to taking actions to mitigate carbon footprint at ports, some have initiated adaptation actions and strategies to counter climate-change impacts in their development, planning and operations. For example, the Port of Melbourne, the largest Australian container port, has initiated climate-change adaptation strategies as evidenced by being a signatory to the World Ports' Climate Declaration, having a corporate climate change policy since 2007, and a climate change strategy since 2009 'inscribing climate change response and mitigation as well as climate change risk and adaption into the port's safety and environment management' (Ng et al. 2013, p. 191). Nevertheless, the adaptation strategy for climate change in the Australian port system tends to be fragmented. Instead of having a specific climate-change plan crossing the entire port system, ports seem to incorporate adaptation strategies and measures in regulatory and operational mechanisms such as workplace health and safety policies, operational plans, environment management policies, risk response/management plans and emergency response plans (McEvoy and Mullett 2013). The following discussion identifies current adaptation measures initiated by Australian ports according to recent studies. What will become apparent is that the adaptation measures used by ports mainly focus on seaside and quayside operations. A potential reason for this is that adaptation actions for multi-modal transport in landside logistics such as road and rail are the responsibilities of State and local government agencies.

Adaptation measures to cyclone tidal surge and sea-level rise flood inundation have been undertaken by ports in Western Australia, Northern Territory and New South Wales. Ng et al. (2013) explain that a weather station to measure the height of waves, the extent of storm surges, changes in the tilt of the tectonic plate and rising sea levels have been established in a port in Western Australia. Ports also incorporate the impact of rising sea levels in their port infrastructure plans, for example a port in the Northern Territory has incorporated cyclone tidal surge and sea-level rise flood inundation mapping as part of its land-use planning. This detail is included in their 20-year port master plan to ensure that new infrastructure will be positioned above the highest predicted tidal surge areas. This process had been occurring over the past few years with relocation of critical infrastructure such as the storage of dangerous goods to areas above the tidal surge zone, for example the repositioning of uranium storage was undertaken to ensure it could not be washed back into the harbor, thus creating serious environmental risks. Similarly, ports in New South Wales have a port plan that ensures future port infrastructure is built higher than the expected sea-level rises over the next 90 years, with reinforced rock walls being constructed to withstand stronger and higher wave surges (Scott et al., 2013).

To manage the risk of erosion as a result of higher sea levels to port superstructure, changes to maintenance practices have been implemented by ports in Western Australia, Northern Territory and Victoria. The changes include incorporating anti-corrosive paints into ongoing maintenance, and using concrete on the wharf and additives to help withstand salt-water inundation (Ng et al., 2013; Nursey-Bray et al., 2013). To manage the risk of heat on the surface of terminal yards, concrete has been used for terminal yards instead of tarmac. In the case of the Port of Melbourne, which was mentioned earlier, the terminal operator converted 15 per cent of a 36-hectare terminal to a concrete surface with a further plan to gradually pave over the entire terminal yard (Chhetri et al., 2012).

A number of measures are being taken to prevent the impact of strong winds on Australian port operations. Ports in New South Wales and Queensland for example, have developed or are developing partnerships with local weather stations to receive forward warning of approaching winds (Scott et al., 2013). In the Northern Territory, ports are enhancing the adaptive capacity of port infrastructure to withstand Category 4 cyclone winds with a minimum requirement of Category 3 to account for the predicted increasing intensity of storms (Ng et al., 2013). Even in the south of Australia, there is a 'specification for a new berth at a regional port in Victoria, being built for a wood-chip carrier, which has taken into account more periods of windy days, increased wind strength and higher bollard pull' (Nursey-Bray et al., 2013, p. 1035).

Several adaptation measures for quayside and terminal operations have been discussed by Scott et al. (2013). For example, in a coal terminal, new coal-stacking regimes (i.e. revised stockpile heights) have been introduced to prevent coal slumping due to heavy rainfalls, and in addition, the dampening of coal stacks is being undertaken to reduce dust being blown over the neighbouring community when the now longer, hotter and drier windy weather conditions prevail. Container terminals are also reacting to the increasing heavy winds by reducing the stacking height of containers, in particular empty containers, to avoid them falling, while general terminals have installed wind monitors on ship loaders. Other measures such as assessing drainage systems to examine whether the capacity is sufficient to handle heavy and prolonged rain events; using active mooring systems when sea and swell conditions deteriorate; and using automated logistics tasks to remove the requirement for cease work during extreme weather events are also employed across a range of Australian ports (Scott et al., 2013).

Workplace health and safety of the port workforce is another area of port operations that is affected by climate risk as they introduce further hazards to an already challenging work environment. The adaptation strategies used by Australian ports in response to an increasing number of hot days are being captured within workplace health and safety policies and industrial agreements with trade unions such as the DP World example illustrated earlier in this chapter (Chhetri et al., 2012; QUT, 2010). Adaptation strategies also include medical screening of port employees for skin cancers, inclusion of the effects of increasing daily temperatures into fatigue management plans, re-evaluating working/break periods, modification of working uniforms to include long-sleeve shirts and long trousers to protect workers from the sun, and undertaking a smart working plan involving more breaks where ice, water and electrical fans are made available to employees (Ng et al., 2013).

Clearly the climate risks are real and current with the potential to add significant costs to operations. Linked to this is the greater need for Australian ports to be working with insurance agencies to manage risk, in particular using protective measures such as flood and/or storm insurance. To reduce the risks, ports have initiated specific response plans to various climate conditions, for example, the implementation of a cyclone response. With an increase in intensity of storms, changes are also being made to emergency response plans, for example, there is growing concern about what may occur if a large vessel loses its moorings and becomes adrift in harbors or grounded and therefore preventing access to other vessels (Ng et al., 2013). The emergency response plan includes evacuation protocols for all ships in the event of severe winds or a cyclone entering the region and is updated in response to the increasing intensity of cyclones.

Table 13.1 summarises the discussions in Sections 2 and 3, including the threats that climate risks impose on the Australian port system and the adaption measures currently undertaken by ports. The following section provides further evidence of the impact of climate change on Australian ports and their supply chains by conducting a case study of a major coal port and its supply chain.

4. Case study: Vulnerability of coal supply chains to climate change

The majority of past studies have tended to focus on climate-change impacts on mainly the port rather than also examining the impacts on the supply chains of which ports are a nodal point. As many of the actors, if not all, will be affected by climate change, this may mean that ports have a duality of impacts, the first related to their own business, and the second as a result of accumulated impacts cascading through the supply chain. A significant supply chain in Australia is the coal supply chain.

Australia is the world's largest exporter of coal, significantly contributing to the national economy. Recent extreme weather events including flooding and cyclones have damaged the coal industry significantly, for example a reduction in coal exports by 25–54 million tonnes occurred as a result of the extensive flooding in 2011 that has led to a revenue loss of 5 to 9 billion Australian dollars (Reisinger et al., 2014). This example demonstrates the vulnerability of coal supply chains to a single event that created large losses of revenue for the coal mining companies, the rail companies, the port and the Australian economy (via royalty loss). The vulnerability in the coal supply chain is extensive as climate change can impact on mining operations, landside transport, the infrastructure that each of these supply-chain actors rely on (such as electricity and water), all the way through to shipping port operations. The issue for coal ports and their supply chain, similar to other ports, is that climate extremes are likely to increase and thus without adaptation strategies, vulnerability is inevitable.

A coal supply chain on the east coast of Australia was chosen as it represents one of the largest exporters of coal. Due to requests for anonymity from the coal port, the coal supply chain is referred to in this chapter as the East Coast Coal Supply Chain (ECCSC) and the coal port as the East Coast Coal Port (ECCP). A highly experienced senior manager from the ECCP agreed to participate in interviews and email communications to explain the climate-change risks, impacts and adaptation strategies related to the major supply-chain actors, including the port.

4.1 The composition of a coal supply chain

A coal supply chain, which is often engaged in the export markets, generally comprises the infrastructure, operations and operating assets required to move coal from a supply point, such as a mine, to ships' holds (the demand source) at an export loading point at a port. There tends to be about six major parties involved in this supply chain:

1 The coal mines, which may be a combination of open cut and underground mines.
2 Above rail operators, which are the competitive rail carriers providing the rail wagons and locomotives to haul the coal.
3 Below rail operators, which are the track owners and users of the electrical systems to run the trains.
4 The port terminal operators, which provide the rail unloading, stockyard and loading of the coal on ships.
5 The customers, who create the demand for the coal and order the ships to carry the coal – these customers are likely to be based in Japan, South Korea and China.

6 Surrounding the above supply chain are also the electricity providers who provide the electrical system for the rail and port operators and other actors.

Coal export supply chains operate within two basic systems:

1 Coal is pushed from the point of supply to a holding point where it is drawn down by a demand stem – this is known as a supply push system and usually exists where the supply capability exceeds the demand.
2 Coal is pulled by the demand stem into the holding point and immediately loaded – this is a demand pull system and usually exists where the demand exceeds the supply or geographical limitations restrict the supply chain's ability to meet demand.

The first system is usually more port capital intensive because of the need to have a larger port-located holding point (stockyard), to balance the excess supply against the demand limitations. In the case of a coal supply chain, the port stockyard needs to be large enough to accommodate different grades of coal and different demands for those grades from the coal buyers. The accompanying transport system used to haul the product from the supply point to the stockyard is usually characterised by an even flow of product. The port stockyard is likely to be large in comparison to its ship queue (demand) and the trains that haul coal from the mines to the port operate so as to balance the rail haulage task across the supply points. This avoids congestion in different parts of the rail system and minimises the requirement for track infrastructure as well as limiting the maximum number of trains and crews required for the haulage task.

The second system (demand pull), which is used by the ECCSC, is a leaner capital port solution because the port stockyard can be smaller. This is because as one product vacates the stockyard by being loaded onto a ship, the next product for the next ship is laid down in that space. As such, the capacity of the system is determined by the velocity in which product moves from the supply point to the demand point (e.g. from a mine to the ship's hold), with the stockyard being a very temporary holding point used to assemble products as dictated by the ship queue. In essence, the higher the velocity, the greater the throughput. The issue with this system is that it must operate like a funnel with more infrastructure capability required to deliver the coal to the port stockyard. Any complications upstream of the port will manifest in delays to the delivery of the coal to the stockyard, which slows the system and decreases the capacity. As a result, the demand source needs to be quite constant so that the supply chain has options in case there is a delay to a particular coal type or ship. In such a case, the waiting ship would be by-passed by another further down the ship queue. This system relies on a large demand stem to maximise opportunity. The ECCP in the case study for example, requires a queue of 15–20 ships at any one time to minimise vacant berth time and maximise upstream supply chain capability.

Table 13.1 Climate-change threats to Australian port systems and existing adaptation measures

Major climate risks	*Threats to port systems*				*Current adaptation measures*
	Approach from sea	*Quayside (Interface)*	*Landside*	*Hinterland*	
Increasing storm intensities and/or frequencies	• Deposition and sedimentation along the port's channels/terminal berth pockets • Coastal erosion at or adjacent to the port • Risks to vessels transiting port channels during high winds • Heightened risk using turning basins and ship manoeuvring during high winds • Storm surge creates difficulties at the berths • Storm-generated wave patterns (short cycle)	• Waves coupled with tidal surge damage port facilities such as berths and handling equipment • Downtime in port operations due to high winds • Decreasing days of operation • Delays in ship loading and unloading • Wave action causes excessive vessel motion when moored damaging fenders or ships' hull	• Floods caused by storms and cyclones damage road and railway lines servicing ports and increase maintenance and replacement costs • Excessive rain overloads environmental systems causing uncontrolled water discharge	• Impact on operations of key industries such as agriculture, mining and tourism • More interruptions to operations of the offshore oil and gas industries • Top soil erosion in agricultural areas	• Weather stations to measure the height of waves and the extent of tidal/storm surges • Incorporation of cyclone tidal surge as part of land-use planning and port master plan • Relocation of critical infrastructure such as the storage of dangerous goods to areas above the cyclone tidal surge zone • Storm/cyclone insurance; flood insurance • Cyclone response plans/emergency response plans • Increased design tolerances for any new equipment/plant
Sea-level rise	• Deposition and sedimentation along the port's channels • Coastal erosion at or adjacent to the port • Coastal recession and land inundation	• High salt-water splash zones for ports' superstructure such as wharfs			• A weather station to measure the changes in the tilt of tectonic plate and sea-level rise • Incorporation of sea-level rise impacts such as flood inundation in port infrastructure or master plans • Relocation of critical infrastructure such as the storage of dangerous goods to areas above the tidal surge zone • Increased and changed maintenance practices to manage the risk of erosion of port superstructure: • Using anti-corrosive paints in ongoing maintenance • Using concrete on the wharf and additives to help withstand salt-water inundation

Major climate risks	*Threats to port systems*				*Current adaptation measures*
	Approach from sea	*Quayside (Interface)*	*Landside*	*Hinterland*	
• Extreme weather: intense rainfall and hot temperature	• Intense rainfalls causing floods and depositing sediment and material in ports, thus reducing channel depths	• Increased vulnerability of cargo handling and equipment • Causing productivity loss and increased operational expenditure, e.g. gusty winds blowing containers over; heatwaves damaging the surface of terminal yards • High temperatures changing the working practice of port workers by increasing breaks or terminating operations when the temperature is over a certain degree	• Floods caused by intense rainfalls damage road and railway lines servicing ports and increase maintenance and replacement costs • Heavier than normal rainfall affecting cargo movement plan to ports • Heatwaves damage to transport infrastructure such as railway tracks	• Some regions have been affected by drier and hotter climate negatively affecting crop and beef production in southern, northern and western Australia, which further impact throughput of ports in those regions • Cruise tourism in particular in Northern Territory and Western Australia may be less attractive as a result of hot temperature and reduced road access to main attractions due to flooding	• Concreting over the surface of terminal yards to avoid heat damage • Overcoming high wind impact on ports: • Partnerships with local weather stations to receive forward warning of approaching winds at ports • Enhancing adaptive capacity of port infrastructure to withstand Category 4 cyclone winds • Building a new berth taking into account more periods of windy days, increased wind strength and higher bollard pull • Using new coal-stacking regimes to prevent coal slumping due to heavy rain • Dampening coal stacks to reduce dust being blown over neighbouring community due to long, hot and dry windy weather • Reducing stacking height of containers to avoid them tumbling during high winds • Installing wind monitors on ship loaders • Assessing drainage systems to exam the capacity of handling heavy and prolonged rain events • Using active mooring systems where sea and swell conditions can be difficult • Using automation of logistics tasks to remove the requirement of ceasing work during extreme weather events • Developing industrial agreements and policies for heat and fatigue management in response to an increasing number of hot days, in addition to medical screening of workers, modification of working uniforms to protect workers from the sun, undertaking smart working plans with more breaks • Flood insurance

Sources: Chhetri et al. (2012); McEvoy and Mullet (2013); McEvoy et al. (2013); Maunsell (2008); Ng et al. (2013); Nursey-Bray et al. (2013); Scott et al. (2013); QUT (2010).

4.2 Subsystems of the ECCSC

Overall, the demand pull system can be broken into four subsystems: (i) the supply source; (ii) the transport mode; (iii) the port; and (iv) the demand source. Each of these components is discussed further in the following subsections in conjunction with the climate risks and adaptation strategies being undertaken by the ECCSC and in particular the ECCP.

The coal supply chain being investigated in the case study is situated on the east coast of Australia in a region that has been subject to increasing adverse conditions including extreme rain events, cyclones and higher wind conditions. To highlight the growing intensity of the rain events for example, the region has been prone to heavy falls over the last four years where on at least three occasions over half a metre of rain has fallen during relatively short time periods in the port region. Although the port has dams to accommodate high rainfall, they were designed in the early 1980s for a one-in-10-year flood of water. However, in the last four years the region has suffered two one-in-100-year events and three one-in-50-year events, thus exceeding the capabilities of the dams' design.

4.2.1 The supply source

In the coal industry on the east coast of Australia, the supply source is the mining operation which utilises trains to transport the coal to the port. Due to the sheer size of the coal seams, the mines are spread over hundreds of kilometres and comprise three different types of coal (for this particular case study): hard coking coal (HCC); pulverised injection coal (PCI); and thermal coal. HCC and PCI are metallurgical coals, meaning they are used in the steel production industry for their carbon content, whereas thermal coal is used as a fuel source to heat boilers for electricity production. Complexity is added to the logistic exercise by adding different types of coal to the supply chain as metallurgical coals cannot be contaminated by thermal coals. As discussed in the port and the demand source subsystems, it is the different types of coal that create additional challenges, particularly when delays are created due to the impact of climate change. For example, because a demand pull system is used due to the coal stockyard at the port not being large enough to have dedicated stockpiles of each grade of coal, the port relies on a direct flow of the coal from the mine to the port. Any delays along any part of the coal supply chain therefore significantly impacts the ships being loaded when there is insufficient delivery of a particular coal grade. For example, excessive rain may convert an open cut mine into an open lake preventing removal of the coal and disrupting loading of the ships at the port. It is here that the accumulating effect of climate-change risks begins because delays created due to extreme storms may also be impacting on other subsystems. As operations may not resume among the subsystems simultaneously, additional delays may occur while waiting for downstream actors to resume. Further disruptions may occur if the managers at the coal mine are not readily communicating the disruptions from the climate-change effects to other actors in the coal supply chain.

4.2.2 The transport mode

In the ECCSC, rail is the primary transport mode which comprises the track infrastructure (below rail) and overhead power delivery system, which in this case study is an electrified system. The trains themselves (above rail), which are required to move coal from the supply source to the port rail inloaders, are a necessary component of the transport system with the coal trains reaching two kilometres in length with up to 10,000 metric tonne payloads. Train scheduling is a critical process that controls the sequence and timely arrival of trains at the port, which in turn must be available and ready to unload the train on its arrival, so as to maintain the coal payload velocity.

One of the biggest climate-change-induced events causing transport delays in the ECCSC is flood inundation. The extreme rain conditions have caused rivers to overflow, and rail tracks and bridges to be washed away. A central Queensland rail system for example, had five kilometres of rail track/bridges washed away after intensive rainfall which besides being expensive to replace, also interfered with operations in terms of transporting coal from the mine to the port.

Extreme heatwaves, which also occur in the region, have an impact as they can warp rail tracks and potentially derail trains, also causing disruption to the transport of coal. As the rail transport is an electrified system, consideration must also be given to the electricity providers and how they may be affected by climate-change effects. Electricity providers also face the same potential for disruption due to the effects of extreme rainfall and heatwaves on poles/lines that may result in fires. Each of these effects can bring down power lines or stop the electrical supply from the power plant. Therefore the electricity operators, although not supply-chain actors, are part of the network as an important infrastructure provider for the ECCSC.

4.2.3 The port

Significant climate-change conditions such as flooding and storm surge impact on the ECCP and can result in possible environmental breaches and reduced operations. As mentioned in Section 4.2, extreme rain conditions have exceeded the capacity of dams constructed to manage water overflows to the extent that a third dam is required. Although the port can deal with flooding by discharging excessive water into the ocean, a greater problem is that in excessive rain events, the discharge includes very fine coal particles picked up with the water as it flows through the coal stockyard. This can impact on the environmental licences granted by the government agencies to the port, resulting in both financial penalties and operational constraints should breaches occur. Breaches of the environmental licence can be significant, with penalties up to the hundreds of thousands of dollars (Australian) occurring for repeat offences. Extra costs incurred by the port due to excessive rainfall from extreme storms and cyclones include constructing of the third dam and annual dredging to remove the fine coal particles. Excessive rain has also impacted on port operations due to flooding within the region cutting off road transport and preventing port employees travelling to work. So, although the port has been capable of operating, due to insufficient workforce operations have been delayed.

When compared to other regions of Australia, high temperatures and humidity are yet to have an operational impact at the ECCP, although the port workforce do have more frequent breaks and the port ensures the workforce are appropriately hydrated. Similarly, although rising sea levels are a concern for the berths and port location in the future, there do not appear to have been an immediate impact.

4.2.4 The demand source

Climate-change impacts on the demand source are generally ship-related and customer-related. The ships that are chartered to load coal from the export port are part of the demand source along with the customers (coal purchaser). Most ECCSB coal mines sell their coal as free on board (FOB), meaning it is their responsibility and cost to get the coal to the ships and into the cargo hold. The responsibility and costs for arranging and providing the shipping is with the coal purchaser. As most steel makers rely on 'just in time' production, the arrival of the ship at the load port will suit the coal buyer's production schedule, not necessarily the ECCSC's requirements for a balanced system. Accordingly, shipping arrivals are uneven which also adds complexity to the logistic task.

In relation to the ships, high winds are a concerning factor, when under navigation in port waters, at berth and at anchor. This becomes more evident when considering that the port's coal berths are situated four kilometres offshore where short sharp swells whipped up by high winds can make it dangerous for ships to remain at berth. This situation worsens when cyclones are declared and the port ceases operations and the harbour master orders ship captains to take the ships out to sea for safe anchorage or sailing, often further up or down the coastline to avoid the cyclone. Periods of up to four days of lost operations have occurred due to ships sailing away from the port to evade a cyclone then having to sail back to the port. As the normal queue of ships waiting to be loaded numbers about 25, this is many ships that must return after a cyclone, lengthening the delays.

A climate change issue that is influencing the customers and the entire coal supply chain is the impact of environmental activism. The mining and export of coal is fraught with environmental concerns, particularly for people and lobby groups interested in mitigating the risk of climate change and reducing global warming. It is ironic that the coal industry, often blamed as a key source of global warming, is also impacted by climate change. The coal supply chain in the case study has been the target of activists who have the goal of creating sufficient delays in the transport of coal from the mine to ship as a means of undermining confidence in the industry. Initially the focus of the activists was on demonstrating in and around the port precinct to delay the coal being loaded onto ships, but it is now expected that the activists will be moving upstream along the supply chain to influence policy and regulation by both demonstrations, court challenges and lobbying, for example regarding dredging and challenging development approvals for mine exploration licences.

The impact of environmental activism is real, effective and increasing. Banks and superannuation organisations are beginning to make moral and ethical decisions about divesting from fossil-fuel-intensive companies, including the coal industry. Recently the Australian National University's council announced it would divest AUD16 million from seven companies following a review of environmental, social and governance (ESG) criteria (Bertini, 2014). This particular event in itself will not change funding outcomes, but demonstrates a concerning trend for the coal industry. Without the funding from the financial institutions, which can be about AUD12–15 billion across a coal supply chain, this is a significant amount of national and regional activity that is being lost, including jobs, purchase of infrastructure, and the royalties that would be returned to the government. Also, consider the situation of a port terminal which has 90 or so years remaining on its long-term lease that no longer is able to export coal, a situation that is quite possible within the next 20–30 years. As one of the senior managers at the port terminal stated, 'There's not much use for a coal port that can't export coal, we don't have another industry to substitute.' The challenge for individuals, governments and associated organisations is to weigh up the considerable short- to medium-term benefits provided by coal against the longer-term challenges of a clearly changing climate. This is not a straightforward comparison. While it is true that leaving the coal from any individual mine in the ground will not significantly affect global climate change, if this argument is prosecuted everywhere, the result is complete inaction and the consensus of the scientific community is that inaction will be more costly than action.

Both government and industry have responded to this with research and development activities such as the carbon capture and storage initiatives of the Department of Energy in the USA through the American Recovery and Reinvestment Act (Recovery Act). The US $1.4 billion investments associated with this Act include industrial-scale activities to sequester carbon underground, and industrial plants attempting to innovate around the reuse of carbon dioxide in manufacturing processes for products such as plastics, biofuels and construction materials (Department of Energy, 2015).

Overall, in the context of this chapter, any delay is critical to each subsystem but even more so in a demand pull system because of the risk to ensuring high velocity. Disruptions from climate-change effects can occur at the subsystem level individually or collectively within the supply chain creating accumulated effects of disruption. Figure 13.3 provides an overview of ECCSC and its subsystems and the current major climate-change impacts. This chapter has not explored what the future may hold in terms of the climate-change impacts. What is evident is that the impacts are already occurring and this will only intensify if the climate-change effect predictions occur.

Although it was clear that the ECCSC is being affected by adverse extreme weather effects and environmental activism that are worsening due to climate change, less evidence is available that explains the adaptation strategies undertaken by the ECCSC actors. Currently, when the climate-change effects occur, this is dealt with in a reactive manner by crisis management approaches by ECCSC actors. A proactive adaptation strategy is yet to be undertaken at individual actor levels or as part of a coordinated supply chain approach.

Conclusion

From the systematic perspective of a supply chain, there are significant interdependencies occurring among supply-chain actors. Of importance, infrastructure independencies (e.g. transport, electricity, water) are critical to the system. For example, ports are a key element of an international supply chain and interdependent on other actors along the chain such as regional shippers using their ports, and transport modes for moving goods such as air, rail, road and sea transport. Any disruption from other actors results in the vulnerabilities of port operations being exposed, while any termination of port operations threatens other actors within the chain such as shippers and consignees. Therefore, as also suggested by Wiltshire (2014), it is critical to understand the interdependence of supply-chain actors in relation to climate-change risk, and thus the necessity to promote a coordinated approach to climate-change adaption, which will be beneficial to the supply-chain actors' adaptive capacity to climate change. However, despite isolated adaption measures being used by port corporations, terminal operators and State road agencies including the ECCSC in the case study, there is still a lack of integrated planning in Australian transport infrastructure for climate change adaption (Climate Institute, 2012). If climate-change adaptation strategies are to be effective, a coordinated approach is required to reduce the cascading effects of the disruption upstream and downstream. Some suggestions on how to implement a coordinated approach are as follows:

1 Conduct an investigation of individual chain actors' awareness of their individual exposure to climate risks as well as potential collective risk interdependencies with other actors upstream and downstream in the supply chain.
2 Undertake a risk measurement and analysis of each chain actor and relevant transport assets, and their risk interdependencies. This requires skills and knowledge to first develop a risk assessment and then develop an analysis model for the supply chain.
3 Develop cooperative adaptation strategies among supply-chain actors to build skills and capacity in effectively managing climate-change risk. The adaptive capacity requires information to flow smoothly and quickly between actors as well as the sharing of information technology, education, skills, infrastructure, and management and leadership to facilitate cooperative adaption strategies to climate risks.

As involving a range of stakeholders to tackle climate risks that may impact on a supply chain network can be challenging due to a variety of ownership and

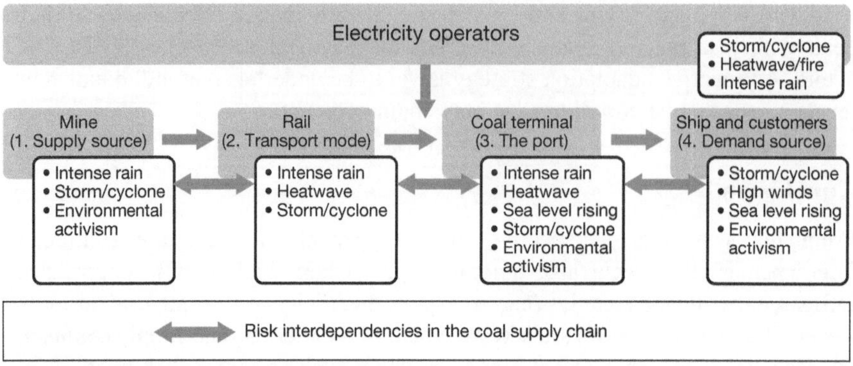

Figure 13.3 Coal supply chain and climate risks.
Source: Authors

management approaches, and/or involving different government levels and the private sector, governance and legislation could provide an important role in coordinating chain actors' involvement. For example, commercial contracts have been used by the private sector when managing and operating transport infrastructure such as rail and port terminals to ensure adequate attention is given to managing climate-change risk (Maddocks et al. 2010; Wiltshire 2014). Government legislation may also be useful for encouraging key chain actors to develop climate-change adaption strategies.

As already highlighted, information sharing is critical for collaboration. An information network for sharing climate-change knowledge among chain actors is critical for collaboration. There are some existing networks which may be able to provide opportunities for supply-chain actors to be collectively involved in climate-change adaption, for example, the Infrastructure Security and Continuity Networks (SCNs) and Trusted Information Sharing Network (TISN). Wiltshire (2014) indicates that these two networks have already been auspiced by the Victorian government, and the Port of Melbourne Corporation and Geelong Port are members of the SCN for Roads, Ports and Freight. Through the networks, infrastructure operators can engage with issues concerning climate-change risk impacts and facilitate support for adaptation measures.

References

Anderson, D. (2010). 'Industry Perspective on the National Ports Strategy', *Proceedings of the Regional Ports Conference 2010 (presentation slides)*, Darwin Port Corporation, Darwin, Australia

Bertini, I. (2014). *Australian National University Divests from Fossil Fuels Amid Government Criticism*. Available at http://blueandgreentomorrow.com/2014/10/13/australian-national-university-divests-from-fossil-fuels-amid-government-criticism/.

Blewett, R. S. (2012). *Shaping a Nation: A Geology of Australia*. Canberra: Geoscience Australia and ANUE Press.

Bureau of Meteorology. (2013). *Special Climate Statement 43–Extreme Heat in January 2013*. Available at: http://www.bom.gov.au/climate/current/statements/scs43e.pdf.

Bureau of Meteorology. (2014). *Climate Classification Maps*. Available at: http://www.bom.gov.au/jsp/ncc/climate_averages/climate-classifications/index.jsp?maptype=kpngrp

Chhetri, P., Hashemi, A., Basic, F., Manzoni, A. and Jayatilleke, G. (2012). *Bushfire, Heat Wave and Flooding – Case Studies from Australia*. Report from the International Panel of the WEATHER project funded by the European Commission's 7th framework programme.

Climate Institute. (2012). *Coming Ready or Not: Managing Climate Risks to Australia's Infrastructure*. Sydney: The Climate Institute.

CSIRO. (2007). *Climate Change in Australia*. Technical Report, CSIRO and Bureau of Meteorology.

CSIRO and Bureau of Meteorology. (2012). *State of the Climate*. Available at: http://www.csiro.au/en/Outcomes/Climate/Understanding/State-of-the-Climate-2012.aspx

Department of Energy. (2015). *Carbon Capture and Storage from Industrial Sources*. Available at: http://energy.gov/fe/science-innovation/carbon-capture-and-storage-research/carbon-capture-and-storage-industrial

Kinrade, P. and Justus, M. (2008). *Impacts of Climate Change on Settlements in the Western Port Region: Climate Change Risks and Adaptation*. Melbourne: Marsden Jacob Associates, CSIRO, Western Port Greenhouse Alliance.

Maddocks, Hassel and Hyder. (2010). *Climate Change and the Transport Sector: Are we Travelling in the Right Direction?* Available at: http://www.maddocks.com.au/uploads/articles/climate-change-and-the-transport-sector-are-we-travelling-in-the-right-direction-update-november-2010.pdf.

Maunsell Australia Pty Ltd. (2008). 'Impact of climate change on infrastructure in Australia and CGE model input', *Garnaut Climate Change Review*.

McEvoy, D. and Mullett, J. (2013). *Enhancing the Resilience of Seaports to a Changing Climate: Research Synthesis and Implications for Policy and Practice*. Enhancing the resilience of seaports to a changing climate report series, National Climate Change Adaptation Research Facility, Gold Coast.

McEvoy, D., Mullett, J., Millin, S., Scott, H. and Trundle, A. (2013). *Understanding Future Risks to Ports in Australia*, Enhancing the resilience of seaports to a changing climate report series, National Climate Change Adaptation Research Facility, Gold Coast.

Ng, A.K.Y., Chen, S. L., Cahoon, S., Brooks, B. and Yang, Z. (2013). 'Climate change and the adaptation strategies of ports: The Australian experiences', *Research in Transportation Business & Management*, 8: 186–194.

Nursey-Bray, M., Blackwell, B., Brooks, B., Campbell, M., Goldsworthy, L., Pateman, H., Rodrigues, I., Roome, M., Wright, J., Francis, J. and Hewitt, C. (2013). 'Vulnerabilities and adaptation of ports to climate change', *Journal of Environmental Planning and Management*, 56 (7): 1021–1045.

Ports Australia. (2014). *Map of Australian Ports*. Available at: http://www.portsaustralia.com.au/assets/Uploads/Ports-Australia-map.pdf

Queensland University of Technology (QUT). (2010). *Impacts and Adaptation Response of Infrastructure and Communities to Heatwaves: The Southern Australian Experience of 2009*. National Climate Change Adaptation Research Facility, Gold Coast.

Reisinger, A., Kitching, R. L., Chiew, F., Hughes, L., Newton, P.C.D., Schuster, S. S., Tait, A. and Whetton, P. (2014). 'Australasia'. In: *Climate Change 2014: Impacts,*

Adaptation, and Vulnerability. Part B: Regional Aspects. Contribution of Working Group II to the Fifth Assessment Report of the Intergovernmental Panel on Climate Change (Barros, V. R., Field, C. B., Dokken, D. J., Mastrandrea, M. D., Mach, K. J., Bilir, T. E., Chatterjee, M., Ebi, K. L., Estrada, Y. O., Genova, R. C., Girma, B., Kissel, E. S., Levy, A. N., MacCracken, S., Mastrandrea, P. R. and White, L. L., eds.). Cambridge, UK and New York, NY: Cambridge University Press, pp. 1371–1438.

Scott, H., McEvoy, D., Chhetri, P., Basic, F. and Mullett, J. (2013). *Climate Change Adaptation Guidelines for Ports*, Enhancing the resilience of seaports to a changing climate report series, National Climate Change Adaptation Research Facility, Gold Coast.

Steffen, W., Hunter, J. and Hughes, L. (2014). *Counting the Costs: Climate Change and Coastal Flooding*. Potts Point, New South Wales: Climate Council of Australia.

Wiltshire, A. (2014). *Governance and Legislative Issues for Critical Infrastructure Adaptation to Climate Change: Case Study – Ports*. Victorian Climate Change Adaptation Research (VCCAR). Available at: http://www.vcccar.org.au/publication/research-paper/case-study-report-governance-and-legislative-issues-for-critical

14 A decision support toolkit for climate resilient seaports in the Pacific region

Darryn McEvoy, Jane Mullett, Alexei Trundle, Andrew Hunting, Daniel Kong and Sujeeva Setunge

Introduction

A changing climate, as represented by a projected increase in the frequency and intensity of extreme events and longer-term changes to climatic variables, will act to amplify the challenges facing societies in the twenty-first century (IPCC, 2014). One of the key societal challenges will be to ensure the ongoing resilience of critical infrastructure assets that provide the essential services that support the modern functioning of communities; ensuring a continuance of economic wellbeing, social fabric and security (see, for example, O'Rourke, 2007). However, although there has been longstanding attention paid to the importance of infrastructure in contributing to social and economic prosperity, consideration of the resilience of critical infrastructure to future climate risks is an altogether much more recent endeavour (US Department of Energy, 2013; Cox et al., 2012; CSIRO, 2006). One such critical system with potential vulnerability to climate change is that of transport. As noted by Ng et al. (2013), there has been very little attention paid to climate risks and adaptation strategies for the transport sector (focusing on Australia in their study), and even less paid to ports, shipping and supply chains. There are however signs that the impacts of the changing climate on seaports are beginning to be considered, particularly in places where ports have been impacted by recent extreme events such as heavy rainfall, storms and storm surge (UK Department of Transport, 2014; Smythe, 2013).

As evidenced in the international academic literature (see, for example, Becker et al., 2011) and in strategic review papers such as the Australian National Adaptation Research Plan for Settlements and Infrastructure (Cox et al., 2012), seaports have been identified as being both vital to future prosperity and considered potentially highly vulnerable to future climate-related impacts due to their often coastal locations. However, while issues relating to sustainability and the reduction of greenhouse gas emissions (mitigation) have gained traction with port authorities in recent times, for example as evidenced through the development of green guidance documents (European Sea Ports Organisation) and the establishment of sectoral sustainability groups (for example, Ports Australia Environment and Sustainability Working Group), as yet the technical detail relating to future climate risks and their integration or mainstreaming into sectoral risk management decision-making processes remains largely absent.

However, an analysis of relevant grey literature indicates a growing awareness within the international ports sector that climatic hazards should be considered as part of a broader spectrum of risks in order to ensure more informed and comprehensive risk assessments and an enhanced resilience of port assets and functions into the future. For instance in the Australian context, the National Ports Strategy (Infrastructure Australia, 2011) emphasises the strategic importance of the sector and for the first time stated that the strategy should involve the development of 'scenarios for the impact of changes such as demography, climate, and energy for planning consideration' (Infrastructure Australia, 2011: 20). The review of relevant literature also provided evidence of changing risk perceptions within the international seaports and shipping industries. For instance, in 2008 the World Association for Waterborne Transport Infrastructure published a paper that examined the potential climate impacts on maritime and inland navigation as well as scoping out possible adaptation options (PIANC, 2008). Other recent relevant documentation includes a series of risk assessments carried out by a number ports in the UK (see, for example, Peel Ports Group, 2011) and reviews of climate-adaptation measures potentially available to seaports (IAPH, 2011; UNCTAD, 2011).

This chapter addresses this contemporary agenda by reporting on recent research activity that was carried out in 2013/14 to tailor a climate-change adaptation decision support tool for seaports in the Pacific region (Suva in Fiji and Port Moresby in Papua New Guinea). This activity, funded by USAID, built on a proto-type toolkit that had been previously developed for the Australian context (McEvoy and Mullett, 2013).1 In line with the assessment carried out by Stenek et al. (2011) on the Cartagena port facility in Columbia, the development of the decision support resource was framed by a system-based and integrated approach which took account of both functions and infrastructural assets according to current-day vulnerabilities and future climate-related risks. The toolkit – *Climate-Smart Seaports (Pacific)* – aimed to support climate change strategic planning by port managers by providing observed and future projected climate data for the port region alongside a series of assessment tools and guidance materials. The development of the decision support toolkit, plus elaboration of its content, forms the basis of this chapter.

The Pacific context

Pacific nations are geographically isolated and have population numbers that are relatively small (see Figure 14.1). The total population for the Pacific Oceania region (Melanesia, Micronesia and Polynesia) was estimated to be approximately 10.5 million in mid-2013, with a land mass around 500,000 square kilometres within a region that covers 30 million square kilometres (ADB, 2013).

Seaports are therefore important nodes in Pacific supply chains. Throughout the region, trade imports are generally of greater monetary value than exports and include imported food and materials necessary to sustain modern life; while

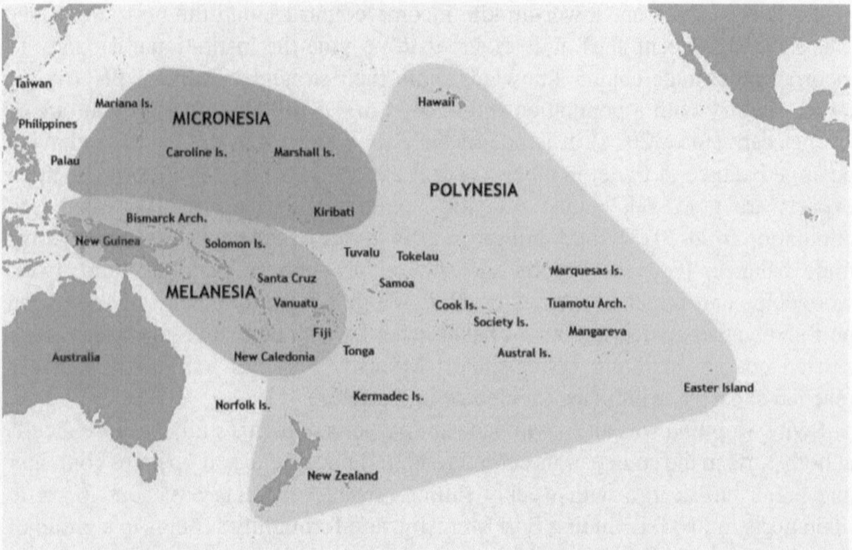

Figure 14.1 Oceania, including case-study countries, Fiji and Papua New Guinea
Courtesy of Kahuroa; http://commons.wikimedia.org/wiki/File:Pacific_Culture_Areas.jpg.

exports focus on agriculture, fishing and mining. Tourism is another major source of income and linked to this the cruise ship industry is growing; leading to an impetus for port upgrades to cater for larger ships and additional customs clearance areas. Most port business is international as inter-island trade is minimal, although there can be significant local transportation of people between islands (see Figure 14.2 for regional vessel movement). The seaports that manage international freight are mainly owned by government with occasional private partners; however there are also single product export ports that are solely privately owned.

Seaports in the Pacific region are currently undergoing a period of transformation. Past import and export operations have been generally inefficient due to a lack of modern equipment and, in many cases, a deteriorated condition of wharves and associated facilities. However, there has been a trend toward port renewal funded mainly by international aid in partnership with government. That said, while infrastructure upgrades are underway for many ports there are still many stressors affecting the sector before factoring in climate change, for example isolation and small volumes of trade which impact the viability of shipping lines that service the islands, lack of access to training which acts as an impediment to progress, and the presence of corruption in some Pacific nations (ADB, 2007).

The two seaport case studies that were selected for the project service the capital cities of Fiji and Papua New Guinea (PNG), Suva and Port Moresby respectively, where the head office for each port authority is also located. Both Suva Port and Port Moresby Port operate a mix of container, bulk and break bulk operations; however other ports around the islands operate the majority of the bulk trade. The two nations are the biggest economies within the Melanesian area of the Pacific, both are lower-middle income countries, and the port authorities thus have significant staff and resources to provide the institutional capacity to incorporate climate change knowledge into their strategic planning. PNG is the larger country with a population (mid-2013) of 7.3 million and a growth rate of 2.3 per cent (SPC, 2014). It is one of the only countries in the Pacific to have a positive balance of trade; in other words it exports more than it imports. Its main exports are gold, oil, copper ore, logs, palm oil, coffee and fish. Fiji has a population (mid-2013) of 0.8 million, a growth rate of 0.8 per cent and a negative trade balance. Its main exports are mineral products, sugar, fish, gold, kava, woodchips and mineral water. Suva Port is Fiji's biggest container and general port and was upgraded in 2005. It has five berths with continuing upgrades being carried out on operating systems. Port Moresby Port has four berths and has reached capacity, with plans to relocate in the future.

Swire Shipping was also approached to participate in this study because it calls at both ports in the course of its commercial trading activities. It operates container and break bulk cargo with weekly sailings from Australian east coast ports to main ports in PNG, including Port Moresby, and fortnightly sailings to a group of Island ports, including Suva. While seaports are all likely to experience impacts from severe storms from time to time, and have an understanding of the broad implications of sea-level rise, the other effects of the changing climate are less immediately obvious. Previous research into the Australian seaport sector (McEvoy and Mullett, 2013) identified that many ports were concerned with potential impacts to the sea-ward side of their operations rather than the land-ward side. Swire Shipping provided valuable insights into some of the practical risk issues from a shipping company perspective.

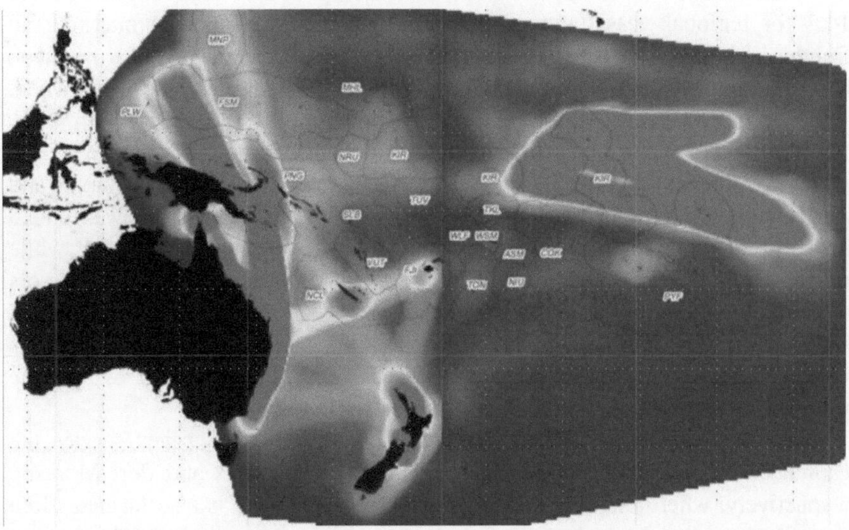

Figure 14.2 Satellite imagery of vessel movement throughout 2013.

Courtesy of SPREP

Climate change is already having an effect on the Pacific region, through the increased intensity of extreme climate events. This observation is confirmed through work undertaken by the Australian Bureau of Meteorology (BoM) and the Commonwealth Scientific and Industrial Research Organisation (CSIRO) in collaboration with Pacific national meteorological organisations over an extended period of research from 2008 to 2014. The most recent report from this project *Climate Variability, Extremes and Change in the Western Tropical Pacific* (BoM and CSIRO, 2014) notes that on a regional scale observations show a persistent mean annual warming trend of 0.18°C since 1961, as well as significantly more warm days and nights2 (BoM and CSIRO, 2014: 4). There has been an increase in both sea-surface temperatures and ocean acidification and in overall sea level. The data were generated using the Coupled Model Inter-Comparison Project Phase 5 (CMIP5) global climate model projections for individual Pacific countries, over four 20-year time periods centred on 2030, 2050, 2070 and 2090, and using four emissions scenarios (very low, low, medium and very high emissions). At the regional scale, climate projections indicate an increase in average temperatures and corresponding increase in extremely hot days and warm nights. In terms of rainfall, there will be high variability as expressed through an increase in annual rainfall, fewer droughts indicated for most areas, coupled with an increase in the number of extreme rainfall events. There are also likely to be fewer cyclones forming, though those that do occur are likely to be at the more severe end of the existing scale for cyclone intensity (ibid.: 5).

A holistic consideration of the elements at risk shows that the network of roads and bridges which connect many ports to their island hinterlands are affected by extreme rainfall and intense cyclones and storms; and it is important to note that recovery from impacts to the transport network can actually take longer than recovery of the port itself (UNCTAD, 2012: 23). The agricultural export trade can also be affected by climate variables such as floods, cyclones and droughts (Morrell and Scialabba, 2009), and the tourism industry is vulnerable to the impact of coral reef degradation from ocean warming and acidification. So far, research has concentrated primarily on future impacts to the agricultural sector and on disaster risk reduction to intense climate events. However, it is important to note that the accumulated impacts of increasing small and medium-size climate-related events over time can be just as important as a single large impact (McEvoy and Mullett, 2014).

Previous research by McEvoy and Mullett (2013) had also identified a gap in the knowledge of small to medium-sized seaports in relation to accessing and prioritising the climate information that was available to them, and then how to operationalise this knowledge. As a result, a prototype decision-support toolkit was developed for Australian conditions, funded by the Australian National Data Service (ANDS), and it is this resource that formed the platform for the Pacific region work.

The Climate Smart Seaports (Pacific) toolkit

The project for the Pacific region, funded by USAID, ran from late 2013 through 2014 and consisted of three main strands of activities: 1) collating and managing the data that are incorporated into the toolkit; 2) further developing the software and structuring the online toolkit processes; and 3) working closely with the case-study stakeholders to ensure that the toolkit design was usable and fit for purpose. The prototype decision-support toolkit, Climate Smart Seaports (CSS), which had already been built for the Australian context, was customised to meet the specific needs of seaports in the Pacific region. The refined toolkit, called Climate Smart Seaports (Pacific) (CSS-Pacific), was developed as open source software.3 It uses the 'Java Development Kit' with Java applications including 'Eclipse' and 'Maven', with the web application organised around Spring Model-View-Controller (Spring MVC). The reporting structure of the toolkit provides the user with a great deal of flexibility, and the design is such that it is possible to add other tools and data in the future according to the evolving needs of the port authority (Figure 14.3).

The content of the toolkit comprises a set of appropriate climate and marine data from the Australian Bureau of Meteorology (BoM) and the Commonwealth Scientific and Industrial Research Organisation (CSIRO), as well as from island

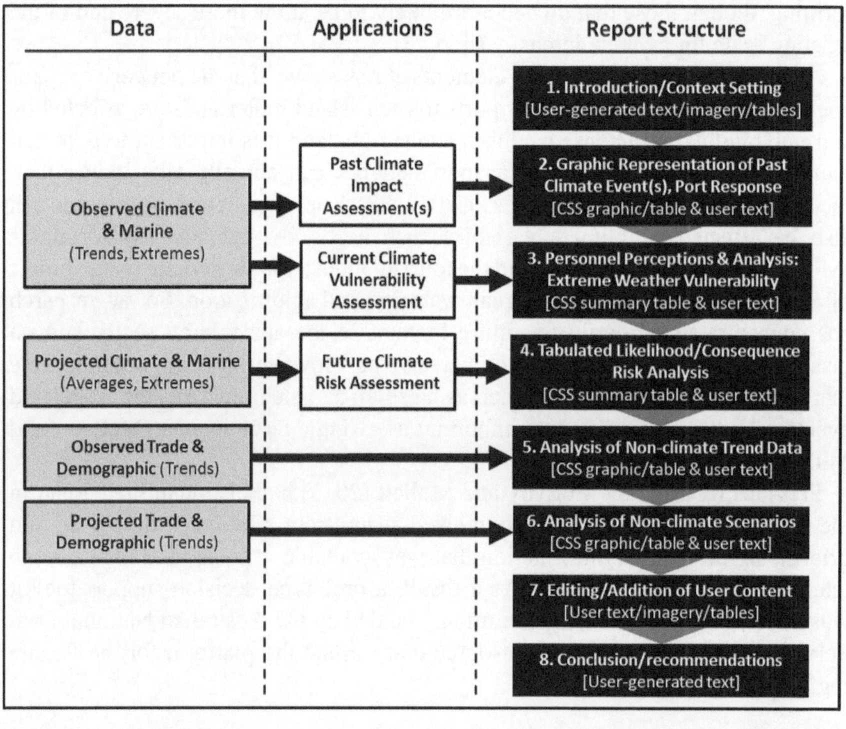

Figure 14.3 Representation of the internal processes of the toolkit informing report production.

nation meteorological services. The climate and marine data are augmented with non-climate data covering trends in trade patterns as well as demographic data. The content was chosen in order to ensure that the decision-support tool carries out an integrated assessment of risks across the whole of the port business, considering ports as systems and as nodes in a logistics supply chain; as well as recognising that climate change is only one of the many drivers affecting the functioning of ports. A framing of integrated assessment as the bringing together of different knowledge – quantitative, qualitative and participatory – therefore underpinned the practice of this project. In terms of the participatory element, iterative feedback from both Fiji and PNG port authorities and from Swire Shipping was gathered through structured workshops in order to feed into the toolkit development throughout the lifetime of the project. Such a 'bottom-up' perspective to problem framing/solving was considered important as there was particular recognition that the expert input and knowledge of the port authorities and other key stakeholders would make an important contribution to toolkit development and its subsequent application.

When using the toolkit for the first time, users choose a nation and then select a particular port. This then opens up a tailored set of relevant data. The data embedded in the toolkit are couched within a fluid interactive framework that ultimately produces a simple formatted report that can be used by port personnel as decision-support for better-informed climate risk management decisions. To achieve this, the toolkit supports the user through a guided process that identifies climate and non-climate stressors by:

- Collating and displaying observed climate and marine data, providing a context of recent climate and marine trends, and identifying a set of plausible future climates and sea levels for a geographical region;
- Providing a strategic context of socio-economic trends, including observed trade trends (where possible) for the case-study port and demographic trends for the defined region;
- Providing a series of hands-on applications that are intended to elicit information from internal workshops, which is then fed back into the tool to inform the final report.

Climate data from the BoM and the CSIRO was sourced from the Pacific-Australia Climate Change Science and Adaptation Planning Science program (PACCSAP), which from 2008 to 2014 worked with Pacific Island meteorological offices to build climate science capacity in the region (Power et al., 2011). This program examined past climate trends and variability and provided detailed regional and national projections for the climate and the ocean in the Pacific region. As a result of access to this rich source of data, the CSS-Pacific project was able to use both the climate data available through the published reports from PACCSAP and compatible climate data provided by the Fiji Meteorological Service for specific variables (such as humidity) that were of particular relevance to the engineering research component of this project (discussed later in this chapter). The data included projected climate variables for temperature and rainfall, and sea-level rise futures for three emissions scenarios (low, medium and high) centred on three time periods (2030, 2055 and 2090; relative to 1990) for Suva, Fiji and Port Moresby, PNG. Corresponding data for observed climate was also included.

Trade and demographic data were provided by the Fiji Statistics Office and the Secretariat of the Pacific Community (SPC). These data are generally too coarse to provide specific insights for each port as they are data collated for the entire nations of Fiji and PNG. Accessing non-climate data was in many ways as challenging as accessing the climate data. As most data is aggregated to the whole island, isolating data to the level of the two ports proved difficult. However, for future use of the tool, each port has individual access to more precise data that can be imported into the tool at any stage in the assessment process.

Participatory tools: co-generation of knowledge

There are a variety of application 'tools' within the toolkit that serve as an intermediary analytical resource between the introduction of scientific and economic 'top-down' data and the production of a report that incorporates 'bottom-up' knowledge and perceptions of the condition of the port infrastructure, its business, and current vulnerability to climate variables. These participatory tools enable port personnel to input their own data in order to build their collective corporate knowledge of the consequences of past climate impacts on the port's structures, workforce and business, to discuss present vulnerabilities to climate variability, and to identify and prioritise potential future risks from a changing climate.

Three such applications were tested at workshops held in Suva and Port Moresby, and were refined as a result of feedback from the participating groups. These include an interactive display of the impact of past climate events (past), a vulnerability matrix that encompasses the range of activities conducted by a seaport, from supply-chain movement through to workforce readiness (present), and a risk assessment format that includes a consequences and likelihood table (future). Matrices can be downloaded to use in a workshop setting and the salient data uploaded to a table that appears in the final report. Figure 14.4 depicts a spider graph which provides a visualisation of people's perceptions of the impact of a particular extreme weather event on port business. This enables comparison of impacts across the different areas of a port's business.

Two workshops were held in each case-study location, separated by several months. Middle management from the port authorities were represented from across a range of departments, including human resources, occupational health and safety, engineering, operations, and finance personnel. Additional briefing sessions were held individually with senior personnel. The first workshop served to introduce climate-change science, and the toolkit and its concepts, in order to get feedback on the issues that were of most importance to the port authorities. Local experts were also invited to present on the climate trends and projections for the port where the workshop was held. The second follow up workshop then

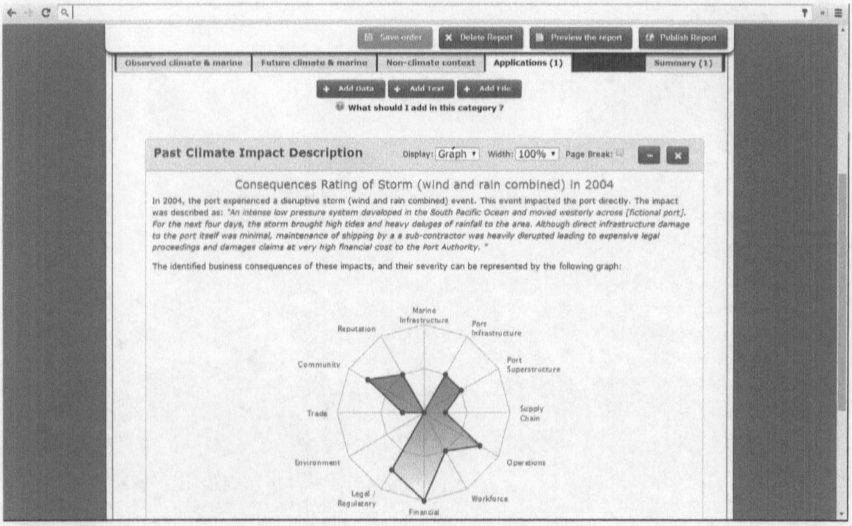

Figure 14.4 Representation of one of the toolkit's functions.

presented the modified toolkit to the participants and facilitated hands-on use of it to ensure that the toolkit was easy to use and 'fit-for-purpose'. In both workshops the three applications previously discussed were presented and made subject to feedback on their applicability.

The Suva Port workshops were shaped by the public–private partnerships that existed. As well as managers from sections within Fiji Ports Corporation Limited (FPCL), managers attended from the private operations body (Ports Terminal Ltd, PTL) and from Fiji Ports Corporation Limited subsidiary body Fiji Ships and Heavy Industries Ltd (FSHIL), which specialises in ship repairs and maintenance. This introduced a strong on-the-ground experience into the Suva workshops. In the case of Port Moresby, the input of the PNG Government Office of Climate Change and Development (OCCD) at the workshops introduced a strong policy-oriented perspective. The participants in both locations were generous with their time and insights (Figure 14.5).

The accessibility of the toolkit benefits from its use of a temporal progression from past (known) climate events to current impacts before then embarking on an analysis of future risks. Also, the applications have been customised – with the aid of the workshops – to better reflect sectoral knowledge, providing a 'known' element, or building block, to offset dealing with the relative 'uncertainty' of climate-change data. In the workshop setting, all groups started off by using the toolkit to develop a set of visual graphs that depicted their combined understanding of a past extreme climate event. Similarly, all groups were familiar with the process of creating risk assessments in their own areas of work, which also added to their comfort level when using the toolkit applications.

While the workshops were invaluable for contributing expert knowledge that could then be incorporated into the toolkit, there were other unintended benefits.

Figure 14.5 Members of PNG Ports and the OCCD (l) and FPCL, PTL and FSHIL (r) work through the toolkit.

Participants from both nations remarked that the workshops had brought managers from across different areas of port business together and that it had been a useful exercise to hear how the climate affected the diverse areas of the port business in different ways. This observation supports the basic tenet of this project, that adapting to a changing climate needs to occur through an integrated 'multi-actor' process. In addition, it was observed at both workshops that the toolkit provided impetus to the need for systematic gathering of climate data at source, and also to build more robust analysis of climate impacts into their daily work.

The first workshops included a presentation on observed climate and future projections. In Suva this was provided by the Fiji Meteorological Service and in Port Moresby by the Office of Climate Change and Development. Surprisingly, there was limited knowledge (on the part of the ports personnel) of the capacities of their local meteorological services. As such, this project facilitated a greater understanding of the capabilities of these services and brokered new lines of engagement between local climate information providers and an important end user community.

Engineering application: accounting for climate change in deterioration models

In addition to the participatory 'hands-on' tools, more specialised technical elements are also capable of being incorporated into the decision support toolkit. One additional application that was developed as part of the project targets the deterioration of infrastructure assets. An engineering group from RMIT University worked on incorporating climate-change variables into existing concrete deterioration models, collaborating with Fiji Ports Corporation Limited to produce detailed data for Suva Port. This engineering 'application' is an example of how more sophisticated tools can also be incorporated into the toolkit as needed to provide additional technical detail in support of decision making.

Seaport structures are subject to various deteriorating agents throughout their lives. The engineering application calculates and presents numerical probabilistic projections on the deterioration rates of concrete assets, allowing the lifespan of a large number of port assets – such as concrete berths, piers or piles – to be analysed. This exercise can be undertaken through the examination of a 'typical' pile within a port precinct assuming common deterioration across neighbouring and similar assets, or alternatively, modelling can be localised down to an individual exposure zone of a particular pile of interest.

The case-study ports are located in significantly corrosion-prone areas given the tropical environment, exposure to salinity and the presence of chlorides (Emmanuel et al., 2012). This combination of the environment and a plentiful supply of chlorides makes the port structures in Fiji and PNG highly susceptible to chloride ingress (Bastidas-Arteaga et al., 2010). This phenomenon is currently costing Western countries around 2.5 per cent of their national annual GDP, or costing the world economy over $1.8 trillion a year (Schmidt, 2009). The case study concentrated on a number of concrete berths at Suva Port. Restricting the analysis to concrete structures only was a practical decision based on the non-availability of design data for wooden or steel assets at the port. Mechanisms for corrosion are typically accelerated by humidity, temperature, carbon dioxide, environmental salinity, and increased wetting and drying cycles (precipitation). As the climate changes, and in many cases becomes more aggressive towards concrete assets (Bastidas-Arteaga et al., 2010), there is a need to better understand the potential for increased deterioration rates of vital assets.

The modelling work carried out for Suva Port builds on the methodology that was developed as part of the Australian seaports study mentioned previously in this chapter. This original work involved a refinement of existing deterioration models to incorporate climate data into modelling runs in order to determine changes to deterioration rates when impacted by changes to climate-related variables. Changes to environmental variables will directly impact the deterioration of concrete, primarily through chloride ingress and carbonation intrusion (Kong et al., 2013, 2012). For this study, the modelling of changes to chloride ingress was used.[4]

The climate data relevant to the Pacific – generated by the CSIRO through their Pacific Futures modelling resource – differs in methodology and timescale to the previous Australian scenarios (to 2090). This is a result of international changes adopted by the scenario modelling community brought about by the introduction of the 5th assessment report of the InterGovernmental Panel on Climate Change (IPCC, 2014). To consider a range of possible climate futures for Suva, the two climate futures (A1B and A2) were analysed under three 'scenarios' – mean, upper and lower – to illustrate the possible ranges of these variables. Thus, the emission scenarios used by the engineering model were different to those used by the other application tools within the *Climate-Smart Seaports (Pacific)* decision-support tool. Although every effort was made to align the data for all tools, because of the complexity of the engineering model it was not practical to rebuild this model completely. This is one of the unfortunate consequences of the speed of changes and improvements to climate data that are derived from evolving global climate models.

While the attention of the engineering analysis was on longer-term changes to climate variables, and their impact on the deterioration of concrete assets over a period of time, it is also worthwhile noting that other 'instantaneous' events such as rehabilitation of the asset, overloading and environmental extremes (storm surges etc.) will also impact on the lifespan of the assets. As the future frequency and magnitude of these events is still uncertain, the modelling of the deterioration relied on modelling average increases due to the long-term effects of climate change.

In order to deliver concise reporting to the port's asset managers, the model required a wide breadth of inputs which involved integrating climate data with the particular requirements of the deterioration modelling (for an overview of some of the challenges facing the translation of climate data for adaptation planning more generally, see McEvoy and Mullett, 2014) and physical sample data. Obtaining accurate local data was problematic; not due to a reluctance to share data by the relevant authorities, but rather a lack of accurate asset portfolio records and adequate tracking of chloride ingress over a period of time. An improvement to data collection and its storage by port operators would assist in asset management in general and improve the accuracy of the results from the model.

Typically, within port infrastructure maintenance operations, when an element reaches a critical trigger (for example based on detected crack width in concrete), a maintenance or replacement regime will commence. The resilience of a seaport asset to climate change is therefore usefully represented by the difference between the deterioration curve of a baseline climate and one that is accelerated by climate change. This can provide a useful illustration of the revised serviceable life of the asset in question. Figures 14.6 and 14.7 below show the quantitative difference in projected corrosion induced by chloride intrusion for a given asset within Suva Port for a given climate future. Interestingly, Figure 14.7 details a significant difference between the baseline and the climate-impacted deterioration curve. For a given probability of a corrosion initiation trigger (for example, 25 per cent) the model indicates a reduction in serviceable lifespan of the asset by over a decade.

Results indicate that existing concrete assets are already experiencing accelerated deterioration due to current changes in climate. As the possibility of more substantial variances in climate futures continue to present themselves, we may see much more profound effects on the deterioration curves. Further, it was found that chloride-induced corrosion is significantly affected by the sea-surface temperatures, with the splash and tidal zones being the most susceptible areas of the asset due to wetting and drying cycles. This is important as the exposure of assets will alter over time due to sea-level rise and increased levels of storm surge as a result of climate change.

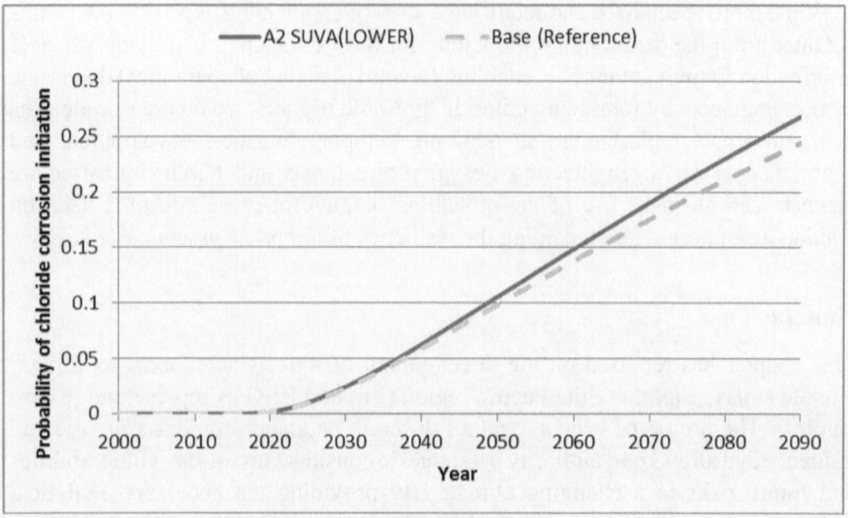

Figure 14.6 1,200mm pile with A2 climate (- extremity).

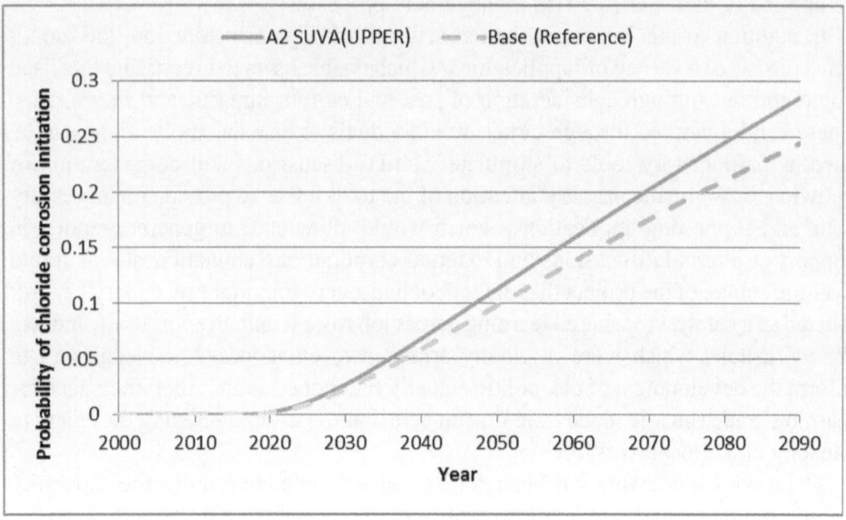

Figure 14.7 1,200mm pile with A2 climate (+ extremity).

Any ambiguity in the collection of data given the variable nature of climate records should not act as an inhibitor or justification for not carrying out adaptation measures. Remedial options for chloride ingress vary from enlarged cover distance and modified concrete mix composition (in the design phase) to retro-fitted cathodic protection and paint-on coatings. The suitability of these remedial options depends significantly on cost and the stage of the deterioration, which demands that more detailed assessments need to take place on a case-by-case basis.

While more specialised and requiring a level of engineering expertise, the results obtained from the deterioration modelling for the Pacific context provide valuable information for port authorities, enabling forward planning of operational downtime by port engineers by foreseeing chloride threshold triggers; from core sample tests to a full wharf replacement. In addition, shipping logistics, development and contracting services, engineering design professionals and public infrastructure planners can all make use of the modelling outputs for more informed decision making in support of strengthening the resilience of important assets.

Conclusion

This chapter has reported on the development of a decision support toolkit for climate smart seaports in the Pacific region (Fiji and PNG as representative case studies). The prototype brings together different data and provides structure and guidance that allows port authority personnel to consider current-day vulnerabilities and future risks to a changing climate. By providing the necessary analytical framework and reporting capability the toolkit not only enables port authorities to access climate and non-climate information more easily but also helps to make sense of the complexity of climate science and how such information can usefully be applied in their existing risk management processes.

In addition to this important data access and management function, the toolkit also consists of a variety of 'applications' which enable users to investigate localised vulnerabilities through consideration of past and current impacts, and future risks. These can be worked through either by individuals or else the applications can be used as participatory tools to stimulate internal discussions and co-generation of knowledge. While the primary intention of the toolkit was to provide the necessary data, and supporting applications, which would allow staff to generate reports in support of internal discussions and external corporate dissemination, it was found over the course of the project that the toolkit had a very important secondary benefit: it acted as a catalyst for shared learning across job roles within the port itself. Indeed, the workshops which were originally arranged to elicit local knowledge and to inform the development of the toolkit, actually functioned as an important 'space for learning' and arguably made a significant contribution to strengthening the adaptive capacity of the local workforce.

The toolkit framework has been deliberately designed to provide the capability to incorporate further 'application' tools into the toolkit in an iterative fashion (or alternatively could be customised to serve other critical sectors, such as energy or agriculture). This chapter has showcased one such technical application, the integration of climate-change variables into deterioration models for concrete; a resource which will enable port engineers to consider risks and adaptation options for seaport assets and how changes may affect maintenance regimes over time. Given the complexity of simulating future climatic conditions for individual assets the task of understanding the direct effects of climate change on seaport performance can be challenging. As such, during the final development phases of the advanced probabilistic modelling, the input methodology was converted into a user-friendly software tool which can be applied locally by port engineers. It has also been designed to ensure that the resource can easily be adapted for different types of future analysis.

Workshop participants identified the potential for the toolkit to be used as a record-keeping device. Two uses suggested by the workshops were, firstly to track climate impacts across all areas of port business and secondly, act as a platform for integrated port planning across all ports managed by the port authority. Participants at both workshops were also keen to investigate the possibility of incorporating their own data into the toolkit, using the reporting function as a framework for regular collection and analysis of climate impacts as well as adaptation initiatives; thus making the resource more comprehensive and multi-purpose (though this is an undertaking that was beyond the scope of the USAID project).

While the climate smart toolkit was successfully tailored for use by both case-study ports, there remains significant potential to further build on its strengths. Firstly, there is interest in extending the scope to provide coverage for all Pacific Island nations and for it to be truly Pacific-wide. However, issues of ownership and longer-term sustainability of the online resource will need to be addressed. Secondly, the range of applications could be added to and refined with greater consideration paid to adaptation decision making (moving beyond risk assessment towards implementation). Thirdly, there is further potential to improve on the data contained in the toolkit, for example extended supply-chain, appropriate wind and wave swell data for shipping, and climate extremes analysis. These remain important knowledge and operational gaps to be addressed in the future.

Acknowledgements

The development of the climate smart seaports toolkit was generously supported by USAID, 'from the American people', under their Climate Change Resilient Development project (activity number CCRDACD005).

The generosity of the management teams at Fiji Ports Corporation Ltd and PNG Ports Corporation Ltd is also recognised.

Notes

1 The original work was supported by the Australian National Data Service. USAID subsequently funded the Pacific version.

2 Countries covered by this area include: Cook Islands, Federated States of Micronesia, Fiji, Marshall Islands, Nauru, Niue, Palau, Papua New Guinea, Samoa, Solomon Islands, Tonga, Tuvalu, Vanuatu and East Timor.

3 A full description of the source coding and documentation of the build process is available at: https://github.com/eresearchrmit/seaports-pacific

4 Chloride is one of the major negative ions in seawater. It assists in the rusting of the internal metal framework of concrete structures. Although this mechanism is not fully understood, it is believed that chloride itself doesn't cause corrosion however it may expedite the process. Chloride penetrates the passive oxide film layer on steel, leaving it vulnerable to corrosion (with the presence of oxygen and moisture). It aggravates the corrosion conditions by catalysing oxidation.

References

ADB. (2007). *Oceanic Voyages: Shipping in the Pacific*. Asian Development Bank.

ADB. (2013). *The Economics of Climate Change in the Pacific*. Philippines: Asian Development Bank.

Bastidas-Arteaga, E., Chateauneuf, A., Sánchez-Silva, M., Bressolette, P. and Schoefs, F. (2010). 'Influence of weather and global warming in chloride ingress into concrete: A stochastic approach', *Structural Safety*, 32 (4): 238–249; doi: http://dx.doi.org/10.1016/j.strusafe.2010.03.002

Becker A., Inoue, S., Fischer, M. and Schwegler, B. (2011). *Climate Change Impacts on International Seaports: Knowledge, Perceptions, and Planning Efforts among Port Administrators*. Climatic Change Online First Online: http://unctad.org/sections/wcmu/docs/ahm2011_2_18_becker_en.pdf

BoM and CSIRO (2014). *Climate Variability, Extremes and Change in the Western Tropical Pacific: New Science and Updated Country Reports*. Pacific-Australia Climate Change Science and Adaptation Planning Program Technical Report. Melbourne, Australia: Australian Bureau of Meteorology and Commonwealth Scientific and Industrial Research Organisation.

Cox, R., McEvoy, D., Allen, G., Bones, A. and McKellar, R. (2012). *National Climate Change Adaptation Research Plan: Settlements and Infrastructure – Update Report*. National Climate Change Adaptation Research Facility, Gold Coast.

CSIRO (2006). *Infrastructure and Climate Change Risk Assessment for Victoria*. Australia: CSIRO.

Emmanuel, A., Oladipo, F. and Olabode, O. (2012). 'Investigation of salinity effect on compressive strength of reinforced concrete', *Journal of Sustainable Development*, 5 (6), 74–82.

IAPH. (2011). *Seaports and Climate Change – An Analysis of Adaptation Measures*. Hamburg, Germany: Port Planning and Development Committee, International Association of Ports and Harbours.

Infrastructure Australia (2011). *National Ports Strategy: Infrastructure for an Economically, Socially, and Environmentally Sustainable Future*. Canberra: Infrastructure Australia and the National Transport Commission, Commonwealth Government.

IPCC. (2014). *Climate Change 2014: Synthesis Report. Contribution of Working Groups I, II and III to the Fifth Assessment Report of the Intergovernmental Panel on Climate Change* (Core Writing Team, R.K. Pachauri and L.A. Meyer (Eds.)). Geneva, Switzerland: IPCC.

Kong, D., Setunge, S., Molyneaux, T., Zhang, G. and Law, D. (2012). 'Australian seaport infrastructure resilience to climate change', *Applied Mechanics and Materials*, 238: 350–357.

Kong, D., Setunge, S., Molyneaux, T., Zhang, G. and Law, D. (2013). *Structural Resilience of Core Port Infrastructure in a Changing Climate*. Gold Coast, Australia: National Climate Change Adaptation Research Facility.

McEvoy, D. and Mullett, J. (2013). *Enhancing the resilience of seaports to a changing climate: research synthesis and implications for policy and practice*. NCCARF report series. Gold Coast, Australia: NCCARF.

McEvoy D. and Mullett, J. (2014). 'Enhancing the resilience of seaports to a changing climate'. In: Palutikof, J. P., Boulter, S. L., Barnett, J. and Rissik, D. *Applied Studies in Climate Change Adaptation: Australian Experiences*. Oxford: Wiley Publishing.

Morell, W. and Scialabba, N. E. (2009). *Climate Change and Food Security in the Pacific: Policy Brief.* Rome: Food and Agriculture Organisation of the United Nations.

Ng, A.K.Y., Chen, S.-L., Cahoon, S., Brooks, B. and Yang, Z. (2013). 'Climate change and the adaptation strategies of ports: The Australian experiences', *Research in Transportation Business and Management*, 8: 186–194.

O'Rourke, T. D. (2007). 'Critical infrastructure, interdependencies, and resilience', *The Bridge*, 37, (1), Spring 2007.

Peel Ports Group. (2011). *Mersey Docks and Harbour Company Ltd. Climate Change Adaptation Report: Report to DEFRA under the Adaptation Reporting Powers*. Bretton, UK: Peel Ports Group.

PIANC. (2008). *Waterborne Transport, Ports and Waterways: A Review of Climate Change Drivers, Impacts, Responses and Mitigation.* Report for Task Group 3: Climate Change and Navigation. Brussels. Belgium: PIANC (World Association for Waterborne Infrastructure).

Power, S. B., Schiller, A., Cambers, G., Jones, D. and Hennessy, K. (2011). 'The Pacific Climate Change Science Program', *Bulletin of the American Meteorological Society*, 92: 1409–1411.

Schmidt, G. (2009). *Global Needs for Knowledge, Dissemination, Research, and Development in Material Deterioration and Control*. New York: World Corrosion Organisation.

Smythe, T. (2013). *Assessing the Impacts of Hurricane Sandy on the Port of New York and New Jersey's Maritime Responders and Response Infrastructure*. Quick Response Report No. 238: Final Report to the University of Colorado Natural Hazards Center. Quick Response Grant Program (National Science Foundation grant CMMI1030670), May 31, 2013.

SPC. (2014). *2013 Population and Demographic Indicators*. Secretariat of the Pacific Community Available at: http://www.spc.int/sdd/index.php (accessed December 1, 2014).

Stenek V., Amado, J.-C., Connell, R., Palin, O., Wright, S., Pope, B., Hunter, J., McGregor, J., Morgan, W., Stanley, B., Washington, R., Liverman, D., Sherwin, H. Kapelus, P., Andrade, C. and Pabon, J. D. (2011). *Climate Risks and Business: Ports-terminal maritimo Muelles el Bosque Cartagena, Colombia.* Washington DC: World Bank.

UK Department of Transport. (2014). *Transport Resilience Review: A Review of the Resilience of the Transport Network to Extreme Weather Events*. UK Department of Transport.

UNCTAD. (2011). *Ad Hoc Expert Meeting on Climate Change Impacts and Adaptation: A Challenge for Global Ports*. United Nations Conference on Trade and Development, September 2011.

UNCTAD. (2012). *Review of Maritime Transport 2012*. United Nations Conference on Trade and Development.

US Department of Energy. (2013). *U.S. Energy Sector Vulnerabilities to Climate Change and Extreme Weather*. US Department of Energy.

15 Canada's Arctic shipping challenge

Towards a twenty-first-century Northwest Passage

Michael C. Ircha and John Higginbotham

Introduction

Climate change and global warming have led to significant melting of the Arctic sea ice, creating long-term commercial shipping opportunities through Canada's Northwest Passage (NWP). For centuries, European explorers sought a shorter sea route to the riches of the Far East across the Arctic Ocean. This almost relentless search for a suitable route over the top of North America ceased in the 1870s with the opening of the Suez Canal, which significantly shortened the distance from India and East Asia to Europe.

Today, climate change has made commercial shipping from Asia to Europe and to the East Coast of North America via the Arctic Ocean a feasible, albeit seasonal, alternative to the major shipping routes through the Panama and Suez Canals. Authoritative estimates vary, but in as little as twenty years Canadian Arctic navigation could be transformed through a combination of melting ice, new icebreakers, ice-tracking technologies, and proactive national investment in marine corridors and safety infrastructure.

Since the 1980s Arctic temperatures have been increasing about 1°C per decade—about twice the global rate. As a result the extent of Arctic sea ice has reduced about 13 percent per decade and its thickness decreased by 42 percent over the past 30 years. The International Panel on Climate Change (IPCC, 2014: 4) predicts that the Arctic Ocean will be nearly ice-free during the summer months within this century. However, ice-free does not mean clear and unimpeded sailing. Arctic ice-free waters may still contain icebergs and floating ice chunks necessitating the use of ice strengthened vessels with extra propulsion power and possible icebreaker assistance (Østreng et al., 2013: 24). The IPCC (2014: 31) predicts that the summer sailing season will increase by a further ten days by 2020 and 20–30 days by 2080. The US Navy's recent *Arctic Roadmap* (US Navy, 2014: 11) predicts that by 2020–25 the NWP will have 200 commercial vessels traversing it per year while Russia's Northern Sea Route (NSR) will annually handle about 450. Canada's Commissioner of the Environment and Sustainable Development is even more optimistic in predicting 300 commercial voyages through the NWP by 2020 (Auditor General of Canada, 2014). Continued sea ice reduction will create opportunities for increased commercial shipping, oil, gas and mineral exploration and extraction, along with cruise, tourism and fishing activities in the north. National measures to facilitate destination shipping will simultaneously support transit traffic through the NWP.

During the past several years, ice-strengthened cargo ships have traversed both the NSR and NWP. In 2010, the ice-strengthened cargo vessel *Monchegorsk* sailed the entire NSR without icebreaker assistance (Østreng et al., 2013: 15). This was followed in 2011 with the 162,000-tonne tanker *Valdimir Tikhonov* carrying 120,000 tonnes of gas condensate from Murmansk to Thailand. The sailing distance on NSR was 40 percent less than the Suez Canal route and reduced the tanker's sailing time by a week (*The Economist*, 2012). In October–December 2012, the ice-class LNG tanker *Ob River* made a round trip through the NSR carrying LNG from Norway to Japan (Kassianenko, 2012). Traversing the NWP in September 2013, the ice-strengthened *Nordic Orion* carried 73,500 tonnes of coal from Vancouver to Finland (McGarrity and Gloystein, 2013). Using the NWP reduced the trip by 1,000 nautical miles compared to the Panama Canal route. In addition, the *Nordic Orion* was able to carry about 25 percent more coal on the NWP trip with its laden draught of 14 meters than it would have been able to carry through the shallower 12-metre deep Panama Canal. The *Nordic Orion* was the first commercial vessel to traverse the NWP. However, on this voyage it required Canadian Coast Guard (CCG) icebreaker escort. In 2014, Fednav's *Nunavik*, a Polar Class 4 vessel had the first commercial, non-icebreaker-assisted voyage through the NWP. It carried 23,000 tonnes of nickel concentrate from Northern Labrador to China. The NWP voyage reduced the full trip length by sixteen days. Unfortunately, ice set in early in the NWP and the *Nunavik* returned via the Panama Canal (Keane, 2014).

In addition to trans-Arctic trips, climate change creates opportunities for increased resource exploration and extraction. This is particularly the case in the NSR with its extensive offshore gas and petroleum drilling platforms. In the Canadian Arctic, mineral extraction is being developed in several locations and increased oil exploration activities are anticipated, particularly in the Beaufort Sea.

This chapter examines the capabilities of the NWP in supporting increased shipping, including ports, ship technology, search and rescue capability, navigational aids, and governance and regulatory concerns. The chapter also considers the shipping services provided in Russia's NSR in comparison with Canada's NWP.

Arctic shipping

Trans-Arctic routes offer shorter voyage distances between Europe and the US East Coast (USEC) and Asia. There are three potential routes across the Arctic Ocean: NSR, NWP and the Trans-polar route (TPP). The TPP crosses the central Arctic Ocean near the North Pole and is not expected to be a commercially viable shipping route alternative for non-icebreaker-assisted vessels within this century. Both the NSR and NWP lie along the southern fringes of the Arctic Ocean where commercial shipping can benefit from seasonal melting sea ice.

The NSR lies between Murmansk and the Bearing Strait along Russia's northern coastline and involves a series of routes varying from 2,200 to 2,900 nautical miles. The route selected depends on seasonal ice conditions (Østreng et al., 2013: 13). The more frequented routes along the southern portion of the NSR extend over shallow water, limiting their use to ships of 50,000 deadweight tons with draughts of 12.5 meters or less (Farré et al., 2014).

The NWP consists of various routes through Canada's northern archipelago from Baffin Bay on Davis Strait in the east to the Beaufort Sea in the west. The shortest, but most northerly route is from Parry Channel north of Baffin Island through the M'Clure Strait to the Beaufort Sea. However, as the M'Clure Strait is normally ice-bound year-round with hard multiyear ice drifting south from the high Arctic, this route will not likely be feasible for commercial shipping for the foreseeable future. The more navigable and commercially favored NWP route is from Parry Channel south and then west between Victoria Island-Banks Island and the northern mainland coast to the Beaufort Sea. The many islands and narrow straits of the NWP have more shore-based ice formation resulting in slower ice clearing during the summer months than on the more open waters of the NSR. As a result, the NSR is currently more attractive than the NWP as a northern passage for commercial vessels.

Canadian Arctic shipping can be divided into two main categories: destination and transit shipping. The most frequent northern commercial shipping is destination traffic. This shipping activity can be subdivided into several subcategories: resupply, mineral resources, fishing, government research, Canadian Coast Guard icebreakers, and cruise ships. Transit shipping from the Atlantic to/from the Pacific is slowly increasing.

Resupply cargo and tanker vessels carry consumer goods, fuel, equipment, and other supplies to remote northern communities. Most Canadian resupply vessels begin their voyages in Montreal and provide twice-a-year service to small northern villages. In many cases, marine resupply is the only option for providing northern communities with heavier, bulk commodities. The lack of suitable marine infrastructure requires resupply vessels to carry a landing barge on board to lighter goods to shore. The exception is Pangnirtung on Baffin Island where the federal government recently built a fixed wharf, sealift ramp, and breakwater to serve the local community (Fisheries and Oceans, 2013a). Landing operations normally take place at high tide on beaches close to the communities and depend on favorable weather conditions. Obviously this resupply method is slow, inefficient, and expensive, further contributing to the already high cost of northern living. Other periodic northern shipping includes government research vessels, coast guard icebreakers, and smaller fishing craft.

Serving mineral resource needs is a second major category of northern shipping. This subcategory includes ships used to support mineral extraction and oil and gas exploration. Typically, purpose-built, ice-strengthened bulk vessels serve northern mines operating from private terminals transporting exports and importing equipment and supplies. For example, Fednav's *Nunavik* is a purpose-built dry bulk carrier transporting copper and nickel concentrate from Deception Bay in northern Labrador to Europe. As pointed out by Tomas Paterson, Fednav's senior vice-president, "The ship is capable of breaking through 1.5 meters of level ice at three knots continuously, and therefore is able to trade into the Canadian Arctic on a year-round basis" (CBC, 2014). The anticipated multi-billion dollar Mary River iron mine on Baffin Island had initially planned on delivering product to Europe with a fleet of ten to seventeen ice-class Capesize ore carriers operating year-round from a new port facility at Steensby Inlet on the west side of Baffin Island (Jordan, 2013). However, recent changed economic circumstances are leading to a more modest mining proposal.

Two-thirds of Canada's Arctic's oil and gas resources are located offshore (Siron et al., 2009: 83). From 1970 to 1990, offshore petroleum exploration was undertaken in the Beaufort Sea. However, offshore exploration experienced extremely harsh environments, which at times forced drill ships to move off station due to heavy ice (LTC, 2013). There has been no offshore drilling since 2006 (Byers, 2013). Several reasons have been cited for this: inadequate infrastructure including the lack of roads and pipelines, poor oil-spill response capability and regulatory and environmental burdens. However, climate change may reduce the challenges facing Arctic offshore oil exploration leading to increased petroleum-based destination shipping. Currently, the Canadian government is constructing an 85-kilometer road from Inuvik to Tuktoyaktuk as the first permanent road linking to Canada's northern coast (Chase, 2014). The new road coupled with the recent placement of an oil-spill response barge at Tuktoyaktuk is an initial step towards increased offshore petroleum exploration.

A growing subcategory of destination shipping is the increasing presence of cruise ships in the Arctic providing adventure tours. A recent television series "The Polar Sea" referred to the NWP as the "new Everest" for both cruise ships and smaller private yachts (TVO, 2014). Their presence in the north create concerns over Canada's ability to provide suitable search and rescue services should a major marine incident occur.

The second category of Arctic shipping is transit connecting the Pacific to/from the Atlantic. There are far fewer of these voyages, but climate change will undoubtedly lead to an increase. The current commercially favored trans-Arctic route is Russia's NSR where 71 crossings were recorded in 2013 (Farré, 2014). In 2014, NSR experienced a steep decline to 31 voyages (North Sea Information Office, 2014a). In contrast, in the NWP in 2013, there were 21 trips comprised of eighteen yachts, two cruise ships and one cargo ship; while in 2014 there were thirteen voyages including ten yachts, two cruise ships and one cargo vessel (Headland, 2014). These northern voyages are a mere miniscule fraction of ships traversing the Panama and Suez Canals.

There have been proposals to develop designated non-Arctic hub ports to trans-ship containers through the Arctic on ice-strengthened, purpose-built, all-season container ships. Arctic container transshipment may reduce transit time and fuel consumption between Asia and Europe and the USEC, lowering costs and minimizing environmental pollution. Iceland is considering a deep-sea port at Isafjordur to trans-ship containers from the Arctic to eastern North America and

Europe (Østreng et al., 2013: 36). Russia's ice-free port at Murmansk has long served as a deep-sea hub port in the eastern Arctic. In the western Arctic, the US Army Corps of Engineers (2013) is evaluating potential deep-sea ports in Nome and Point Clarence Alaska. Along with Dutch Harbor, these ports could serve as western Arctic hub facilities (Arctic Portal, 2014). However, considerable challenges lie ahead for any container trans-shipment system. Container shipping depends on tight schedules, something that is difficult to predict with variable and seasonal ice conditions in the north. Further, double handling at trans-shipment facilities will contribute to further delay and costs that may offset the advantages of a shorter trans-Arctic voyage (Farré et al., 2014).

Arctic shipping challenges

Despite retreating sea ice, shipping in Canada's north faces extraordinary challenges, including the lack of ports, minimal marine charts and navigational aids, the need for trained and experienced northern seafarers and ice navigators, the use of ice-strengthened ships and icebreaker support, slow search and rescue services, and Arctic security.

Ports

One of the major challenges facing the development of commercial shipping in the NWP is the almost complete lack of ports and places of refuge for ships in distress. The only northern deep-sea port is Churchill in Hudson Bay located a considerable distance south of the NWP. Although not on the NWP, Churchill offers a potential place of refuge for ships in the Arctic. Churchill has never fulfilled its original potential as a marine access port to middle North America. However, climate change may lead to its greater role as an off ramp from the NWP.

In contrast, as shown in Table 15.1, the NSR has several well-distributed ports that may be able to provide some services, shelter, and search and rescue capabilities for larger ships traversing the route. Although these ports are available, many are currently in a deteriorated state (Young et al., 2013). Russia with its partner Teekay and China LNG is constructing a new deep-sea port at Sabetta on the Yamal Pennisula to support NSR shipping. As part of the project, the

Table 15.1 Major Russian Northern Sea Route ports

Port	*Location*	*Depth*	*Comments*
Murmansk	Kola Bay on Barents Sea	10 m	Year-round, ice-free; ship repair
Dikson	Kara Sea near Yenisey River	14 m	Year-round with icebreaker support
Dudinka	230 miles up Dudinka River	11 m	Year-round for Norilsk Mining
Khatanga	Laptev Sea at Khatanga River	5 m	Seasonal
Tiksi	Laptev Sea near Lena River	6 m	Two offshore moorings
Pevek	East Siberian Sea	13 m	Seasonal
Providenyia	East side Chukotka Peninsula	10 m	Six offshore moorings; ship repair

Source: Northern Sea Route Information Office; http://www.arctic-lio.com/ (accessed April 7, 2014).

consortium is building sixteen ice-breaking LNG tankers to export gas year-round to Europe and Asia (Farré et al., 2014, Doyle, 2014).

Marine charts and navigation aids

A second major challenge to NWP shipping is the lack of suitable marine charts. Only about 1 percent of the northern waters are charted to modern standards and about 25 percent of paper charts are considered "good" (Auditor General of Canada, 2014). Improving marine charts in the Arctic is a priority for the Canadian Hydrographic Services. In contrast, much of Russia's NSR waters have been charted to modern standards.

Navigational aids for larger ships in the NWP are also problematic. The aids currently provided by the Canadian Coast Guard (CCG) are seasonal and aimed at access to local harbors rather than supporting transit shipping. In addition, GPS use is limited in the Arctic due to high levels of electronic precipitation in the northern ionosphere coupled with low elevation angles of geostationary satellites (Jensen and Sicard, 2010). The Russian GLONASS system has recently been restored providing enhanced GPS service for high altitude use (GLONASS, 2015). Many GPS equipment suppliers can access this system to improve coverage in the north. Currently, new Canadian satellites are being deployed to improve Arctic ice management, search and rescue, and sovereignty-related activities.

There is also a lack of adequate communication systems in the NWP. The CCG only provides a seasonal marine broadcast service for ships. In addition, CCG icebreakers leave the Arctic at the end of the summer sailing season for redeployment in southern waters. To further compound the NWP's ice challenges, Canada's Auditor General found the CCG is decreasing the presence of icebreakers in the Arctic (Auditor General of Canada, 2014). In contrast, Russia's NSR provides communication and icebreaker services year-round.

Ships' crew and ice navigators

Operating in the extreme conditions of the Arctic requires the use of seafarers with northern training and experience. The International Maritime Organization (IMO) has provided *Guidelines for Ships Operating in Polar Waters* (International Maritime Organization, 2010) to the shipping community on the need for training and hands-on experience for seafarers and ice navigators. More recently, the IMO adopted a more detailed and specific "Polar Code" for ship operations in Arctic and Antarctic waters (International Maritime Organization, 2014). Regulations in both the NWP and NSR require the use of certified ice navigators for ships sailing in ice-covered waters. There is an estimated shortage of 4,000 ice-trained seafarers (Østreng et al., 2013: 232). Thus, there is a clear need for appropriate northern marine training for suitable candidates. A particular group that could be considered for such training is Inuit youth. The Centre for Marine Simulation at Memorial University in Newfoundland offers a variety of Arctic-related courses including ice navigation. It may play a crucial role in training northern seafarers.

Northern shipping technology

As discussed above, the NWP is predicted to be ice-free with floating ice until at least mid-century. In this case, appropriate classed ice-strengthened ships will be required for NWP voyages. This implies thicker steel hull plates and more powerful propulsion systems increasing the ship's weight. In turn, this reduces the ship's cargo-carrying capacity and increases construction and operating costs. For example, the LNG tankers being built for the project at Sabetta cost $100 million each or about twice as much as normal vessels (Doyle, 2014). Such purpose-built vessels are effective in ice-covered waters, but are less efficient in other trade routes in more moderate climates. These added costs and potential seasonal operating inefficiencies, along with other factors such as darkness, ice cover, inadequate search and rescue, and extreme weather conditions can lead ship owners to question the viability of northern operations.

Other technological advances are being developed for northern shipping. The use of rotating azimuth thrusters replacing static propeller systems for northern vessels increases their position agility, particularly important in maneuvering in ice-covered waters. New double-ended ships coupled with azimuth propulsion allow them to break ice in either direction. Innovative icebreaker ship designs are being developed, such as a new Finnish oblique icebreaker. It not only operates ahead and astern but also sideways to widen the cleared path through the ice for broad beam ships (Almeida, 2014).

The NSR formerly had a user pay fee for ship inspections prior to entering the NSR and for mandatory icebreaker escort. However, in 2013, Russia changed its rules to set the NSR fee based on the vessel's ice-capability, time of navigation, and prevailing ice conditions, and eliminated the requirement for mandatory icebreaker escort (Farré et al., 2014). In Canada, CCG icebreaker escort services are provided at no charge and are not mandatory.

The Russian icebreaker fleet is the largest in the world with 36 heavy and medium icebreakers (Østreng et al., 2013: 202). The Russians are building five new icebreakers and have plans for eight more. Canada operates eight icebreakers, two heavy and six medium sized; with six icebreakers deployed to the NWP for seasonal operations. The CCG's icebreaker fleet is aging, having been constructed in the late 1970s and early 1980s. Canada is planning the construction of a new heavy polar icebreaker, although issues with design and cuts to government funding have delayed construction until 2022 (McCague, 2014).

Search and rescue and oil-spill response capability

A further challenge for commercial shipping in the NWP is inadequate search and rescue (SAR) capability. For most of the NWP, SAR is provided by fixed wing and helicopter services from CFB Trenton located on Lake Ontario about 3,600 kilometers from Cambridge Bay on the NWP. Thus, it takes considerable time to respond to a SAR incident by air. In the eastern portion of the NWP, SAR support is provided from Halifax; again, a considerable distance away. The CCG supplements SAR operations with its helicopters and icebreakers.

In contrast, in Russia's NSR, Maritime Rescue Coordination Centers (MRCC) are located in the major ports of Murmansk, Dikson, Tiksi, Prevek, and Provideniya (North Sea Information Office, 2014b). Each MRCC has multifunctional rescue vessels (including having an oil-spill response capability) and aircraft. The Russian government has committed to expand the NSR's SAR coverage by developing an additional ten stations (Byers, 2013).

Oil spillage is a major northern environmental problem as cold water reduces the capabilities of oil-eating bacteria. Canada pioneered Arctic shipping pollution concerns with its Arctic Waters Pollution Prevention Act of 1970 and the Arctic Shipping Pollution Prevention Regulations of 1978. The latter outlines the requirements for ice-strengthened ship design, construction, and operations in various zones in the Canadian north. Currently, the federal government's Tanker Safety Expert Panel is determining the requirements for appropriate oil-spill response in the Arctic.

Arctic security and international maritime disputes

With climate change and the growth of northern economic development, there are increasing security concerns about unauthorized entry into Canadian, US and Russian waters. The Canadian military presence in the north is limited to occasional winter and summer exercises. Locating CF18 jet aircraft in the north was considered, but extreme weather conditions discouraged this deployment. Canada's larger naval vessels are not ice-strengthened and thus only venture into clear northern waters in summer months. However, the government has announced a program of building six to eight Arctic offshore patrol ships with icebreaking capability to reinforce Canada's sovereignty. In time, they may operate from the proposed deep-water naval supply station at Nanisivik. However, the construction of these ships has been delayed and budget issues surround these small but expensive "made in Canada" vessels.

The Russians have boosted their military presence in the NSR region with the deployment of two additional Arctic brigades supported with new tanks and specialized Arctic personnel carriers. The Russian navy is also increasing its NSR presence in support of shipping and oil and gas extraction (Zhalko-Tytraenko, 2014). Despite this northern military build-up, Russia's ambassador at large for the Arctic has stated that this is not a threat as Russia is "solely concerned with defending its own vast northern regions, which are becoming more vulnerable due to climate change" (Blanchfield, 2014).

The US is also expanding its naval operations in the Arctic reflecting increased commercial shipping activity. By 2020, the US Navy expects to significantly increase northern trained personnel and expand transits through the Arctic by submarines, surface ships, and aircraft to better understand the evolving northern operating environment (US Navy, 2014). Tensions with Russia may accelerate US plans for building ice-strengthened surface vessels for deployment in the Arctic. This reinforces the argument for the joint development of a North American NWP as a strategic hedge against the Russian NSR as these northern passages become important sea routes.

Canada has ongoing border disputes with the US in the Beaufort Sea and with Denmark over the ownership of Hans Island between Canada's Ellesmere Island and Greenland. These border disputes, particularly in the Beaufort Sea, have led to investment uncertainty and likely deterred more extensive oil and gas exploration in that region. Canada and the US are also at odds over the designation of the NWP. Canada claims it as territorial waters within the UN Conference of the Law of the Sea (UNCLOS), whereas the US argues the NWP is an international strait. The difference is significant in that the Canadian position allows it to regulate and control ship traffic through the NWP while the US stance opens the straits for the innocent passage of all ships. Providing effective safety and security of ships requires the cooperation of the three Arctic countries, Canada, US, and Denmark. Resolving these longstanding disputes will promote future NWP economic development. Recently, the US State Department indicated a willingness to come to an agreement with Canada on the Beaufort Sea maritime boundary (Gardner and Shalal-Esa, 2014).

Despite these minor border and interpretation irritants, there is considerable cooperation amongst the eight countries that lie along the Arctic and their indigenous peoples through their active membership in the Arctic Council. The Council has established international cooperation agreements for aeronautical and marine search and rescue in 2011, and in 2013 oil pollution preparedness and response. Again however, issues with Russia may chill future substantive cooperation by Arctic nations until the Ukraine situation is resolved.

Canadian Arctic ports

The only Arctic facility with a dock, sheltered waters and adequate draught is the former Nanisivik mining terminal in Northeast Baffin Island. This facility is being converted to a Royal Canadian Navy refueling station. However, delays in environmental cleanup by the former mining company and the high costs of northern construction are delaying development by the military (Canadian Press, 2013). There are various small community harbors available, but typically these consist of beaches for barge operations from resupply ships anchored offshore. They do not offer port services for larger trans-Arctic and destination ships.

Despite the lack of NWP ports, Canada does have the potential to develop port facilities to serve evolving shipping needs. Table 15.2 outlines existing and potential NWP ports.

There are many smaller communities along the NWP that may provide some support to transiting vessels. The government's recent development of a small harbor facility in Pangnirtung on the east side of Baffin Island could be the first of many similar facilities in other northern communities. Community harbors provide sheltered water for smaller vessels and support local fishing and marine traffic. These facilities could be developed under the auspices of Fisheries and Oceans Canada.

Table 15.2 Potential Northwest Passage deep-water ports

Port	*Location*	*Depth*	*Comments*
Tuktoyaktuk	Beaufort Sea near Mackenzie River	4–6 m	New road link to northern coast; hub for oil and gas exploration; airport; small shallow harbor requiring dredging
Cambridge Bay	South Victoria Island on Queen Maud Gulf	Anchorage: 4m	Sheltered water; airport
Bathurst Inlet	South of Cambridge Bay, up the Bathurst Inlet	Serve ships up to 50,000 tons	Proposed private port for several mining companies; ship zinc concentrate to Europe; sheltered water, airport
Resolute Bay	Cornwallis Island on Barrow Strait	Anchorage: 5–6 m	Central portion of NWP; sheltered water, airport
Nanisivik	Northern Baffin Island	10+ m	RCN vessel supply facility; place of refuge; airport
Iqaluit	Frobisher Bay, southeast Baffin Island	11 m with development	Capital city of Nunavut; plans prepared for deep-water port; high tidal range; airport
Churchill	West side of Hudson Bay	10 m	Rail link south; major port; place of refuge

Conclusion

Despite climate change leading to sea ice retreat, many other factors militate against the rapid growth of commercial shipping in Canada's NWP. These factors include: the lack of ports; inadequate marine charts coupled with relatively poor GPS service (although new navigational technologies may resolve this concern); lack of navigational aids to support commercial shipping; slow response time for search and rescue; extreme weather, including intermittent fog; the need for ice-strengthened and more powerful ships; seasonal shipping through the NWP; a lack of trained and experienced seafarers and ice navigators; and, higher marine insurance costs.

These infrastructural and institutional problems have led the government to claim that the growth of NWP shipping will be slow in the short and medium term. The Minister of Transport, Lisa Raitt, stated that Arctic shipping remains a distant dream. As she put it, "I'm passionate about it. But I don't think it's a panacea, and I don't think the Panama Canal or the Suez Canal … have any worries about competition from the Northwest Passage right now" (Panetta, 2014). However, Minister Raitt's comment contrast with those found in the Auditor General's report, "Environment Canada has estimated that new mining projects in the eastern Arctic could result in about 300 new voyages per year by 2020, nearly doubling current traffic levels. The Canadian Coast Guard expects that as resource development projects move forward in Canada's Arctic region, the size and variety of ships will increase, as will the demand for services" (Auditor General of Canada, 2014).

On the other hand, the shipping industry seems reluctant to invest in northern shipping. In an earlier study, Lasserre and Pelletier (2011) found, "In a survey of shipowners' motivations to engage in the Arctic, eight out of 15 businesses favor the [NSR], which has better infrastructures, more local ports to service and more mining and oil and gas operations [than the NWP]." It seems clear that there is a significant gap in leadership in the provision of northern marine infrastructure between Canada on one side and Russia on the other. Russia's NSR has better facilities to support shipping, less sea ice, and is likely a decade or more ahead of Canada's NWP in providing maritime infrastructure along with navigational and search and rescue services for trans-Arctic shipping.

In response to a growing interest in Arctic shipping, Transport Canada and the CCG adopted as a 2013–14 operational priority, the development of a plan for a "Northern Marine Transportation Corridors Initiative" under the auspices of the government's "Northern Strategy" (Fisheries and Ocean, 2013b). This plan will support safe navigation in Arctic waters by improvements to Canadian Hydrographic Services charting and other steps to support economic and resource development in the north.

Canada's Northern Strategy should be renewed to clarify and consolidate the many initiatives being undertaken by federal and territorial departments, agencies, and their private-sector partners involved in Arctic maritime issues. This strategy includes many interlinked elements including developing appropriate transportation infrastructure to underpin regional economic and social development and by supporting destination and trans-Arctic shipping with investment in community harbor facilities, providing navigation aids, supporting training for northern seafarers and ice navigators, and developing required surface infrastructure.

On the international front, Canada's Northern Strategy must aim to resolve outstanding border and sovereignty concerns with the US and Denmark. Further, in cooperation with the US and Denmark, the Northern Strategy should designate the NWP and its marine approaches as a tri-national transportation and trade corridor. A safe, secure, and efficient NWP transportation corridor regulated and managed by the three nations would ensure this key shipping route would have: a geographic distribution of ports; modern and accurate marine charts; real-time vessel traffic monitoring; appropriate aids to navigation; comprehensive coverage for search, rescue, and oil-spill response; and, tighter regulation of itinerant marine traffic to improve safety and security.

Canada's Northern Strategy should adopt, with the US, a robust cooperative bi-national arrangement tailored for the Arctic that deals with energy, border control, infrastructure development, and other pressing issues. Such an arrangement can be built upon existing innovative bi-national institutions such as: Canada–US St. Lawrence Seaway Development Corporation, International Joint Commission on the Great Lakes, Beyond the Borders Action Plan, North American Aerospace Defence Command, and the North American Free Trade Agreement.

Canada should continue to take a pro-active role on the Arctic Council to maintain its focus on responsible marine, resource, economic, and community development as well as promoting national and tri-national shared goals among the Arctic states.

Developing Canada's North West Passage as a viable and active shipping route is essential for ensuring the country's sovereignty, through appropriate understandings with our Arctic neighbors and Western security by providing a strategic alternative to Russia's Northern Sea Route.

Over the past two centuries, the development of Canada's national economy has been based on significant government investment on east–west transportation infrastructure including national railways, ports, bridges, and highways. This same level of national public infrastructure development is now needed to support Canada's third coast symbolized by the North West Passage.

Acknowledgments

The authors wish to acknowledge the continued support from SSHRC's Borders in Globalization Partnership Program, the Centre for International Governance Innovation, and Carleton University.

References

Almeida, R. (2014). *Building the Baltika – The World's First Oblique Icebreaker. gCaptain:* http://gcaptain.com/building-baltika-worlds-first-oblique-icebreaker/ (accessed January 13, 2015).

Arctic Portal. (2014). *Arctic Hub Ports Future Trajectory Predictions*: http://portlets.arcticportal.org/hub-ports (accessed April 4, 2014).

Auditor General of Canada. (2014). Marine Navigation in the Canadian Arctic. Chapter 3. *2014 Fall Report of the Commissioner of the Environment and Sustainable Development.* Ottawa: Queen's Printer of Canada.

Blanchfield, M. (2014). *Putin's Arctic Envoy Defends Russian Military Buildup*: http://www.thestar.com/news/world/2014/01/30/putins_arctic_envoy_defends_russian_military_buildup.html (accessed April 7, 2014).

Byers, M. (2013). *The (Russian) Arctic is Open for Business*. Canadian International Council. http://opencanada.org/features/the-think-tank/the-russian-arctic-is-open-for-business/ (accessed April 6, 2014).

Canadian Press. (2013). *Ottawa's Arctic Port Plan Mired in Delays*: http://www.cbc.ca/news/canada/north/ottawa-s-arctic-port-plan-mired-in-delays-1.1399519 (accessed April 7, 2014).

CBC. (2014). 'MV Nunavik The Newest Icebreaker to Hit Arctic Waters', http://www.cbc.ca/news/canada/north/mv-nunavik-the-newest-icebreaker-to-hit-arctic-waters-1.2583861 (accessed April 8, 2014).

Chase, S. (2014). 'Nationalism, northern style', *The Globe and Mail,* January 18.

Doyle, A. (2014). 'High Arctic costs deter business despite thaw', Reuters: http://www.reuters.com/article/2014/10/27/business-arctic-idUSL6N0SJ3YW20141027 (accessed December 26, 2014).

Farré, A. B., Stephenson, S. R., Chen, L., Czub, M., Dai, Y., Demchev, D., et al. (2014). 'Commercial Arctic shipping through the Northeast Passage: Routes, resources, governance, technology and infrastructure', *Polar Geography*, 37 (4): 298–324.

Fisheries and Oceans Canada (2013a). *Government of Canada Opens First Small Craft Harbor in Nunavut. Ottawa*: http://www.dfo-mpo.gc.ca/media/npress-communique/2013/hq-ac49-eng.htm (accessed April 6, 2014).

Fisheries and Oceans Canada (2013b). *Report on Plans and Priorities 2013–14:* http://www.dfo-mpo.gc.ca/rpp/2013-14/rpp-op-po-eng.html (accessed June 12, 2014).

Gardner, T. and Shalal-Esa, A. (2014). 'White House releases plan to make Arctic shipping safer', Reuters*:* http://www.reuters.com/article/2014/01/31/usa-arctic-whitehouse-idUSL2N0L501O20140131 (accessed April 10, 2014).

GLONASS. (2015). http://www.glonass-center.ru/en/ (accessed January 14, 2015).

Headland, R. K. (2014). Transits of the Northwest Passage to the End of the 2014 Navigation Season: http://www.americanpolar.org/wp-content/uploads/2014/10/NWP-2014-X-5-layout-for-PDF.pdf (accessed December 26, 2014).

International Maritime Organization. (2010). *Guidelines for Ships Operating in Polar Waters*. http://www.imo.org/Publications/Documents/Attachments/Pages%20from%20E190E.pdf (accessed April 7, 2014).

International Maritime Organization. (2014). *Shipping in Polar Waters*: http://www.imo.org/MediaCentre/HotTopics/polar/Pages/default.aspx (accessed January 14, 2015).

IPCC (.2014). *Climate Change 2014: Impacts, Adaptation, and Vulnerability*, Fifth Assessment Report, 38th Session. Yokohama, Japan: International Panel on Climate Change.

Jensen, A.B.O. and Sicard, J. P. (2010). 'Challenges for positioning and navigation in the Arctic', *Coordinates*, VI (10): 10–13; http://mycoordinates.org/challenges-for-positioning-and-navigation-in-the-arctic/ (accessed January 13, 2015).

Jordan, P. (2013). 'Baffinland Iron Mines sharply scales back Mary River project', *The Globe and Mail,* January 11.

Kassianenko, V. (2012). 'Gazprom successfully completes the world's first LNG shipment through the Northern Sea Route', *Blue Fuel,* 5 (4): 21.

Keane, T. (2014). *Industry Update*. Ottawa: National Marine Advisory Board, Canadian Coast Guard.

Lasserre, F. and Pelletier, S. (2011). 'Polar super seaways? Maritime transport in the Arctic: An analysis of shipowners' intentions', *Journal of Transport Geography*, 19 (6): 1465–1473.

LTC Consulting and Salmo Consulting. (2013). *Oil and Gas Exploration and Development Activity Forecast: Canadian Beaufort Sea 2013–2028.* Beaufort Regional Environmental Assessment. Ottawa: Aboriginal Affairs and Northern Development Canada. http://www.beaufortrea.ca/wp-content/uploads/2013/06/NCR-5358624-v4-BREA_-_FINAL_UPDATE_-_EXPLORATION_AND_ACTIVITY_FORECAST-__MAY_2013.pdf (accessed April 7, 2014).

McCague, F. (2014). 'Arctic shipping routes develop at top of the world,'*Cargo Business News*, 92 (4): 18–22.

McGarrity, J. and Gloystein, H. (2013). 'Northwest Passage crossed by first cargo ship, the *Nordic Orion* heralding a new era of Arctic commercial activity', *National Post*, September 27.

North Sea Route Information Office. (2014a). List of NSR transit voyages in 2014 navigational season. http://www.arctic-lio.com/docs/nsr/transits/Transits_2014.pdf (accessed December 26, 2014).

North Sea Route Information Office. (2014b). Search and rescue. http://www.arctic-lio.com/nsr_searchandrescue (accessed April 7, 2014).

Østreng, W., Eger, K. M., Floistad, B., Jorgensen-Dahl, A., Lothe, L., Mejlaender-Larsen, M., et al. (2013). *Shipping in Arctic waters: A Comparison of the Northeast, Northwest and Transpolar Passages*. Chichester, UK: Praxis Publishing.

Panetta, A. (2014). 'Arctic shipping remains a distant dream for now, Transport Minister says', *The Canadian Press,* March 25.

Siron, R., VanderZwaag, D. and Fast, H. (2009). 'Ecosystem-based ocean management in the Canadian Arctic'. In: Alf Håkon Hoel (ed.), *Best Practices in Ecosystem-Based Oceans Management in the Arctic.* Tromso: Norwegian Polar Institute, pp. 81–100.

The Economist (2012)' Short and sharp: Arctic sea routes have been a long-standing dream; now they are becoming a reality', *The Economist,* June 16.

TVO. (2014). *The Polar Sea: How Climate Change Is Changing Us*: http://tvo.org/story/polar-sea-unprecedented-look-northwest-passage (accessed January 13, 2015).

US Army Corps of Engineers. (2013). *Alaska Deep-Draft Arctic Port System Study*. Alaska District, Jber, Alaska. http://www.poa.usace.army.mil/Portals/34/docs/AKports/1ADDAPSReportweb.pdf (accessed April 8, 2014).

US Navy. (2014). *U.S. Navy Arctic roadmap for 2014 to 2030*. Washington, DC: Department of the Navy. http://greenfleet.dodlive.mil/files/2014/02/USN-Arctic-Roadmap-2014.pdf (accessed April 6, 2014).

Young, O. R., Kim, J. D. and Kim, Y. H. (2013). *The Arctic in World Affairs. Part I The Future of Arctic Shipping.* Korea Maritime Institute. http://www.eastwestcenter.org/publications/the-arctic-in-world-affairs-north-pacific-dialogue-the-future-the-arctic-2013-north-pac (accessed December 26, 2014).

Zhalko-Tytraenko, A. (2014). 'Russian Arctic expansion: What does it mean for Canada?', *Diplomat and International Canada*, 25 (1): 42–46.

16 Arctic transportation and new global supply chain organizations

The Northern Sea Route in international economic geography

Jerome Verny

Introduction

Over the past two decades, the acceleration of economic globalization has revolutionized the interaction between all countries in the world (Hammami et al., 2008; Robertson and Scholte, 2006). Therefore, the new organization of flows reflects the increase in the international mobility of goods. This mobility is based on a few maritime routes, using passages now becoming gradually saturated, such as the canals of Suez and Panama or the Straits of Gibraltar and Malacca. Indeed, the main centers of global production and consumption correspond to the new extended triad (Europe, North America and North East Asia). So the flows of goods remain essentially confined to the Northern hemisphere. In this context, arises the question of the opening of new international trade routes.

However, global warming—which causes the gradual melting of the Arctic ice cap—suggests the possibility of a growth of the maritime traffic along the Arctic region (Figure 16.1): the Northern Sea Route (NSR) between Europe and Asia; the passage of the North-West between the two sides of North America; and the "bridge" between the coasts of White and Barents Seas and those of Hudson Bay. A route that should be acknowledged is the transpolar route passing through the North Pole, but it should remain impassable in the foreseeable future. Thus, this chapter focuses on the NSR, which is a potential route for linking Western Europe to East Asia via the Arctic Ocean. The objective is to analyze the development of this Arctic route for international trade. But beyond the purely commercial question, we can ask about the possible effects of the increased commercial traffic on the Russian territories bordering the Arctic Ocean, especially in seaport development. Russian Arctic territories were hitherto relatively marginalized in the global economy. But the development of merchant shipping along the NSR could have positive effects on their integration. In order to provide some information, we assume that the passage of merchant ships cannot, by itself, constitute a factor of local development, but it can be an item for the strategies adopted by public and private stakeholders. Indeed, it has been demonstrated in numerous research works that seaport infrastructures do not generate direct local

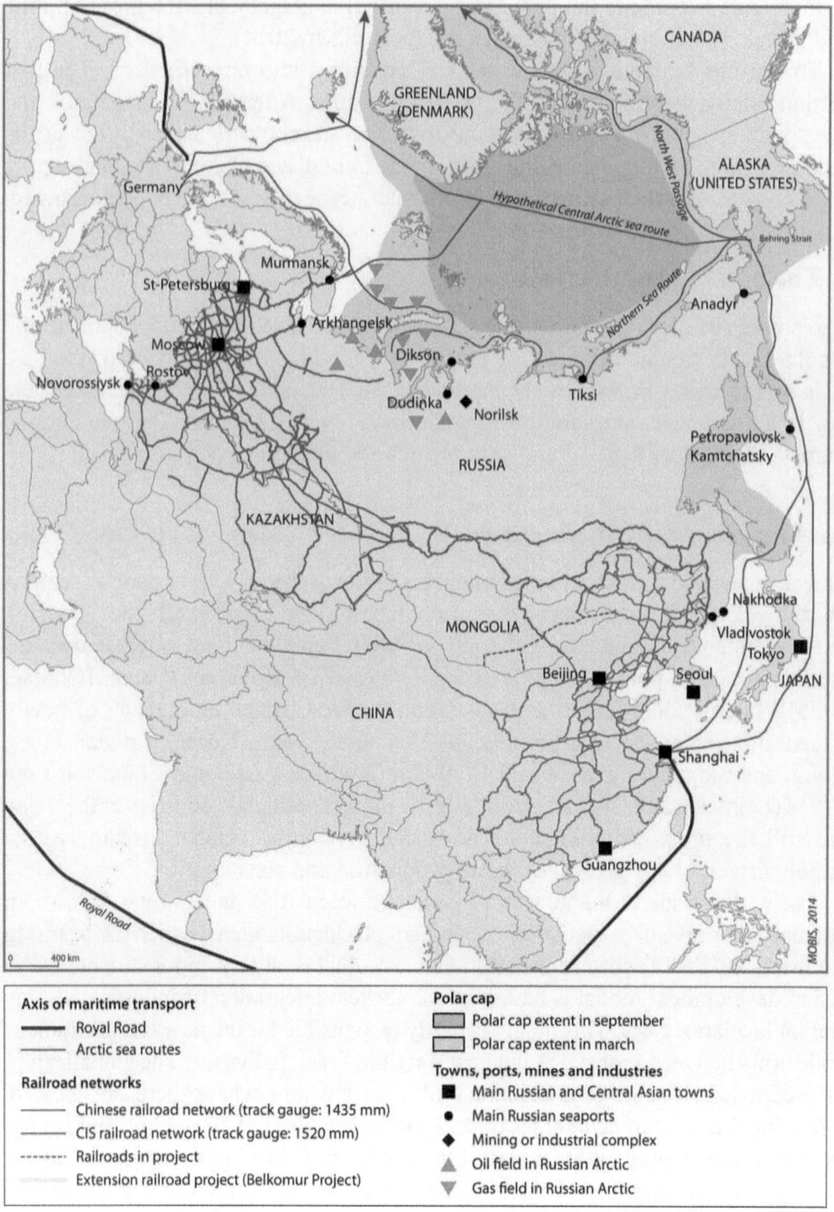

Figure 16.1 The Northern Sea Route and the main railroad networks.

economic development, but that it is the result of management strategies of public and private stakeholders (Song and van Geenhuizen, 2014).

To test this hypothesis, this chapter is organized into three sections. The first section briefly describes the NSR for freight traffic. After that, the strengths and constraints generated by this maritime route for stakeholders are analyzed in the second section. Finally, in the third section, the importance of the new global supply chains organization in the near future for Russian Arctic seaports will be demonstrated.

1. The potential of the NSR

The process of global warming regularly revives the idea of the establishment of a regular line of freight traffic during part of the year, linking Western Europe to East Asia by bypassing Eurasia by the Arctic Ocean. The value of this alternative route lies in the looming saturation of the main road—which connects the two ends of Eurasia by the Suez Canal—and in the search for an alternative transit route.

1.1 A new geography of global trade

The current trend—towards economic integration of the various economic regions of the world—results from the globalization of trade and the growing interdependence of national economies. After being held for several decades around the three poles of the Triad (US, Western Europe, and Japan [Ohmae, 1985]), international trade has been revolutionized by the emergence of newly industrialized countries, especially in Asia: first, South Korea, Taiwan, Hong Kong, and Singapore in the years 1970–80; South East Asia and China since the 1990s (Leinbach and Capineri, 2007). If the main developed countries of the Triad are still the most technologically advanced economies, global growth is now largely driven by the growth of Asian production and consumption.

Today, the trade in manufactured goods between Asia and Europe follows an unbalanced movement between the Asian production area (which is a strong exporter) and the European consumption market that is largely import-oriented. The Asian development model is based on the vertical integration of industrial sectors and on low labor costs. This model is partly responsible for the new organization of trade flows between East and South East Asia and Western Europe. The globalization of production systems uses resources of each territory where there are located, following the logic of economies of scale, reducing costs of production and supply as well as developing optimization of the potential of local labor (Li et al., 2012). Finally, the improvement of international freight transport promotes thinning and increases flows of goods, thanks to increasing vessel size, increasing frequency through the expansion of fleets, and the spread of containerization.

The combination of these factors explains the weight of Asia in international trade. Asian imports have indeed increased by 6 percent between 2005 and 2012, and Asian exports by 7 percent over the same period (WTO, 2013). In this context, the role of the main maritime shipping routes is essential, especially the traditional "Royal Route" via the Suez Canal.

1.2 The bottlenecking of the Royal Route

Since the opening of the Suez Canal in 1869, the major maritime shipping routes between Asia and Europe have not changed much. Ships pass through almost exclusively the Suez Canal, the Straits of Calais, Gibraltar, Bab el-Mandeb, and Malacca (Figure 16.2). On this maritime shipping route, the Suez Canal and the Straits are strategical, both politically and economically, due to the concentration of vessels resulting from the traffic intensity (Mostafa, 2004). In recent years, the increased flows of goods have caused congestion for some strategic passages. This is the case of the Suez Canal, which is used by the majority of maritime traffic each year between Asia and Europe: more than 17,200 vessels, including 37 percent (6,332) of container ships, and 948 million tons of freight in 2012 (Suez Canal Authorities, 2012).

The increase in vessel size does not solve congestion easily insofar as it inevitably reduces the number of ships in each convoy. In this case, the waiting time should then increase and affect the overall journey time of ships. While shipping companies choose to go through the Suez Canal to save time, the bottleneck caused by this infrastructure might harm and disrupt the momentum of trade between Asia and Europe. So, the gradual saturation of the Royal Route produces calls to address the opening of alternative maritime shipping routes (Liu and Kronbak, 2010). But keep in mind that in any event, the Suez Canal will always stay for the next decades the main maritime trade route between North East Asia and North-West Europe.

On the Asia–Europe axis, the NSR—which runs along the Siberian and Norwegian coasts—is a potential shipping route in addition to the Royal Route

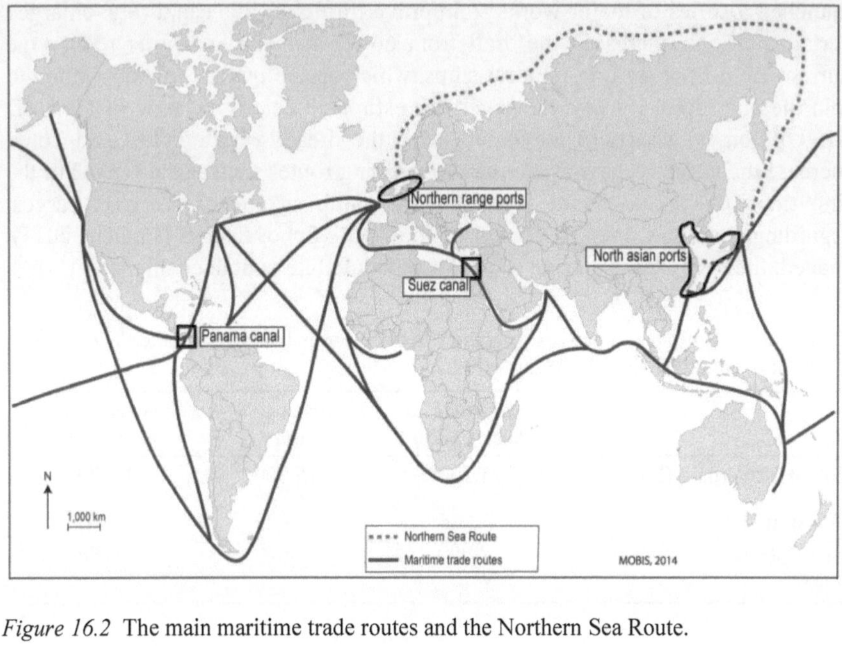

Figure 16.2 The main maritime trade routes and the Northern Sea Route.

which is progressively congested. It would connect the ports of North East Asia to the European ports of the Northern Range via the Arctic Ocean, subject to a significant improvement of navigation conditions (whether physical, technical, or administrative) in the coming years (Buixadé Farré et al., 2014).

1.3 The advantages of the NSR in the new geography of flows

In theory, the Arctic routes—particularly the NSR—are economically viable on the Asia–Europe axis, for seaports ideally at latitude above 30°N.[1] Indeed, one of the advantages of the NSR over traditional maritime shipping routes via the Suez Canal and the Panama Canal is the shorter physical distances (Table 16.1).

The example of the flows of goods between Tokyo and Rotterdam shows the shipping route along the Russian coast turns out to be shorter by a third (in km) than via the Suez Canal. Therefore, the transit time is reduced by about 17 days: 19 days by the NSR against 36 days via the Suez Canal. The passage through the NSR generates significant gains in terms of cost (one of the most important positions) with the fuel consumption (in tons) generating a saving of approximately 50 percent for the NSR. However, one should not forget that the NSR is opened only three to four months per year, approximatively between July and September–October. For the remaining time, the accessibility is very restricted due to the development of sea ice.

Nevertheless, besides the distance, travel time, and fuel costs, other gains are also possible during these four months, such as the economy of the Suez Canal tolls, estimated to average USD $290,000 per ship in 2013. Crossing the Suez Canal, with its total length of 193 km, takes a half-day (12–16 hours of navigation). However, in 2014, Egypt, which manages this international transport infrastructure, launched a series of major works to improve traffic on this canal (e.g. enlarged certain portions, increased the draft from 66–72 ft). The goal is to reduce the current eight-hour waiting time for ships, which are obliged to anchor waiting to join the only daily convoy. However, the estimated cost of these works is USD $100 billion, which would, inevitably, affect the already very high toll, and would increase the attractiveness of alternative shipping routes such as the NSR. For the first time, this could create interest from tramp steamers, like bulk cargos, regarding the short opening access of the NSR (Schoyen and Brathen, 2011). Indeed, the tramp trade has not a regular schedule like container ships.

Table 16.1 Maritime alternatives shipping routes between Tokyo and Rotterdam with a 6,000 TEU container ship (in August)

	Royal Route (via Suez Canal)	*Northwestern Sea Route*	*Northern Sea Route (NSR)*
Tokyo–Rotterdam (km)	21,200	15,700	**13,500**
Transit time (days)	36	22	**19**
Fuel consumption by way (tons)	2,900	1,600	**1,400**

Source: Various industrial sources.

The above shows that the NSR has numerous advantages over the Royal Route, both in terms of distance, transit time, and cost. However, it is clear that the NSR has not received the favor of shipping companies, which continue to prefer the Royal Route, even though it is saturated. For this reason we will seek to understand the reasons for this, at least temporary, lack of interest.

2. Global warming and improving diplomatic relations: new perspectives, but significant constraints

For several decades, global warming and improving diplomatic relations—as a consequence of the end of the Cold War—provided the groundwork for the massive opening of Arctic shipping (Verny and Grigentin, 2009). Despite the interest in the NSR for the transportation of goods in terms of distance, time, and cost in trade between Asia and Europe, its extensive use in the near future is unlikely (Hong, 2012). Indeed, weather, technical, and geopolitical constraints presented by Northern Siberia and the Arctic Ocean are basic limitations to the development of maritime traffic in this part of the world (Furuichi and Otsuka, 2013).

To understand why ship owners and marine insurers are reluctant to use the NSR, we will successively analyze the geographic, technical, and geopolitical conditions for the transit of goods along the Russian Arctic coast.

2.1 Extreme weather conditions for navigation

As its name implies, the Arctic Ocean is covered by the polar ice cap for much of the year. But currently, the global warming phenomenon suggests an increase in the period during the year in which navigation is possible—three or four months—because of the rapid melting of ice at the end of the polar cap followed by a too slow recovery of the ice. However, the first problem faced by ship captains when they navigate on the NSR is the result of extreme weather in the region (Smith and Stephenson, 2013).

The United Nations Conference on the Environment (UNCE), held in Stockholm in 1972, was the first to mention the existence of global warming. Then, many studies validated this hypothesis, including the reports of the Intergovernmental Panel on Climate Change (IPCC). One of the immediate effects of this warming is the gradual melting of the polar ice cap that covers the Arctic Ocean. According to experts, the Arctic sea ice has already declined by 75 percent and the most pessimistic forecasts predict the disappearance of the ice by 2030/2040 (Sakhuja, 2013; ISEMAR, 2014). Thus, the navigation season would be extended from 20 to 30 days to reach about 120 days. According to forecasts by the ACIA (Arctic Climate Impact Assessment), the navigable season could even be extended to 170 days.

However, shipping on the Arctic routes is theoretically possible for three to four months in the year (July to September–October), although, in practice, the NSR is actually free of ice (when weather conditions allow) for 30 days, from August 15 to September 15 (Marchand, 2014). Indeed, the polar ice cap naturally evolves seasonally with the formation of ice in winter and its extension to the coasts, and the melting of part of the ice in the summer. In addition, an exact date for the melting ice is difficult to predict, which represents a major constraint for stakeholders in global supply chains, especially logisticians and ship owners who need to plan routes in advance. This seasonal temporality of the opening of the NSR prevents the establishment of regular lines, which significantly reduce its profitability, particularly for container traffic.

Moreover, when the seas are clear or almost icy, ships may also face all kinds of risks (Stephenson et al., 2013). Icebergs can continue to drift and cause damage to ships, aggravated in cases of heavy storms. Similarly, under the influence of winds and currents, ice sheets can form ridges and constitute walls that even an icebreaker should not cross and would have to pass around. Among the vagaries of weather, cyclones and polar lows, whose trajectories are difficult to predict, may be formed in a short time. Their effects are even accentuated by the almost closed contour of the ocean, reinforcing swell and wave power. In fact, eight-meter-high waves may occur in a very short time. Finally, fog, especially ice fog,[2] is more significant than in the oceans of temperate and tropical zones, which adds an extra difficulty to navigation (Berglund et al., 2007). Only a few ships—with reinforced hulls and navigational tools—can transit in these conditions without the assistance of an icebreaker. However, their draft should not be too deep because of the low depths in some areas in the Arctic Ocean, such as along the NSR with the 13-metre-deep Sannikov Strait.

2.2 Significant technical constraints

In addition to harsh weather conditions are technical constraints linked to the geographical characteristics of the 2,500 nautical miles of the Siberian coast, located between the Bering Strait and the port of Murmansk. On this coast—where no major port is developed—no call is possible. This prevents the optimization of the productivity of shipping lines, according to the same model as the Royal Route which has a combination of important calls at ports well-connected to their respective hinterlands thanks to quality services along dense transport networks—both inland (roads, railways, waterways) and maritime (feeder) networks. In Russian Arctic seaports, their hinterlands are relatively small. Their traffic is exclusively organized around the supply of the local population and the export of raw materials.

In addition, the limited number of seaports along the NSR and their low equipment currently limit the services linked to technical problems related to extreme weather conditions. However, the development of international maritime trade traffic on the NSR requires sailors to have assistance in a minimum of time, along with replacement ships, the help of icebreakers, and specific handling equipment. To this is added the extra costs of insurance, still prohibitive, certainly justified by weather constraints, but mainly due to the still limited knowledge of shipping routes on the Arctic. These costs would then affect the clients of shipping companies' portfolios, which currently operate on the Royal Route and would be likely to transfer some of their traffic to the NSR.

Currently, the isolated nature of the territory bordered by the NSR can be of benefit to hazardous materials only. In order to optimize the presence of ships on these Arctic shipping routes by navigating during the margins of the summer season, the ships must meet several technical requirements, including having reinforced bow and hull, round and smooth hulls in order to avoid hooking ice (e.g. icebreakers). In addition, natural conditions involve the use of smaller vessels than those encountered in the Suez Canal, contrary to the maritime economic logic where massification is required.

2.3 A new geopolitical context

The improving diplomatic relations between the countries bordering the Arctic Ocean, following the collapse of the Soviet regime in 1991, substantially amended the terms of use of the NSR (Chaturvedi, 2012). Indeed, if the NSR had already been established by the USSR, its commercial use by capitalist countries has really only become possible since the 1990s (Brubaker, 1999). However, initially, the collapse of the Soviet system and the deepening economic and social crisis facing the country since the 1980s had a negative impact on the use of the NSR, with the infrastructure deteriorating for lack of maintenance. Thus, ports, icebreakers, and scientific stations were abandoned for many years. Maritime transport in the vicinity of the Arctic Ocean therefore considerably reduced during the 1990s (Bennett, 2014). It was not until the turn of the century that steps were taken to revive the project as a regular trade route.

In addition, beyond the economic issues related to maritime traffic, the regional development of the Arctic is, for the Russian authorities, a way of asserting their presence and control in a region where the stakes are high. Thanks to the development of maritime traffic on the NSR—inbound, outbound, and transit—the geopolitical balance in this part of the world evolved in Russia's favor (Wegge, 2015). The NSR is both a new international maritime route and a new large mineral reserve to explore. So, the control of this part of the world reinforces the power and influence of the Russian government on Europe and Asia. The international economic geography is evolving. In Europe, the historical industrial regions in Western Europe are gradually losing influence to Eastern Europe. In Asia, many emerging industrial regions are now located in the western part of China—about 500–2,000 km from major seaports. The eastern coast of China is becoming too expensive for the development of industry. The distances between these new industrial regions in Europe and Asia are shorter via Russia than via the maritime route by the Suez Canal. Therefore, in the huge market between Asia, more precisely China, and Europe, the influence of Russia is becoming significant.

The first way is energy-related, as the basement of the Russian Arctic, as well as that of the Kara Sea, contains significant oil and gas fields (Figure 16.1), that Gazprom (largely controlled by the Russian state) has already begun to explore and exploit (Harsem et al., 2011). However, the exploitation of resources in the Arctic, like the passage of merchant ships, presents the problem of the relationship between economic development and environmental protection. If the current

Russian government chooses economic development, several non-governmental international organizations in which media power is strong (e.g. Greenpeace, WWF) are considering action against this policy.

The second issue is human nature, namely the settlement and equipment of a quasi-desert region due to its physical hostility. In Soviet times, the Russian settlement was the result of *zeks*, the Gulag prisoners sent to labor camps to exploit the mineral resources of the north of Russia or to build transportation infrastructure to link the northern regions to the rest of the country (a canal from the Baltic to the White Sea; 501 and 503 railway lines between Vorkuta and Igarka, which were never commissioned). Soviet concentration policy has strengthened the Russian settlement, next to the more ancient Nordic peoples (e.g. Evenki, Nenets, Dolgan, etc.). But since the end of communist rule, the population of the north of Russia has mostly dropped, raising the question of control of the territory by the Russian government (Marchand, 2014). Local economic development, desired by the Russian government for several years, could then be a way to repopulate this hostile region.

The third challenge is geostrategic. Indeed, the Arctic was, during the Cold War, a bone of contention for the USSR and the USA. During the Cold War, military bases were built on the shores of the Arctic Ocean, and submarines of the two great powers met in the Arctic Ocean. Today, if the ideological conflict between the two superpowers has disappeared, the territorial dispute remains alive and well, to the point that some media sources do not hesitate to talk about a "new Cold War" (Ciuta and Klinke, 2010). Indeed, the Arctic is subject to the claims of the five riparian countries (Canada, Denmark, USA, Norway, and Russia) seeking to expand their territorial waters in order to ultimately exploit its natural resources. Russia is particularly active, with a strong expansion of its exclusive economic zone on the claimed Lomonosov Ridge (where the geographic North Pole is located), but challenged by Danish Greenland. The Russians also sent an underwater expedition in 2007, aiming firstly to collect rock samples with the aim of proving the geological continuity between the Eurasian continental shelf and the Lomonosov chain, thereby supporting their land claims; but also with objective to strengthen Russian patriotism, with the installation of the national flag on claimed territory (Dodds, 2010; Dittmer et al., 2011).

In this context, in 2000, Vladimir Putin expressed his wish to see shipping traffic on the Arctic Ocean along the Siberian coast and on whole or a part of the NSR in the coming decade.[3] To implement a development strategy for the NSR, the Russian authorities want to act on the technical and technological aspects. This is why Russia is now investing in the renovation of its fleet of icebreakers. Six new nuclear icebreakers should be built by 2030, three new conventional icebreakers delivered by 2015, and 60 polar-capable ships by 2020 (Petterson, 2012a). In terms of seaworthiness, heavy investments are already planned (survey vessels, four new research and rescue centers, four new satellites for meteorology, climate change and real-time navigation [Petterson, 2012b]).

However, the administrative management of navigation in the territorial waters of the Russian Northern Sea Route Authority (NSRA) can be an additional burden for ship owners. Indeed, the requirement for ships to use Russian icebreakers to make the transit is considered a major obstacle to increased traffic on the NSR (Verny, 2013). In 2010, only four ships had made the transit between Asia and Europe (door-to-door). By 2014 31 had passed, plus around 100 ships which made only part of the transit. More than 240 requests for permission to navigation were made in 2013. Therefore, facing a growth of traffic on the NSR, the number of icebreakers will be insufficient. One possibility for solving this problem is to consider a passage in convoy, like on the Suez Canal, with a minimum of two icebreakers. Beyond the geopolitical aspects, both technical and Russian, technological voluntarism is therefore a key to the possible development of a regular trade route along the Siberian coast. But, so that the NSR can assert itself as a new international maritime trade route, it is still necessary that shipping companies are interested in it.

3. New logistics organizations for the development of Arctic seaports

Although regular lines on the NSR will not be established for at least twenty years,4 traffic on the NSR will promote Russian Arctic seaport activities. This section aims to analyze the conditions for the development of seaport activities along the NSR and the forms these activities may take.

3.1 The valuation of Arctic seaports thanks to mineral, oil, and gas exports

The first direction of development of Russian Arctic seaports through the NSR is based on mineral, oil, and gas exports. This line of development is conventional and takes its inspiration from Russia's current economic policy, which is based on exports of raw materials. An economy based on energy and mining incomes is in line with the traditional Soviet economy in which development relied on heavy industries. The weight of mining and energy industries in the Russian economy increased after the collapse of the Soviet industrial system, which became uncompetitive in the global economy (Kumpula et al., 2011). In this context, only the mining and energy sectors were able to maintain profitability for export.

For Russian Arctic seaports, export of raw materials has always been, and still is, the main activity, with the supply of local populations. Two cases can be distinguished. The first is mineral exports. Central Siberia is indeed the first mining region of Russia. We can find nickel, cobalt, platinum, niobium, and rare earths required by the computer industry. The emerging economies of Asia are increasingly consumers of these minerals exported from Russian seaports. The NSR is the shortest shipping route in terms of physical distance. Therefore, bulk cargos could reinforce the attractiveness of the NSR but mainly for inbound and outbound traffics. The door-to-door from Asia to Europe by bulk shipping will stay for the next years an exception. Actually, the transit by the NSR should not be increased as quickly as the inbound and outbound traffics because the loading location is in the Arctic Ocean, with these natural ressources. However, in terms of local development, user industries of these minerals are located abroad. So their transit through Russian Arctic seaports should not generate local industry. Therefore, we have a dynamic process on the NSR with this terriroty being regarded as a new maritime trade route and at the same time like a huge new mineral reserve to explore.

The second case is that of hydrocarbons. Most of the reserves are located in the West Siberian plains, including the oil fields of Baku III, operated since the Soviet era, and Vankor since a couple of years ago. We also have an oil field in the basin of the Pechora but it is smaller. Regarding the main gas fields, the largest are concentrated in the peninsula of Yama, and near the Pur and Taz. Eight-nine percent of Russian production is concentrated in these three gas fields. In recent years, explorations have led to the detection of very important potential offshore gas fields in the Kara Sea and Barents Sea—for example, more than 10,000 billion m^3 are estimated in the Kara Sea. However, these new fields are not yet fully exploited.

Yet if oil and gas in the north of Russia can be exported by ship, thanks to oil and LNG ships, competition with pipelines will remain strong. Europe is well-connected to the Siberian and Arctic oil and gas fields thanks to pipelines. But the situation is quite different to that in the emerging Asian countries, especially China (Hong, 2014). These countries are still dependent on the use of ships for hydrocarbon supply, despite the recent construction (in 2011) by the Chinese branch of the pipeline ESPO (East Siberia-Pacific Ocean), which connects the fields of Central Siberia and Vankor to the north of China.

In the near future, the Russian pipeline networks will be developed thanks to new axis between Siberian fields and South Korea and Japan. The steady growth of Asian demand for oil and gas could also be developed as an important traffic of ships on the NSR. Therefore, Russian port activities could grow thanks to new wharfs for this specific industry. Also, one can imagine a diversification of oil and gas activities in Russian Arctic seaports in a similar way to that which happens in the Middle East. In fact, we could have in these Russian seaports new industries due to the transition from simple storage of crude oil to refining and exporting oil products with high added value. But we have two problems. Firstly, the possible reluctance of companies to refine oil products before export due to the increasing risk of losing an added value product, for example, during a shipwreck in the Arctic Ocean where navigation is still uncertain. Secondly, there needs to be a proactive Russian policy in order to build such oil and gas production facilities as well as seaport infrastructures.

3.2 Russian Arctic seaports: simple places of transit or future areas of industrial and logistics activities?

The second approach to the development of Russian Arctic seaports focuses on the evolution of their traditional function, which is the transit between the foreland and their hinterland to production activities and services better integrated into the global economy. This is to affirm the place of Russian seaports in the changing geography of the value chain. Thanks to the massification of goods in seaports, it could be possible to develop a new value chain with the implantation of logistics platforms (Ng and Liu, 2014). The development of traffic on the NSR may become a factor in local economic development due to the integration of port activity as a step in the manufacturing process of semi-finished products. These patterns are now developed in some Western European seaports. For example, the postponement strategy can create jobs in port logistics platforms before shipping finished goods to markets for consumption.

The postponement strategy maximizes benefits and minimizes risks thanks to the optimization of the value chain. This is allowed due to delay in the last stages of production near the consumer market areas (e.g. assembly, co-packing, quality control, etc.), adding value on the product mitigates the cost of the transshipment during the cross-deck operation. However, the transformation model and adding value that tends to develop in Europe is based on semi-finished products imported in containers. In the short term, consumer goods container traffic on the NSR should not be the most consistent in terms of volume because of low population density. But we can imagine the development, in the medium term, of a specific business in these Russian Arctic seaports, such as in Mauritius with its diversification of its economic sectors during the 1970s. For example, the automotive industry could be a new economic sector in this part of the world. This sector is increasingly globalized and the number of places involved in the production of cars has increased during the last decades. Russia is an emerging market for automobiles, far away from the production countries located in Asia or Europe. Russian Arctic seaports could be a step in the operations management of the automotive industry between the production market in Europe and Asia and the consumer market in Russia. In this postponing strategy and with the internationalization of automotive supply chains, Russian Arctic seaports could take a new place in order to finish the production of the automobiles (e.g. customization) for the Russian marketplace. One problem in this scenario is the lack of suitable port terminals and port areas where industrial and logistics activities could be developed. To attract public–private partnerships in these new organizations of production and distribution methods, knowledge of markets in the medium term is necessary. Beyond this, a stable geopolitical environment is requred along with a real proactive policy to integrate these visions into the development of Russian territory plans. However, it is necessary to adopt measures in order to attract labor to these remote territories. For example, in the region of Norilsk and Dudinka, at the mouth of the Yenisei (Figure 16.1), 300km north of the Arctic Circle, there are ten months with negative temperatures, nine months with snow and six months with polar nights. Under these weather conditions, material and financial benefits must be offered by either companies or the Russian authorities to attract labor, as in Soviet period.

Therefore, the effects of increased traffic along the NSR on Russian seaport activities could likely be limited (Lee and Song, 2014). Nevertheless, an exception may be the coastal cities located at the ends of the NSR, namely Arkhangelsk and Murmansk on the Barents and White Seas and Nakhodka, Vladivostok, and Petropavlovsk-Kamchatsky in the Russian Far East (Figure 16.1). In fact, these ports are ice-free all year and better equipped than the northern Siberian ports.

They are already beginning to attract foreign investment and to emerge as issues of regional planning for the Russian authorities.

3.3 Multimodality in international supply chains via the NSR: the example of the Belkomur Project

The third option to be considered for developing Russian Arctic seaports is to strengthen the connection of coastal cities with the mainland. The issue here falls within the geopolitical and geo-economic strategy of the Russian state. Indeed, the main seaports of Russia in Europe, excluding the White and Barents Seas, are located on the Baltic Sea (St. Petersburg), the Black Sea (Novorossiysk), and the Sea of Azov (Rostov-on-Don). However, from these seas, oceans are accessible only via straits controlled by members of NATO: the Bosphorus Strait and the Dardanelles (Turkey), and Kattegat and Skagerrak Straits (Denmark and Norway). In the context of Russian-American diplomatic relations, likely to reach a conflict situation at any time, the development of Russian seaports along the NSR becomes strategic, both militarily and commercially.

Economic valuation of White and Barents Seas ports (Murmansk, Arkhangelsk) depends on proactive policy and public–private partnerships. This is the case for the Belkomur5 Project. The objective is to build a railway track between the port of Arkhangelsk and Perm, in the southern Urals, where it will be connected to the Trans-Siberian railway network with an extension to Murmansk in the Kola Peninsula (Figure 16.3). The project, submitted in 1932 and discontinued in 1953, was revived in 2008 after two unsuccessful attempts in 1996 and 2002, which failed due to lack of funding. This regional development project is integrated into Russia's transport strategy for 2030 and into the strategy for Russian railway transport in 2030. The Belkomur Project is widely financed by Chinese funds.

The length of this railway route project is 1,252km, with 705km still to be built. The estimated cost for the project is around 15 billion Euros, 80 percent of which will come from public–private funds and/or foreign funds, in particular from the China Civil Engineering Construction Company (CCECC). This corridor between China and Arkhangelsk could provide an opportunity for Chinese carriers to access a new gateway to the sea via Mongolia. This will be a new axis in order to increasingly export Chinese manufactured goods abroad. Indeed, from Arkhangelsk, it is possible to develop loops, all year round, with Western European seaports and also North American seaports via the North East West Corridor. In this case, the Russian Arctic seaports of the White and Barents Seas are included in the Northern Transportation Corridor linking East Asia to North America through Central Asia and Europe.

New services can be developed around this infrastructure and integrated into a new global supply chains organization. In Arkhangelsk and Murmansk ports, transshipment between trains and vessels provides a good opportunity to work on freight flows in order to add value to the products. This postponement strategy could be an achievement for these developing territories, such as introduced in the last part of Section 3.2. For example, the assembly of spare parts could be carried

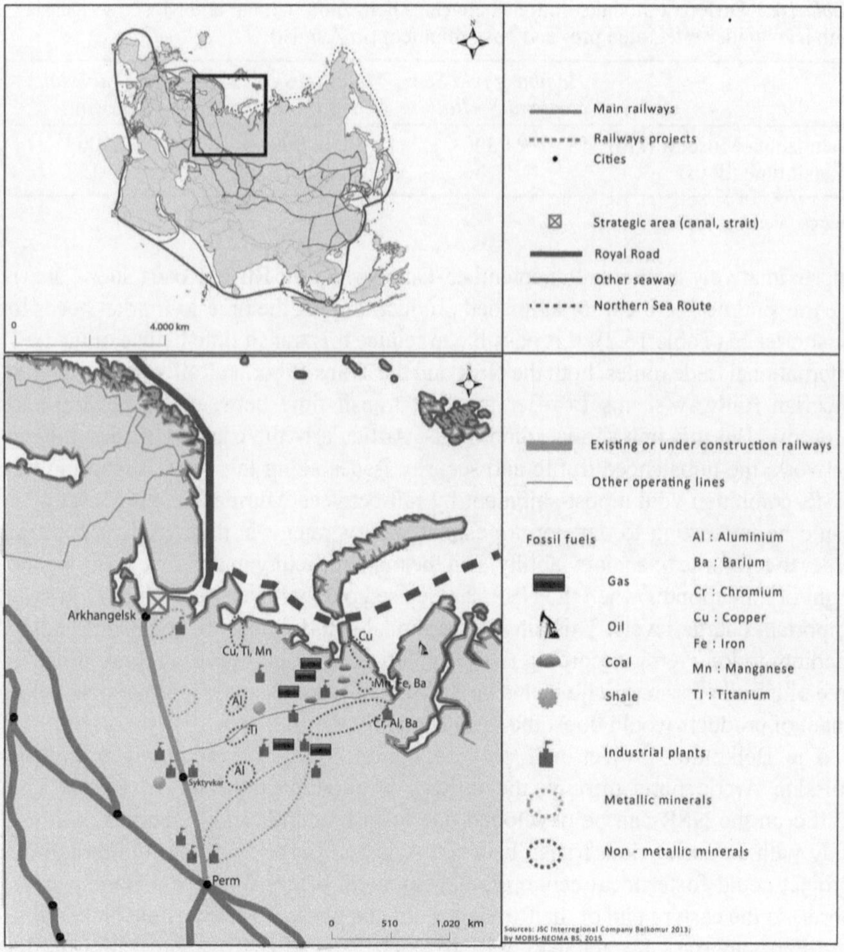

Figure 16.3 The Belkomur Project.

out in the industrial areas of ports. This activity could lead to the development of a local industry, not only in assembly but also in production, with some parts coming from East Asia and others manufactured locally or in the hinterland of these Russian ports. This organization could be relevant for mechanical and/or automotive industries where CKD (Complete Knock Down) and SKD (Semi Knock Down)6 are common practices: the spare parts would be made in Asia and then sent to Russia where they would be assembled, and vehicles would be shipped to Western Europe and North America. According to modal solutions, the transit time between Zhengzhou (China) and Moscow (Russia) is really different. Why send all finished products like automobiles from Zhengzhou to Moscow by the Royal Road (Suez Canal) when it is possible to use alternative emerging routes? This comparison is only relevant during the opening period of the NSR,

Table 16.2 Different modal solutions between Zhengzhou (China) and Moscow (Russia) with a container integrated pre- and post-shipment (in August)

	Hamburg (via Suez Canal) – Rail	*Murmansk (via NSR) – Rail*	*Trans-Siberian Railway*
Zhengzhou–Moscow (km)	24,800	14,900	7,200
Transit time (days)	40	17	12

Source: Verny (2015).

approximatively from July to September–October. By the Royal Road (Suez Canal), the transit time is critical for a finished product. In fact, the time-to-market needs to be shortened (Table 16.2). It is possible to reduce this transit time by use of the new international trade routes, both the NSR and the Trans-Siberian Railway. The Trans-Siberian Railway seems to offer the best transit time between Zhengzhou and Moscow. But this link shows some limits, particularly the capacity of the railway network, the unbalanced traffic and security issues along this way. Regarding the NSR, combined with a post-shipment by rail between Murmansk and Moscow, it could be interesting to develop a postponement strategy in this Arctic seaport in order to reinforce financial viability. But the main difficulty in this case is during the eight or nine months when the NSR can be used only with an icebreaker, with very important charge levels. This alternative modal solution may be reasonable in the medium to long term, according to the evolution of global warming, only if it was free of ice all year long. The inclusion of Russian seaports in the international value chain of products would boost the local economy.

The Belkomur Project will provide opportunities for the revitalization of Russian Arctic ports through the railway hinterland connected to China. The traffic on the NSR can be developed due to to inbounds and outbounds and not only with a door-to-door transit between Asia and Europe. Indeed, the Belkomur Project could foster local economic development of the White and Barents Seas ports. At the eastern end of the Russian route, the ports of Vladivostok, Nakhodka, and Petropavlovsk-Kamchatsky will also benefit from a better integration in the East Asian economy as a result of increased traffic on this part of the NSR.

Conclusion

The ongoing development of freight transport on the Arctic Ocean, especially along the NSR, could have several advantages during the opening period (from July to September–October). Globally, the NSR may be, proportionately, an alternative to the Royal Route. This alternative route does not call into question the supremacy of shipping for international trade via the Suez Canal. But depending on the value of the products, the time sensitiveness, the seasonality, the strategies of shippers, freight forwarders, and other logistics providers, the reliability and efficiency of the relationship, and the value of the logistics services, the NSR could be a new option for clients looking for an alternative to shipping via the Suez Canal, especially if suppliers and/or consumers are located in Eastern Europe, Russia, Central Asia, or western China. So it depends on the capability of stakeholders to develop different logistical innovations in addition to the construction of transport and logistics infrastructure such as wharfs, railways, and warehouses. One of these innovations could be the postponement strategy, in response to the revitalization of the Russian industrial sector. At the local level, the traffic on the NSR could revitalize Russian Arctic ports, which have greatly declined since the collapse of the USSR. It seems obvious that new inbound and outbound traffics along the NSR could provide huge opportunities for the development of northern Russian territories. This northern part of Eurasia could integrate with economic globalization.

Beyond economic issues, the NSR could also be a way for Russia to occupy an important strategic position in world trade. This could be part of the response to recovery initiated by the current Russian government. It points, moreover, to the emergence of a new economic geostrategy combining Russia and China—and to a lesser extent South Korea and Japan—whose strategic interests combine to establish investment partnerships on the NSR. These Russian and foreign investment partnerships will be both commercial (exploitation of the raw materials and natural resources of the Arctic) and logistical (development of seaport terminals and areas of industrial and logistics activities along the NSR).

However, the problems of navigability on the NSR currently represent fundamental limitations on its development, particularly in terms of regularity and efficiency for all-year freight traffic. The potential of this maritime trade route could increase with the enhancement of global warming, but the mechanization of traffic—more liner shipping—is not possible in the next couple of decades. Today, the possibilities for the development of Russian Arctic ports gravitate towards other corridors, such as the Arctic Bridge or the North East West Corridor for the White and Barents Seas ports and, for Russian Far East ports, to inclusion in the economic system of North-East Asia. Nevertheless, the combination of proactive policy and public–private partnerships with foreign investments is the key development for the Russian ports along the NSR. The challenge for the Russian authorities is to facilitate foreign investments while strengthening their control over sparsely populated areas where the geostrategic issues may harden in the coming years.

Acknowledgment

The author would like to thank the anonymous reviewers for their constructive comments and advice which have enhanced the scientific quality of the chapter.

Notes

1 In Asia, the major seaports are: Dalian, Qingdao, Shanghai, Tianjin, Busan, Kobe, Tokyo; in Europe, the major seaports on the Northern Range are: from Le Havre to Hamburg.

2 Haze of reddish-brown, reducing visibility to a few meters.

3 This statement came during a meeting in the Russian port of Murmansk on the Barents Sea, a port accessible all year round.

4 Nunatsiaq online, *Commercial Arctic shipping a long way off, Maersk boss says. Head of global shipping firm not keen on Northern Sea Route, Northwest Passage*, http://www.nunatsiaqonline.ca/stories/article/65674commercial_arctic_shipping_a_long_way_off_maersk_boss_says (accessed July 29, 2014).

5 Beloyo More-Komis-Ural (White Sea–Komis –Oural).

6 CKD is a method of organizing production and distribution, widely used in the automotive industry. The objective is to export vehicle parts, from their place of production to the place of assembly, usually countries where vehicles are sold. This organization aims to avoid paying high taxes on manufactured goods imports, like vehicles. For its part, SKD is a similar form of organization to CKD, except that the parts are already partially assembled.

References

Bennett, M. (2014). 'North by Northeast: Toward an Asian-Arctic region', *Eurasian Geography and Economics*, 55 (1): 71–93.

Berglund, R., Kotovirta, V. and Seina, A. (2007). 'A system for icebreaker navigation and assistance planning using spaceborne SAR information in the Baltic Sea', *Canadian Journal of Remote Sensing*, 33 (5): 378–387.

Brubaker, D. (1999). 'The legal status of the Russian baselines in the Arctic', *Ocean Development and International Law*, 30 (3): 191–233.

Buixadé Farréa, A., Stephenson, S. R., Chen, L., Czubd, M., Dai, Y., Demchevc, D., Efimov, Y., Graczyk, P., Grythe, H., Keil, K., Kivekäs, N., Kumar, N., Liu, N., Matelenok, I., Myksvoll, M., O'Leary, D., Olsen, J., Pavithran, A.P.S., Petersena, E., Raspotnik, A., Ryzhov, I., Solski, J., Suo, L., Troein, C., Valeeva, V., van Rijckevorsel, J. and Wighting, J. (2014). 'Commercial Arctic shipping through the Northeast Passage: Routes, resources, governance, technology, and infrastructure', *Polar Geography*, 37 (4): 298–324.

Chaturvedi, S. (2012). 'Geopolitical transformations: Rising Asia and the future of the Arctic Council'. In: Axworthy, T. S., Koivurova, T. and Hasanat, W. (Eds.), *The Arctic Council: Its place in the Future of Arctic Governance.* Toronto: Munk-Gordon Arctic Security Program, pp. 226–260.

Ciuta, F. and Klinke, I. (2010). 'Lost in conceptualization: Reading the "new Cold War" with critical geopolitics', *Political Geography*, 29: 323–332

Dittmer, J., Moisio, S., Ingram, A. and Dodds, K. (2011). 'Have you heard the one about disappearing ice? Recasting Arctic geopolitics', *Political Geography*, 30: 202–214.

Dodds, K. (2010). 'Flag planting and finger pointing: The law of the sea, the Arctic and the political geographies of the continental shelf', *Political Geography*, 29 (2): 63–73.

Furuichi, M. and Otsuka, N. (2013). 'Costs analysis of the Northern Sea Route (NSR) and the conventional route shipping', *Proceedings of the IAME 2013 Conference*, Marseille, France, July 2013.

Hammami, R., Frein, Y. and Hadj-Alouane, A. B. (2008). 'Supply chain design in the delocalization context: Relevant features and new modeling tendencies', *International Journal of Production Economics*, 113 (2): 641–656.

Harsem, Ø., Eide, A. and Heen, K. (2011). 'Factors influencing oil and gas prospect in the Arctic', *Energy Policy*, 39: 8037–8045.

Hong, N. (2012). 'The melting Arctic and its impact on China's maritime transport', *Research in Transportation Economics,* 35 (1): 50–57.

Hong, N. (2014). 'Emerging interests of non-Arctic countries in the Arctic: A Chinese perspective', *The Polar Journal*, 4 (2): 271–286.

ISEMAR. (2014). *Arctique/Antarctique. Les enjeux des usages polaires*, Note de Synthèse 164, April, p. 4.

Kumpula, T., Pajunenb, A., Kaarlejärvib, E., Forbesb, B. C. and Stammlerb, F. (2011). 'Land use and land cover change in Arctic Russia: Ecological and social implications of industrial development', *Global Environmental Change*, 21 (2): 550–562.

Lee, S. W. and Song, J. M. (2014). 'Economic possibilities of shipping though Northern Sea Route', *The Asian Journal of Shipping and Logistics*, 30 (3): 415–430.

Leinbach, T. and Capineri, C. (2007). *Globalized Freight Transport: Intermodality, e-Commerce, Logistics and Sustainability*. Transport Economic, Management and Policy Series. Cheltenham: Edward Elgar.

Li, L., Dunford, M. and Yeung, G. (2012). 'International trade and industrial dynamics: Geographical and structural dimensions of Chinese and Sino-EU merchandise trade', *Applied Geography*, 32 (1): 130–142.

Liu, M. and Kronbak, J. (2010). 'The potential economic viability of using the Northern Sea Route (NSR) as an alternative route between Asia and Europe', *Journal of Transport Geography*, 18 (3): 434–444.

Marchand, P. (2014). *Géopolitique de la Russie. Une nouvelle puissance en Eurasie*. Paris: PUF, Major, p. 182.

Mostafa, M. M. (2004). 'Forecasting the Suez Canal traffic: A neural network analysis', *Maritime Policy & Management*, 31: 139–156.

Ng, A.K.Y. and Liu, J. J. (2014). *Port-Focal Logistics and Global Supply Chains*. Basingstoke: Palgrave Macmillan, p. 264.

Ohmae, K. (1985). *Triad Power: The Coming Shape of Global Competition*. London: Free Press.

Petterson, T. (2012a). 'Three new nuclear icebreakers in the pipeline', *Barents Observer*. Available at: http://barentsobserver.com/en/arctic/three-new-nuclear-icebreakers-pipeline-14-11 (accessed May 11, 2015).

Petterson, T. (2012b). 'Arctic emergency center opens in Murmansk in 2013', *Barents Observer*. Available at: http://barentsobserver.com/en/arctic/arctic-emergency-center-opens-murmansk-2013 (accessed May 11, 2015).

Robertson, R. and Scholte, J. A. (2006). *Encyclopedia of Globalization*. London: Routledge.

Sakhuja, V. (2013). *India and the Melting Arctic*. New Delhi: The Institute of Peace and Conflict Studies.

Schoyen, H. and Brathen, S. (2011). 'The Northern Sea Route versus the Suez Canal: Cases from bulk shipping', *Journal of Transport Geography*, 19 (4): 977–983.

Smith, L. C. and Stephenson, S. R. (2013). 'New Trans-Arctic shipping routes navigable by midcentury', *Proceedings of the National Academy of Sciences*, 110: 4871–4872.

Song, L. and van Geenhuizen, M. (2014). 'Port infrastructure investment and regional economic growth in China: Panel evidence in port regions and provinces', *Transport Policy*, 36: 173–183.

Stephenson, S. R., Brigham, L. W., and Smith, L. C. (2013). 'Marine accessibility along Russia's Northern Sea Route', *Polar Geography*, 37 (1): 111–133.

Suez Canal Authorities (2012). *Detailed Yearly Statistical Report*, Suez Canal traffic statistics. Available at: http://www.suezcanal.gov.eg/TRstat.aspx?reportId=3 (accessed May 11, 2015).

Verny, J. (dir.) (2015) (in progress). *International Trade Routes between Asia and Europe and Value Added Logistics Strategy*. European Research Project, EC—French Ministry of Research—Normandy Region, Rouen.

Verny, J. (2013). 'European perspective on potential Arctic shipping'. In: Young, O. R., Kim, J. D. and Kim, Y. H. (dir.), *The Arctic in World Affairs: A North Pacific Dialogue on Arctic Marine Issues*. Honolulu: KMI-EWC, 349 p.

Verny, J. and Grigentin, C. (2009). 'Container shipping on the Northern Sea Route', *International Journal of Production Economics*, 122: 107–117.

Wegge, J. (2015). 'The emerging politics of the Arctic Ocean: Future management of the living marine resources', *Marine Policy*, 51: 331–338.

WTO (2013). *Statistics on International Trade 2013*. Geneva: WTO Publications, p. 261.

17 The state of climate adaptation for ports and the way forward

Austin Becker

Human civilization in the twenty-first century depends on ports for jobs, the transport of energy supplies, raw materials, and finished products. Ports help societies meet their basic needs of food and shelter and they enhance quality of life. Indeed, civilization as we know it would be very different without the ports and maritime trade that serves as the backbone to both global and local development. But global warming and the resulting changes to environmental conditions on the coast require that we rethink the future of this vital infrastructure that serves the needs of the international community.[1] This rethinking will be a monumental task for decades, if not centuries, to come. At the time of writing this volume, we are only in the very first stages of rising to this challenge.

Climate scientists project that sea levels will rise dramatically and storms like Hurricane Sandy, Hurricane Katrina, and Super Typhoon Haiyan will only become more frequent as the planet warms.[2,3] Yesterday's sea levels and precipitation patterns served as the benchmark for most coastal infrastructure, and engineers designed ports based on these historical conditions. Thus, many ports will find themselves exposed to new higher levels of risk as probability of flooding increases for some areas, drought frequency increases in others, and extreme heat events occur more frequently in yet others. As businesses, ports already have strong incentives to build resilience as the climate changes, as high costs result from operational delays, repairs, and cleanup. A survey conducted in partnership with the International Association of Ports and Harbors (IAPH) and the American Association of Port Authorities (AAPA) found that port operators feel very concerned about these climate-change challenges. In fact, 80 percent of survey respondents agreed that the port community needs to address this problem head on.[4]

What will happen as sea levels rise and storms become more intense and more frequent? What types of impacts should we expect and upon whom? What can be done to build resilience? Though studied by scientists for decades, practitioners and policymakers have only recently begun to explore the potential implications of climate change on society.[5,6,7] The rapid rate of sea-level rise expected toward the end of the twenty-first century, coupled with changes in storm patterns and intensities, will likely produce a complex set of new conditions within which ports will continue to operate if ours remains an economy based on maritime trade. Due to their exposed locations along coasts and waterways, many ports must ultimately face the climate-change challenge and opportunities head on. Through case studies from around the world, this volume introduces a range of perspectives on this topic. Taken together, this collection indicates an emerging area of concern, some initial steps underway, and the critical need for future research.

Historically, ports have considered planning for storm resilience in a relatively siloed fashion: engineers and consultants developed construction guidelines for structures, while the local coast guard or emergency management agency worked with the port to ensure safe navigation and protection of lives. Most ports have some kind of evacuation and storm preparation procedures designed to minimize damage. But there was no need to plan for long-term changes in sea levels or for storms unlike those seen before. Ports developed their level of resilience in line with historical conditions. However, for many ports, achieving an appropriate level of resilience for future conditions may be beyond their means.

2. The adaptation process

This concluding chapter contextualizes the contributions to this volume by applying a commonly used framework for adaptation, defined here as "any adjustment in natural or human systems in response to actual or expected climatic stimuli or their effects, which moderates harm or exploits beneficial opportunities".[5] Though the book largely follows a geographical framing that explores adaptation issues across the continents, when taken as a whole the volume may be thought of as an indication of "the pulse" of the global port community on this matter. Thus, this concluding chapter tracks a framework that follows an iterative approach to adaptation along five broadly defined steps as seen in Figure 17.1, all of which require a strong stakeholder engagement component.[8] We discuss each of these steps in more detail by drawing from the works represented in this volume, indicating gaps and opportunities for future research.

2.1 Defining the problem

The process of adaptation generally begins with identifying the problem, including understanding impacts, risks, opportunities, and vulnerabilities.[8] This is the subject of a great deal of research in the emerging area of climate adaptation[9,10] and is the focus of much of this volume, which contributes to a better understanding of the nature of the climate-change problem for ports and their stakeholders (Figure 17.1).

Based on a previous chapter, we can understand the concept of risk as:

> The likelihood over a specified time period of severe alterations in the normal functioning of a community or a society due to hazardous physical events interacting with vulnerable social conditions, leading to widespread human, material, economic, or environmental effects that require immediate emergency response to satisfy critical human needs and that may require external support for recovery.[6]

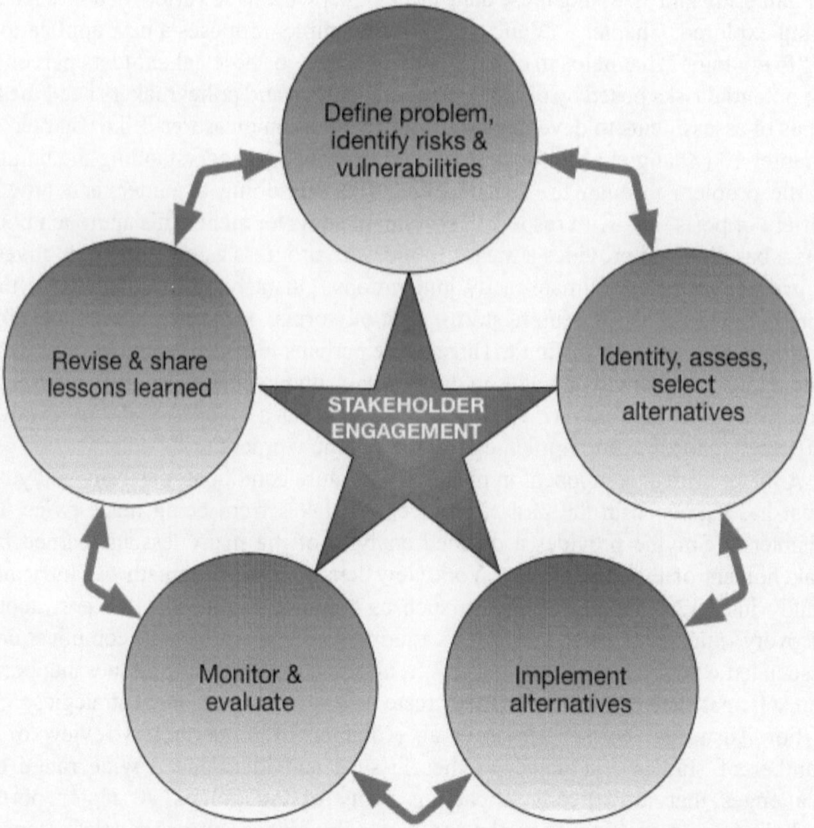

Figure 17.1 The process of adaptation to climate change impacts (after Moser and Ekstrom, 2010).

To simplify, "risk" is the product of the probability of an event and the damage consequences that result. For some ports, this process has easily reduced through reducing damage consequences or vulnerable social conditions. Here, vulnerability is defined as, "the propensity or predisposition to be adversely affected … including the characteristics of a person or group and their situation that influences their capacity to anticipate, cope with, resist, and recover from the adverse affects of physical events."[6]

Though the process is iterative, problem and solution identification generally takes place early on. For ports, this process has recently begun[11,12,13], but in general a clearer definition of the anticipated issues and solutions for ports is still needed.[1,14] Through the chapters herein, this book well represents the difficulty inherent in understanding the nature of the risks, vulnerabilities, and opportunities. Across the globe, policymakers, practitioners, and academics grapple with the complexities of uncertainties in climate projections, high upfront costs, and unquantifiable benefits of action. As yet, there are no generally applied methods for gathering and assessing these data and we thus see a wide variety of approaches being explored. Chapter 3 (Yang et al.), for example, proposes a new application of "fuzzy logic" that helps to quantify how a variety of port stakeholders perceive the potential risks posed by climate change. Planners and policymakers need these types of assessments to develop appropriate adaptation measures. Like Chapter 3, Chapter 10 (Zhang et al.) proposes a novel approach to understanding the nature of the problem through the use of Evidential Reasoning to understand how a variety of ports rank with respect to sustainable development. This approach both sets a baseline and provides a means to measure progress as ports begin to invest in greener and more climate-ready innovations. Quantifying in dollar terms the potential impacts from a future storm, a set of storms, or the rise in sea level (for example) remains very difficult. Thus, while perhaps less satisfying than a precise calculation of potential damages in dollar terms, understanding the perceptions of experts directly involved with port planning serves as a foundational step towards ultimately choosing and implementing the best new practices.

Another critical component in planning for future conditions, carefully studying what has happened in the past can prevent mistakes from being made twice. In Chapter 6, Smythe provides a detailed analysis of the many lessons learned by stakeholders of the port of New York/New Jersey in the aftermath of Hurricane Sandy in 2012. Unexpected issues, such as communicating with the port about recovery, managing personnel in the aftermath of the storm, and complications associated with the loss of fuel and power, all stand as examples of issues that ports can anticipate and prepare for as they create new storm management strategies.

For Australian ports, Cahoon et al. (Chapter 13) undertook a review of a number of studies conducted on these issues and identified a wide range of challenges that are already impacting many of Australia's 70 major ports, including sedimentation of berths and channels, higher salt-water splash zones, tropical cyclones, and melting tarmac. Many of the issues that ports may face result from disruptions along other portions of the supply chains (e.g. coal mines and coal rail operators) and ports must consider these potential disruptions all along this chain, not just at the port itself.

In Chapter 8, Cusano et al. discuss the current state of port planning and climate change policy in Italy. Like many of the other examples in this book, Italian ports have only recently begun to undertake problem identification and have not yet moved into the phase of strategy assessment or implementation, though as discussed in Section 2.2 of this chapter, they are beginning to make strides in this direction.

Though many ports have focused on understanding the significant threats due to increases in storm intensities[15] and SLR,[16] the increasing probability of draught poses a particular challenge for the Panama Canal Authority. Draught can lead to disruptions of operations in the Canal that result in local, national, and global economic consequences. In Chapter 12 (Acciaro), an extensive study of these consequences quantifies their economic impacts and suggests the study and implementation of adaptation measures that could increase the drought tolerance of the Canal.

Although climate change could bring significant challenges, opportunities may also present. In Chapter 11 (Stenek et al.), the authors discuss a wide variety of both potential threats and benefits from climate change. Overall, the study in Muelles El Bosque (Cartagena, Columbia) found only a small number of threats to the port's commercial success (e.g. reduced vehicle movements due to more flooding, decrease in global grain shipments, and degradation of mangroves around the access way to the port). However, since the port does not experience tropical storm activities thanks to its location near the equator, its relative resilience compared to ports closer to hurricane belts could give it a market advantage. The study notes other potential benefits, such as increases in rainfall reducing the generation of dust. Chapter 16 (Verny) illustrates more examples of opportunities, such as the potential economic advantages for shippers and for Russian Arctic ports as the Northeast Passage becomes more navigable due to melting sea ice. New routes such as this and the Northwest Passage could cut down on vessel transit times, reduce fuel costs, and offer the chance to develop new ports for trans-shipment of freight and passengers. However, as Ircha and Higginbotham caution in Chapter 15, a number of daunting challenges stand in the way: there is currently a lack of ports and places of refuge, minimal navigational aids and nautical charts, a dearth of experienced ice navigators, and a complex geopolitical field. Though for some shippers, the Arctic represents exciting possibilities, they must work with territorial departments and agencies to first develop and implement a comprehensive northern strategy. Chapter 4 (Slack and Comtois) also find some good news for ports along the Great Lakes, where projections indicate a longer navigation season due to an absence of ice on both locks and navigation routes. Unfortunately, however, the good news comes with increased costs from dredging as lake levels overall fall due to decreases in regional precipitation, and threaten the navigability of the St. Lawrence River-Great Lakes navigation system and access to local river ports (e.g. Montreal).

All of these chapters break new ground in helping port stakeholders to better understand the nature of the climate-change challenge. This lays the critical groundwork for the important work of identifying and selecting strategies and alternatives to best position ports for these new environmental conditions.

2.2 Identifying and selecting strategies

With a firmer sense of the nature of the challenge faced by port stakeholders, the adaptation process can move to identifying and selecting potential solutions.[17,18,19,20] In this book, a number of chapters address this shift from understanding the nature of the problem, to selecting and implementing strategies toward adaptation. This work can be particularly vexing, as many of the strategies require significant investment today and a payoff that may not be realized for several more decades. In the Great Lakes, for example, Slack and Comtois (Chapter 4) discuss the implementation of structural changes, such as deepening channels or constructing new retention dikes in anticipation of lower lake levels. They also discuss a number of non-structural measures, such as loading ships to draft capacity in one port and "topping off" in a second, shifting certain freight transport modes from ship to rail, or even relocating ports entirely. Lastly, they suggest technological changes, such as designing ships with wider beams and shallower drafts or using a greater number of barges. Though the authors point to a number of potential solutions, they rightly acknowledge the difficulties inherent in implementing such changes in a system fraught with uncertainties with respect to climate, market, and technological shifts.

Japanese ports, in particular, face a unique set of challenges due to their exposure to strong typhoons as well as increasing rates of sea-level rise. Engineers designed much of the existing coastal infrastructure based on historical records of sea levels and wave conditions. In Chapter 9, Esteban et al. suggest consideration of a number of new strategies to prevent an increase in port downtime associated with future climatic conditions. For example, breakwaters could be strengthened or new storm barriers could be constructed for storm-vulnerable areas such as the entrance to Tokyo Bay.

Challenges to ports do not, of course, begin and end at the boundaries of the terminal facility itself. In Canada, for example, the whole supply chain faces an emerging issue of rail freight service disruptions that result from the thawing of the permafrost layer underlying the tracks. To address this, Wang et al. (Chapter 5) propose two basic, albeit expensive, strategies to shore up the tracks as warming continues over the coming century. However, though these both may be feasible, implementing them would be expensive and likely only a temporary fix.

In Chapter 8, Cusano et al. share one example of implementing a strategy in Italy, where the National Government appointed responsibility for coastal zone management to regional authorities. Similar to systems in other countries, this policy does not in and of itself serve to adapt Italian ports to climate change, but it does clarify responsibilities for addressing climate change concerns. This, for example, led to the construction of the MOSE barrier system that may protect Venice from a 60 cm sea-level rise.

For the most part, we see that ports around the world are only at the beginning stages of understanding the full range of potential solutions to the climate-change challenge. Once identified, the arduous phase of weighing costs and benefits, securing necessary funding, design selection, environmental permitting, and actual construction (of new structures) or facilitation (of drafting new policies and plans or other "softer" solutions) can begin. Most ports and port communities have not yet reached this phase.

2.3 Implementing strategies, monitoring and evaluating, and sharing lessons learned

Moving from understanding the problem and potential responses becomes all the more challenging. Most adaptation measures result in both winners and losers, as decision makers invest resources now even when the benefits may not be realized for decades. Thus examples of strategy implementation, monitoring, and revision still remain few and far between.[21] However, as seen through this volume, some ports around the world have begun to implement initial steps. Perhaps leading the charge in port adaptation, a number of Australian ports have implemented seaside and quayside measures such as repositioning storage tanks, designing new structures that account for sea-level rise projections, and using new heat-resistant surfaces in terminal yards (see Table 13.1 and 13.2 in Chapter 13 for a complete list). Some areas have begun to incorporate sea-level rise projections into design specifications.[22,23] In the US, planners at the port of San Diego worked closely with consultants and stakeholders to identify and implement adaptation strategies to protect port areas from projected sea level rise. As discussed in Chapter 2 (Messner et al.), this proved very difficult due in part to the relatively shorter planning horizons utilized by many of the port tenants and the uncertainties around long-term trends for sea-level rise.

For now, many of the "lessons learned" consist of successful methods to help ports and port communities to understand the nature of the problem and the range of solutions. In the coming decades, more and more port communities will no doubt be implementing a wide variety of these strategies to cope with rising seas, draughts, and changing storm patterns. As adaptation planning and implementation evolve, the global port community will benefit from continued dissemination of progress, lessons learned, and methods to evaluate the effectiveness of various alternatives.

2.4 The critical role of the stakeholder network

Throughout, the chapters in this book all highlight the importance of including perspectives of stakeholders from multiple sectors in all phases of the adaptation process, on every scale, from specific organizations[24] to nations as a whole.[25] Adaptation will involve multiple actors, policies, and practices and require communication and public engagement to ensure success.[21] Chapter 14 (McEvoy et al.) provides an overview of a stakeholder-driven process for adaptation planning in Fiji and Papua New Guinea, where port planners participated in facilitated workshops to apply a new tool designed to help discover potential climate-related problems.

In Chapter 7, Osthorst argues that adaptation must go hand-in-hand with the other strategic choices made by many of the stakeholders. In the case of the port of Bremerhaven, adaption governance depends heavily on the voluntary and network-based processes of the individual participants in the stakeholder system. Market pressures, the costs of normal infrastructure maintenance programs, and ecological challenges must be integrated with adaptation strategies in ways that benefit the various actors in the system.

Chapter 5 (Wang et al.) also emphasizes the importance of involving multiple stakeholders (e.g. government, industry, private-sector firms, environmental groups) in the decision-making process so as to ensure buy-in, representation, and the opportunity to balance trade-offs to maximize the positive outcome for all parties. A "partnership" approach can create an integrated knowledge base and management system, adding value to port stakeholders' efforts and building the capacity to address climate-change impacts. Partnerships can bring together key researchers, policymakers, industrial practitioners, interest groups, and other port stakeholders to achieve a common objective, i.e. enhancing the understanding and reducing the uncertainties of adaptation to climate change. The formation of this broad partnership can serve as the pioneer step to develop a transnational collaborative network between port stakeholders around the globe and encourage mutual trust among stakeholders from different sectors, countries, and regions, while facilitating knowledge and information flows.

In addition, partnerships can establish "common stories" that lead to the gradual development of international best practices—a necessity to effectively address a growing global problem, with strong local perspectives and interests. Information exchange through partnerships allows policymakers, practitioners, and academics to compare their local approaches to international practices. For example, assessment tools can demonstrate to the broader port, transportation, logistical, and supply chain communities how well particular elements are functioning. Finally, partnerships can also lead to unified data collection, alleviating the burden of filling in the same data for diversified purposes.

Conclusion

This book provides numerous case studies from around the world which, taken together, indicate that many ports have begun the adaptation process through initial analysis of the challenges and opportunities posed by climate change. It provides a few examples of ports that have identified possible alternatives to reduce their vulnerability and/or take advantage of new opportunities. Fewer, however, have successfully implemented strategies specifically designed to address climate change.

During the United Nations Conference on Trade and Development's (UNCTAD) *Ad Hoc* Expert Meeting on climate change and port adaptation (held in Geneva, Switzerland, September 29–30, 2011), participants highlighted the importance of obtaining reliable information that provides clear insight on how climate-change risks would impact port facilities and operations, and thus the quality of a port's strategic planning.[26] As we find throughout this volume, policymakers and practitioners need knowledge, innovative tools, ideas and solutions to deal with the climate-change challenges for ports, transportation, and supply chains. Simultaneously, individual researchers often found it difficult to access appropriate personnel for data and information collection. Major barriers include: multiple languages, diversified local cultures, finding the right channels to the appropriate people, and a lack of data at the appropriate scale. Innovative partnerships comprised of researchers, policymakers, industrial practitioners, interest groups, industrial alliances, and other maritime/port stakeholders can help overcome these barriers.

This book calls on policymakers, practitioners, and academics to collaborate to the address the risks that climate change will bring to their facilities and operations. We strongly believe that an international partnership approach can address these obstacles and challenges, adding value to effective adaptation to climate-change risks to ports. We invite relevant individuals and organizations from different corners around the world to engage intellectual leadership toward the creation and dissemination of knowledge on this emerging topic.

Notes

1 Becker, A. et al. (2013). 'A note on climate change adaptation for seaports: A challenge for global ports, a challenge for global society', *Climatic Change*, 120 (4): 683–695.

2 Rahmstorf, S. (2010). 'A new view on sea level rise', *Nature Reports: Climate Change*, pp. 44–45.

3 Knutson, T. R. et al. (2010). 'Tropical cyclones and climate change', *Nature Geoscience*, 3: 157–163.

4 Becker, A. et al. (2012). 'Climate change impacts on international seaports: Knowledge, perceptions, and planning efforts among port administrators', *Climatic Change*, 110 (1–2): 5–29.

5 IPCC (Intergovernmental Panel on Climate Change), *Climate Change 2007: Impacts, Adaptation and Vulnerability*. Contribution of Working Group II to the Fourth Assessment Report of the Intergovernmental Panel on Climate Change (eds. Parry, M. L., Canziani, O. F., Palutikof, J. P., van der Linden, P. J. and Hanson, C. E.). Cambridge, UK.

6 IPCC (Intergovernmental Panel on Climate Change) (2012). 'Managing the risks of extreme events and disasters to advance climate change adaptation. Special report of the Intergovernmental Panel on Climate Change'. In: *SREX*, eds. Field, C. B. et al. Cambridge, UK, and New York, NY, USA.

7 NRC (National Research Council) (2010). *America's Climate Choices: Adapting to the Impacts of Climate Change*, ed. Wilbanks, T. J. et al. Washington, DC.

8 Moser, S. and Ekstrom, J. (2010) 'A framework to diagnose barriers to climate change adaptation', *Proceedings of the National Academy of Sciences*, 107 (51): 22026.

9 Bierbaum, R. et al. (2013). 'A comprehensive review of climate adaptation in the United States: More than before, but less than needed', *Mitigation and Adaptation Strategies for Global Change*, 18 (3): 361–406.

10 Preston, B. L., Westaway, R. M. and Yuen, E. J. (2010). 'Climate adaptation planning in practice: An evaluation of adaptation plans from three developed nations', *Mitigation and Adaptation Strategies for Global Change*, 16 (4): 407–438.

11 McEvoy, D., et al. (2013). *Understanding Future Risks to Ports in Australia*. Enhancing the resilience of seaports to a changing climate report series. Australia National Climate Adaptation Research Facility, Gold Coast, Australia.

12 Stenek, V. et al. (2011). *Climate Risk and Business: Ports*. International Finance Corporation.

13 Port of San Diego (2013). *The Climate Mitigation and Adaptation Plan of the Port of San Diego*. Available from: http://www.portofsandiego.org/climate-mitigation-and-adaptation-plan.html (accessed June 10, 2013)

14 EPA (United States Environmental Protection Agency) (2008). *Planning for Climate Change Impacts at U.S. Ports. White Paper prepared by ICF International for the USEPA*.

15 Grinsted, A., Moore, J. C. and Jevrejeva, S. (2013). 'Projected Atlantic hurricane surge threat from rising temperatures', *Proceedings of the National Academy of Sciences*, 110: 5369–5373.

16 Rahmstorf, S. (2007). 'A semi-empirical approach to projecting future sea-level rise', *Science*, 315 (5810): 368–370.

17 Moser, S. (2009). 'Whether our levers are long enough and the fulcrum strong? Exploring the soft underbelly of adaptation decisions and actions', *Adapting to Climate Change: Thresholds, Values, Governance,* pp. 313–334.
18 Adger, W. N. et al. (2007). 'Assessment of adaptation practices, options, constraints and capacity', *Climate Change*, pp. 717–743.
19 Travis, W. R. (2009). 'Going to extremes: propositions on the social response to severe climate change', *Climatic Change*, 98 (1–2): 1–19.
20 Kates, R. W., Travis, W. R. and Wilbanks, T. J. (2012). 'Transformational adaptation when incremental adaptations to climate change are insufficient', *Proceedings of the National Academy of Sciences*, 109 (19): 7156–7161.
21 Moser, S. C. and Boykoff, M. T. (2013). *Successful Adaptation to Climate Change: Linking Science and Policy in a Rapidly Changing World.* Routledge, London, England.
22 USACE (United States Army Corps of Engineers). (2014). *Comprehensive Evaluation of Projects with Respect to Sea-Level Change.* Available from: http://www.corpsclimate.us/ccaceslcurves.cfm.
23 USACE (United States Army Corps of Engineers). (2013). *Incorporating Sea Level Change in Civil Works Programs*, in Regulation No. 1100-2-8162.
24 Berkhout, F., Hertin, J. and Gann, D. M. (2006). 'Learning to adapt: Organisational adaptation to climate change impacts', *Climatic Change*, 78 (1): 135–156.
25 USGCRP (United States Global Change Research Program). (2013). *National Climate Assessment (DRAFT)* [cited July 1, 2013]. Available from: http://ncadac.globalchange.gov/
26 UNCTAD (United Nations Conference on Trade and Development). (2012). "Climate Change Impacts and Adaptation: A Challenge for Global Ports, *Ad Hoc* Expert Meeting (September 2011)." UNCTAD/DTL/TLB/2011/3. 29–30 September.

Index